Giant Sloths and Sabertooth Cats

Giant Sloths
and Sabertooth Cats

Extinct Mammals and the Archaeology
of the Ice Age Great Basin

Donald K. Grayson

With Animal Reconstructions by Wally Woolfenden

THE UNIVERSITY OF UTAH PRESS
Salt Lake City

The Defiance House Man colophon is a registered trademark of the University of Utah Press. It is based on a four-foot-tall Ancient Puebloan pictograph (late PIII) near Glen Canyon, Utah.

20 19 18 17 16   1 2 3 4 5

LIBRARY OF CONGRESS CATALOGING-IN-PUBLICATION DATA

Names: Grayson, Donald K. | Woolfenden, Wally, 1941–
Title: Giant sloths and sabertooth cats : extinct mammals and the archaeology of the Ice Age Great Basin / Donald K. Grayson ; with animal reconstructions by Wally Woolfenden.
Description: Salt Lake City : The University of Utah Press, 2016 | Includes bibliographical references and index.
Identifiers: LCCN 2015033646| ISBN 9781607814696 (pbk. : alk. paper) | ISBN 9781607814702 (ebook)
Subjects: LCSH: Extinct mammals—Great Basin. | Extinct mammals—North America. | Glacial epoch. | Great Basin—History.
Classification: LCC QL707 .G73 2016 | DDC 599.168—dc23
LC record available at http://lccn.loc.gov/2015033646

Printed and bound by Sheridan Books, Inc., Ann Arbor, Michigan.

To Heidi
Who Gave Me Back My Life

CONTENTS

FIGURES

TABLES

PREFACE

Dinosaurs are cool, but except for the ones in *Jurassic Park* they've never done much for me. The last dinosaurs were lost about 66 million years ago. I am incapable of imagining that kind of time depth. My perception of the past has a certain logarithmic aspect to it. I have no problem imagining what ten years ago means, feel comfortable picturing 100 years ago (my mother lived to be 103), have a pretty good grasp of 1,000 years, and have spent enough time working with things that are 50,000 years old that I don't have much problem with that either. Take me back 66 million years, however, and I am lost.

I'll take the Ice Age, or Pleistocene, any day. It dates from 2.6 million to 10,000 years ago, and I can make myself believe that I have an intuitive feel for how long ago 20,000 or 30,000 years ago was. I will also take the Ice Age because it was full of what are, to me, some of the most interesting animals that ever walked the face of the earth or flew through the air. Imagine giant ground sloths with the bulk of elephants, giant cats with saber-like teeth, and giant beavers the size of black bears. Imagine pronghorn antelope with four horns and imagine that the smallest of them couldn't peer over the top of a healthy sagebrush. Imagine giant scavenging birds with 18-foot wing spans. If you can imagine these, you can imagine what some of the Ice Age animals of North America were like. Even more enticing, we know that people encountered many, and perhaps most, of these animals.

Then imagine a landscape of stark beauty, composed of long, broad, flat valleys separated by massive mountain ranges. Imagine that the valley bottoms are covered by shrubs and grasses and that as you walk upslope, the mountains become covered with trees that can live 1,000 years or more, until the mountains became so tall that they can't support trees at all. Imagine that in the valley bottoms, you encounter birds that impale their prey on thorns—they are called shrikes—and that if you walk high enough into the mountains, you hear the flute-like calls of thrushes, as if you were in New England. Imagine that the water that falls in this entire region has no way to escape unless it evaporates and is carried away on the wind. Imagine an area that is still so isolated that it can be quiet enough to let you hear your own ears. If you can imagine all this, then you can imagine the Great Basin.

We can still see the Great Basin, we can still live in it, work in it, play in it, and hear our own ears in it. Many of the Ice Age animals that this area supported, however, are gone, having failed to get any closer in time to us than about 10,000 years ago.

That's what this book is about—Ice Age animals, and in particular the Ice Age mammals of the Great Basin—and the people who might have seen them. These were some of the coolest animals ever to have lived, in one of the most remarkable places on earth, along with the people who may have interacted with them. What a lucky guy I am to have spent a substantial part of my career studying these animals and people, and working in this part of the world.

Dinosaurs. Humph.

ACKNOWLEDGMENTS

A list of the people who made this book possible would include everyone I have ever met who has worked in the Great Basin or with any of the world's now-extinct Pleistocene mammals. I restrict myself to thanking those who were so generous with their time when I bothered them about specific issues discussed in the pages that follow or who provided much-needed help in other ways. For all this help, thanks to Iván Alarcón Durán, Ellen Alers, Rebecca Alexander, Joaquin Arroyo-Cabrales, Charlotte Beck, Monica Buhler, Bill Cannon, Mike Cannon, David Charlet, Ken Cole, Peggy L. Corson, Tom Deméré, Steve Emslie, Elmer Finck, Kay Fowler, Amy Gilreath, Gene Hattori, Suzann Henrickson, Bryan Hockett, Wendy Hodgson, Vance Holliday, David Huckaby, Lori Hunsaker, Tom Jones, Renee Kolvet, Heidi Lennstrom, Lisbeth Louderback, Spencer G. Lucas, Ernie Lundelius, David Madsen, E. K. McDaniel, Greg McDonald, Jim Mead, Dave Meltzer, Constance I. Millar, Jim O'Connell, Jim Patton, Bruce Pavlik, Dave Rhode, Diana Rosenthal, Alan Simmons, Greg Smith, Kathy Swain, Dave Thomas, Arnold (Jerry) Tiehm, Glen Whorton, and Wally Woolfenden.

Some more detailed acknowledgments are also needed.

The reconstructions of extinct mammals found in chapter 3 are the work of Wally Woolfenden, as formidable an artist as he is a Quaternary scientist. The projectile point illustrations in chapter 6 are the work of Peggy L. Corson, whose art has graced my work before. Some of the photographs I have used were provided by Iván Alarcón Durán, David A. Charlet, Kenneth L. Cole, Heidi A. Lennstrom, David B. Madsen, and James F. O'Connell. My friends have helped keep this book within the Great Basin family and I am deeply fortunate to have been able to work with them and to have them in my life.

The University of Washington's Quaternary Research Center provided funding for fieldwork that was essential to completing this book; special thanks to Eric Steig for facilitating that support. The National Science Foundation has supported my research in this realm for decades (especially grants BNS88-03333, SBR98-04692, BCS02-03278, and BCS04-04510); I am extremely grateful for their support as well.

Gene Hattori, E. K. McDaniel, Warden Greg Smith, and Kathy Swain arranged a visit to the visible trackways at the Nevada State Prison. Rocky Koch and Ed Papke were ideal tour guides during our visit there; Glen Whorton provided crucial information on the recent history of the tracks after our visit was over. I am deeply indebted to all of them for this.

I am equally indebted to Bryan Hockett for providing a valuable tour of Mineral Hill Cave and for being helpful in so many other ways. We are lucky to have him in the Great Basin.

Chapter 6 was spurred by a question Paul Martin asked me and that I talk about in that chapter. Paul was a close friend and valued scientific sparring partner. I am in deep intellectual debt to him and wish that he were still here to point out, as he so often did, where I might have gone wrong.

An early version of chapter 6 was commented on in detail by four scientists marked by an intimidating combination of great intellect and broad knowledge. David Charlet and Connie Millar made me think more deeply about conifers, Dave Rhode made me think about plants shorter than I am, and both Dave and Wally Woolfenden made me worry about deeper plant histories. In these and many other ways, they did their best to keep me out of trouble with the arguments I make in that chapter. To the extent that I failed, it was because I did not follow their advice.

Heidi Lennstrom, Ernie Lundelius, and Dave Rhode were kind enough to read an entire draft of this book for me. The three of them, individually and together, provided deeply insightful comments that helped me improve that draft substantially. Had I known that I was going to have such knowledgeable and critical readers, I might have thought twice about launching this project. By the time I got those comments, it was too late. It would not be possible to thank them enough for all the help they gave me.

Very much the same can be said of Laurel K. Anderton. Her thoughtful editing made this a much better book than it otherwise would have been. I am deeply indebted to her for the care and patience she brought to the task.

Heidi Lennstrom knows how grateful I am to her. This book would not have been written without her, but you should not blame her for that.

A TINY BIT OF BACKGROUND

DATING

My prime focus in this book is on the extinct mammals that lived in the Great Basin between about 50,000 and 10,000 years ago, and on the people who may have encountered them toward the end of this time span. Most of the ages I give in this book were provided by radiocarbon dating, the gold standard for figuring out the ages of things that were deposited in archaeological and paleontological sites during the past 50,000 years or so. I explain how this dating method works, and how it can go wrong, in chapter 4. Here, I just need to mention that radiocarbon years are not the same as real, calendar years. How different they are depends on the age of whatever it is you are dating. Something that has a radiocarbon date of 3,000 years ago, for instance, is actually about 3,200 years old. Depending on your goals, that may or may not be a big deal. Something that has been radiocarbon dated to 10,000 years ago is actually about 11,500 years old, which is a bigger deal. Something that has been radiocarbon dated to 20,000 years ago is, in real years, about 23,900 years old. That's a huge difference. For reasons discussed in chapter 4, I use radiocarbon years throughout this book. In appendix 1, I provide an easy way to see the relationship between the two.

BIOLOGICAL NAMES

One of the advantages of living in Washington State is that you don't need the Internet to get Canadian television (one of the disadvantages is the weather; another is the Seattle Mariners). Recently I watched a show called *Wild Canada*. In the midst of impressive images of wolverines, moose, and beaver, up popped a Gray Jay. Except they didn't call it a Gray Jay. They called it a Whiskey Jack.

There is nothing wrong with either term. Gray Jays are jays that are gray. They are also jays that are bold, pesky, and happy to steal your food. The name Whiskey Jack is derived from an Algonquian word for a mythological trickster. Both names capture essential aspects of the bird.

This is the problem with common names. Different people call the same animal different things. Scientific nomenclature solves that problem, since by agreement, the same animal gets the same scientific name. The bird I was just talking about is *Perisoreus canadensis* no matter what common names it might get.

In this book, I have used a series of standard sources for the common and scientific names of existing plants, birds, and mammals. For plants, my prime source was the US Department of Agriculture PLANTS Database (http://plants.usda.gov), with the second edition of *The Jepson Manual* (2012), edited by B. G. Baldwin and his colleagues, not far behind. Following standard practice, I have given both the common and scientific names of plants at first mention in the text, but often only one or the other after that. For those who want to be reminded of these pairings, appendix 2 provides a list of them. For mammals, I have used the third edition of D. E. Wilson and D. M. Reeder's *Mammal Species of the World* (2005): http://www.vertebrates.si.edu/msw/mswcfapp/msw/index.cfm. For birds, I have used the World Bird Database (http://avibase.bsc-eoc.org).

Now let's go to the Nevada State Prison.

1

A Sloth in Prison

"Footprints of Monster Men" read the headline in the *New York Times* for August 18, 1882, reporting an astonishing discovery at the Nevada State Prison.[1] "Mysterious Tracks in Stone" had appeared four days earlier, but that was a far more innocuous title that might have been overlooked by the casual reader.[2] "Footprints of Monster Men" is hard to ignore.

Carson City's prison came into being in 1862. Two years later, Nevada became a state and the prison became the Nevada State Prison. In 1867, the original buildings burned to the ground and construction began anew, using sandstone quarried by convicts on the site itself.[3] It is not clear when the first fossil bones were uncovered, but it is clear that no particular attention was paid to their potential significance until William Garrard became warden in 1881. By then, the quarry had not only yielded fossils but had also revealed the presence of something far more remarkable: the footprints of ancient animals captured in layers interbedded within the sandstone. It was these tracks that captivated Warden Garrard.

W. J. Hanks, the sheriff of Storey County, called in the scientists. During a visit to San Francisco's California Academy of Sciences in June 1882, Hanks told the academy's curator of mineralogy, Charles Drayton Gibbes, that he had heard of "footprints, both of men and animals having been found at the Nevada State Prison."[4] He promised Gibbes that he would examine them personally, and this he quickly did.

On June 26, he wrote to Gibbes to report that the prison quarry was rich in the tracks of ancient animals. Among these were the tracks of men, 22 inches long, along with "bird-tracks, woman-tracks, and many other fossil remains."[5] This is, he said, "the most wonderful formation in the world,"[6] and he urged

Gibbes to get someone knowledgeable to examine what the prisoners had uncovered.

With Warden Garrard's assistance, Gibbes soon arranged a visit to the prison, bringing Harvey Harkness and J. R. Scupham with him. Harkness had been a member of the California Academy of Sciences since 1871 and was a physician and eminent mycologist.[7] J. R. Scupham had been a member of the academy since 1873 and, as an employee of the Southern Pacific Railroad, had helped arrange the railroad's California exhibit at the Centennial Exhibition in Philadelphia in 1876.[8]

These were not the only people Garrard invited to the prison. On July 2, he told Gibbes that he expected the eminent geologist Joseph Le Conte and his University of California colleague W. B. Rising to visit as well.[9] All arrived at the prison between July 20 and 22, 1882.[10]

Impressed by what they saw, all but Rising prepared reports to present to the California Academy of Sciences upon their return. It was this visit, and those reports, that led to the intense newspaper coverage that the prison footprints were to receive.

Harkness, the expert in fungi, was the first to give his paper at the academy on the evening of August 7. Speaking to "an unusually large number of interested auditors"[11] and accompanied by illustrations of the fossil tracks, he reported that he had seen the fossil footprints of "the mammoth, the deer, the wolf, of many birds, of a horse, and most important of all, the imprints of the sandaled foot of man," all intermingled with one another on the flat floor of the prison quarry.[12] The human footprints, he said, averaged 18.5 inches long, with a stride of 2 feet 3 inches; the impressions of the left and right feet were about 18 inches apart measured from center to center. The feet themselves might not have been that big, however, because they seemed to have been protected by wooden sandals. In one spot, he emphasized, the footstep of a mammoth had obliterated a human footstep, leaving little doubt that the two had visited this spot very close to one another in time. "FOOTPRINTS OF MONSTER MEN" provided an accurate summary of what Harkness said that evening.

We know the *New York Times* got it right because Harkness published his talk.[13] The world-famous paleontologist Edward Drinker Cope saw that published version and was convinced by what he read. In the pages of the *American Naturalist*—then, as now, an important scientific journal—Cope observed that Harkness had presented compelling evidence for the existence of what was probably "the ancestor of existing man."[14]

Scupham gave his view of the footprints on the same evening, but since his paper was never published and may no longer exist, it is not possible to know all that he said. Fortunately, some of his comments appeared in the *Popular Science Monthly* in 1883. Those comments make it clear that Scupham had an impressive imagination, making it even more unfortunate that his full presentation is not available. Scupham, the *Popular Science Monthly* tells us, described

> confused tracks of a man and some large animal. Only two or three steps of the man are distinct; then the confusion that appears to mark a struggle, and then the impression where a great body has fallen. After the struggle the great crane has waded about over the spot; its tracks winding in and out as if it had been avoiding with care the deep impressions made by the combatants, till at last, stumbling into one, it rises in startled fright.[15]

For Scupham, some of the tracks showed a prehistoric human hunter successfully pursuing his large prey with a frightened wading bird as a survivor.

On August 27, geologist Joseph Le Conte gave a far more cautious interpretation of what the prison quarry had revealed. The tracks, he agreed, included those of mammoth, large wading birds, deer, and something else—the "supposed human tracks."[16] The origin of these was not at all clear to him. They did, he agreed, bear a remarkable resemblance to human footprints but, at 18–20 inches long and 8 inches wide, seemed far too large. He had no problem with the stride, which, at 2.5 to 3 feet, he compared to that of an average man walking briskly. More problematic, the 18–19-inch distance between the right and left feet, or straddle, seemed inconsistent with the tracks being of human origin. Perhaps Harkness was right and the human makers of these tracks had worn large wooden sandals, but the rounded nature of the base of the tracks did not support this idea, and the straddle, *"as great as that of the contemporaneous elephant,"*[17] simply did not support a human origin for them.

If not a human, then what could have created them? Le Conte had two suggestions: a bear, or, more likely, the extinct giant ground sloth known to scientists as *Mylodon* (now referred to as *Paramylodon* and talked about in detail in chapter 3). A careful scholar, Le Conte concluded that "the judicious mind should hold itself in suspense awaiting more evidence."[18] He made it clear, though, that he very much doubted that these tracks had a human origin, and he made it equally clear that he thought that they had been made by a quadruped, and that the quadruped was most likely a giant ground sloth.

Soon after Le Conte presented his paper, two things happened.

First, the contents of Le Conte's talk made the popular press from San Francisco to New York. As often happened at the time, and as still happens today, some of the renditions of Le Conte's talk ignored both his caution and his conclusion that the tracks in question had likely been made by a sloth. Instead, they made it seem as if he agreed that the footprints were, in fact, human.[19] Even the paleontologist Edward Drinker Cope found that Le Conte's paper did not differ in any important way from the one Harkness had given,[20] although other syntheses of Le Conte's presentation were impressively accurate,[21] including Cope's initial discussion of that presentation.[22] The tracks of giant sloths might have been interesting, but the tracks of ancient people were far more so.

Second, Le Conte had been in contact with the superb paleontologist O. C. Marsh and discovered that Marsh agreed with him that the supposed human footprints had been made by a giant sloth. Le Conte wrote an addendum to his paper to point this out. That paper, however, was not published until long after the talk had been given, and it was the version that he had given on August 27 that made the news.

Charles Gibbes was the last to present the results of his work to the academy, on September 4. In some ways, this was the most important of these presentations, because it was the most empirical. After Harkness, Scupham, Le Conte, and Rising had left Carson City, Gibbes had stayed behind to make careful measurements, maps, impressions, and photographs of the tracks (figs. 1.1–1.4). Gibbes's list of the animals that had left their tracks behind included mammoth, birds, elk, a large cat, and, of course, those that might have been human. The large cat, he suggested, might have been the sabertooth (*Smilodon*; see chapter 3). The tracks that had made the Nevada State Prison so famous, though, were clearly of human origin. There were, he said, at least 114 of those (other counts suggested there were hundreds), in at least six separate series. Not only did his own experience in tracking both game and people tell him that these were of human origin, but "many frontiersmen of great experience in tracking Indians, and also bear and other game, have witnessed these foot-prints, and all give their judgment without a question in favor of their human origin."[23]

Of the four people who had visited the prison and given presentations at the California Academy of Sciences in 1882—Harkness, Scupham, Le Conte, and Gibbes—only Le Conte expressed skepticism that this site contained the footprints of ancient people intermingled with those of other animals, including species that were now extinct. The popular press stressed the human origins of these tracks, and it was that press that spread the word most widely. People ate it up, just as they so often do today with such things.

The California Academy of Sciences did not publish its *Proceedings* between 1877 and 1884, meaning that the papers that were presented at meetings

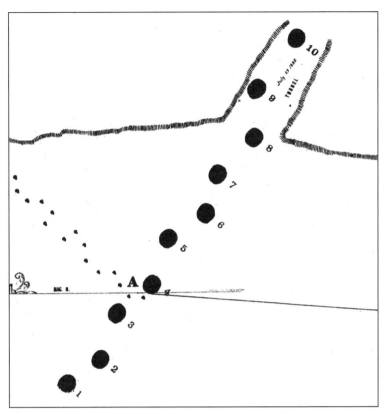

FIGURE 1.1. Mammoth tracks in the Nevada State Prison quarry.
Charles Gibbes's 1882 drawing.

during those years did not automatically have a formal outlet. Instead, those
who wished to have their work published had to pay to have it done. Harkness,
Le Conte, and Gibbes did so; Scupham did not.[24]

It apparently took some months for these papers to appear. Lester Stephens,
Le Conte's biographer, indicates that they appeared sometime in 1883 and suggests that Le Conte was annoyed at the amount of time required for publication. As a result, Stephens concludes, Le Conte wrote a separate paper and
submitted it to the very prestigious scientific journal *Nature*.[25]

This is only partially true. While the academy's *Proceedings* had not appeared, and never did, Le Conte's paper had already been privately printed
and distributed.[26] In addition, Le Conte's paper in *Nature*, which appeared
on May 31, 1883, included only three paragraphs on the Carson City footprints
and was in response to a paper that had appeared in *Nature* on April 19, by
George Douglas Campbell, the 8th Duke of Argyll.

Long interested in the subject of human antiquity, Argyll had written
"Primeval Man, An Examination of Some Recent Speculations," which had

FIGURE 1.2. Mammoth tracks in the Nevada State Prison quarry. From Charles Gibbes's 1882 photograph.

appeared in 1869 in response to the discovery, in Europe, that our ancestors had coexisted with such extinct beasts as mammoths and woolly rhinos.[27] In 1883, his son, the 9th Duke of Argyll, was the governor-general of Canada. Knowing of his father's interest in such things, he sent him a copy of the paper that Gibbes had presented at the California Academy of Sciences on September 4.

Argyll's point was simple. To him, Gibbes's illustrations "leave no doubt on the mind of anyone who sees them" that the footprints in question were

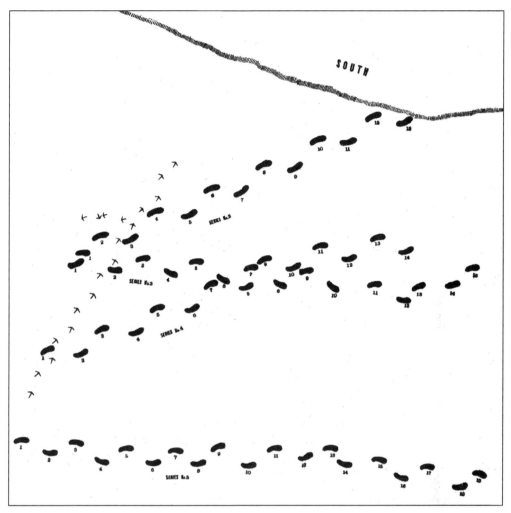

FIGURE 1.3. The supposed human footprints in the Nevada State Prison quarry. Charles Gibbes's 1882 drawing.

human. Deeply religious and no fan of Darwinian evolution, Argyll was fully aware that our ancestors had been on earth far longer than the traditional biblical chronologies allowed, and he was not surprised to learn that human footprints had been found intermingled with those of extinct mammals in western North America. He did, however, note that if this discovery were confirmed, it would provide support for Scripture and, in particular, for the statement in Genesis 6:4 that "there were giants on the earth in those days."[28]

Because Argyll had accepted the human origin of the Carson City tracks, Le Conte wrote to *Nature* to provide a very different interpretation. While admitting that the origin of the supposed human tracks was not known with certainty, Le Conte repeated his core conclusion: these tracks had probably been

FIGURE 1.4. The supposed human footprints in the Nevada State Prison quarry. From Charles Gibbes's 1882 photograph.

made by a large quadruped, and the giant ground sloth *Mylodon* provided an obvious candidate. He noted that others, including O. C. Marsh, agreed with this conclusion.

Marsh made his agreement clear in an article he contributed to the *American Journal of Science* for August 1883. Here, he provided an illustration of the hind foot of *Mylodon* arrayed next to an illustration of one of the Nevada State Prison tracks taken from Harkness's paper (fig. 1.5). The match, he observed, was excellent. In addition, and just as Le Conte had observed, both the size of the footprints and the straddle matched those of *Mylodon*, not those of

1.

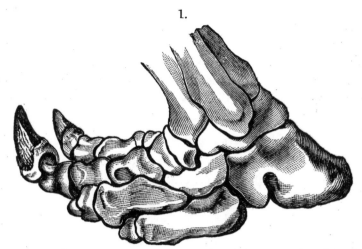

Figure 1.—Left hind foot of *Mylodon robustus* (after Owen).   One-sixth natural size.

2.

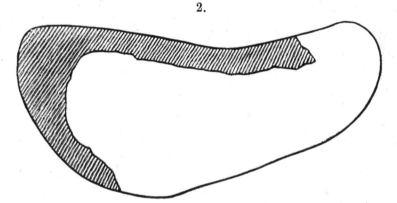

Figure 2.—Left foot-print at Carson (after Harkness).   One-sixth natural size.

FIGURE 1.5. O. C. Marsh's comparison of the hind foot of *Mylodon* with H. W. Harkness's drawing of a "human" footprint from the Nevada State Prison quarry (from Marsh 1883).

people. The tracks, he concluded, were made by a large sloth. Edward Drinker Cope, on reading this paper, found Marsh's conclusion so compelling that he retreated from the support he had earlier given to Harkness's interpretation of those footprints.[29]

Although geological and paleontological opinion was turning staunchly against the interpretation of any of the Nevada State Prison footprints being of human origin, Harkness stuck to his original opinion. In July 1883, he announced that he had discovered a second set of human footprints "some distance from the Carson State Prison quarry,"[30] and that he was endeavoring to buy the land to allow the scientific potential of the site to be explored. If anything came of this, I am unaware of it.

On August 6, 1883,[31] Harkness presented a second paper to the California Academy of Sciences on this topic. As with Scupham's earlier presentation, his second offering went unpublished, and only secondhand accounts exist.[32] These, fortunately, are consistent in the details they provide and tell us that Harkness was as convinced as ever that not only were the disputed Carson tracks those of sandal-shod people, but that their makers deserved their own species name, *Homo nevadensis*. The size of the tracks reflected not the size of the foot, but the size of the wooden sandals; the great width of the straddle reflected the fact that people wearing such sandals in mud would naturally walk with their feet far apart, much as snowshoers do.

Harkness was not the only person to present a paper on the Carson footprints to the academy that night, but the other papers, by Gibbes and by George W. Davidson, failed to gain the attention that Harkness's gained. This is unfortunate, since Davidson's paper was a model of clarity and reason, much as Le Conte's had been. Fortunately, however, Davidson's paper was published.[33]

George Davidson served as president of the California Academy of Sciences from 1872 to 1886, a span that included the Carson footprint years. On July 2, 1883, he requested support from the academy's Board of Trustees to visit the Nevada State Prison, and to take Gibbes and the distinguished geologist William Phipps Blake with him. The trustees agreed, the three visited, and Davidson presented his thoughts on the prison site in tandem with Harkness's paper.[34]

Davidson described the geology of the site and then carefully reviewed the tracks he had seen, of mammoth, birds, a large cat, horse, bison, dog, and, finally, the supposed human footprints. Unlike the cautious Le Conte the year before, Davidson left no doubt about what he thought about these. "The so-called man is a quadruped,"[35] he concluded, the evidence for sandals a misinterpretation of the nature of the impressions themselves. Whether the beast that produced them was a sloth or a bear he left for the paleontologists to figure out, but he was clear on the most important issue: "the animal was a quadruped."[36]

Davidson's paper was separately printed and obscure, but the same was not the case for W. P. Blake's interpretation of the site.[37] As befit a scientist at the top of his profession, he published his observations in one of the world's leading scientific outlets, the journal *Science*. His conclusions were akin to those of Le Conte, Marsh, Cope, and Davidson. The tracks themselves were not consistent with having been formed by people wearing sandals, nor were the length of the unshod footprints or the straddle. Marsh, he concluded, had it right: these were most likely formed by one of the great sloths.

During the 1880s, Davidson and Harkness became antagonistic toward one another,[38] and it is easy to believe that the evening of August 6, 1883, might have had something to do with it. Harkness was on the losing side in the Carson City affair. When, in 1885, the distinguished anthropologist E. B. Tylor reviewed the status of American anthropology, including evidence for a deep human antiquity in the Americas, he simply dismissed "the sandaled human giants, whose foot-prints, twenty inches long, are declared to have been found with the foot-prints of mammoths, among whom they walked, at Carson, Nevada."[39] Although, in 1890, Hubert Howe Bancroft still referred to the "stone quarry marked by the footprints of primeval man" at Carson City,[40] no geologist or paleontologist still accepted them as such.

No matter what animal had provided the contentious footprints, the state of Nevada was understandably proud of them, proud enough to exhibit them at the famous World's Columbian Exposition, held in Chicago in 1893. Along with award-winning honey, flour, apples, pears, butter, cheese, decorated china, and a broad variety of minerals, the state's contribution to this huge World's Fair included "the supposed human footprints and fossils from the quarry at the Nevada State Prison":

> In order to illustrate the exhibit, the platform raised six inches from the floor was covered with canvas, upon which was reproduced the tracks of the supposed man, elephant, horse, tantalus (or wading bird) and saber-tooth tiger. This reproduction was from the reports of Professors Harkness, Le Conte and Davidson, with surveys and maps by Drayton Gibbs [*sic*].... The entire exhibit was surrounded by a railing three feet high, from which were suspended large bromide pictures that fully illustrated the position of the tracks, tunnels, and height of the surrounding walls. Small maps illustrating the course and number of each series were on the west wall, and quotations from eminent scientists were freely posted, giving their views in full regarding this interesting exhibit.[41]

The exhibit was a success, winning its own award and becoming coveted by museums. The $285.80 that the state had provided for this display had been well spent.[42]

But what animal had created the footprints that had caused the ruckus? Akin to an earlier suggestion that the footprints were very recent[43] was Mark

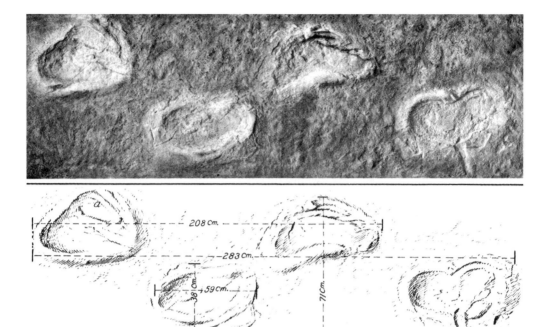

FIGURE 1.6. Chester Stock's illustration of casts of *Mylodon* footprints from the Nevada State Prison quarry (from Stock 1925).

Twain's suggestion in 1884 that the footprints had been created by the members of the first Nevada Territorial Legislature in their rush to the nearest saloon.[44] Geographically, Twain had it right. The first Territorial Legislature met in 1861, in the buildings that were to be transformed into the Nevada State Prison the following year.[45]

It turns out, though, that Le Conte's comparison of the footprints to those of the giant ground sloth *Mylodon* was correct. In 1917, paleontologist Chester Stock published a detailed study of the structure of the foot of the extinct ground sloth *Mylodon harlani*. This analysis led him to agree with Le Conte and Marsh, that the supposedly human footprints from the Nevada State Prison were most probably created by this animal, though he could not rule out the possibility that another sloth had been involved.[46]

Stock felt even more certain of the *Mylodon* identification in 1920, when he reported that he had visited the prison and discovered that fossil bones found at the site belonged to this creature.[47] In 1925, he went one step further and, on the basis of casts that he had made of some of the footprints (fig. 1.6), announced not only that the tracks matched the size of a *Mylodon* foot but also that he had detected traces of a sloth claw in some of them.[48] He summarized all of this in a popular article in 1936, dismissing all claims for a human origin of the once-controversial footprints, "either in the fossiliferous or legislative sense,"[49] and simply declaring that they had been made by the giant Ice Age ground sloth *Mylodon*.

No one disagrees, but there are three important codas. Recently, the world's leading expert on extinct sloths, Greg McDonald, examined existing casts of the sloth prints.[50] He did not do this to verify that they had been made by *Paramylodon*, since Stock had made this quite clear. Instead, he did it to combine our considerable knowledge of the anatomy of extinct American sloths with what the sloth tracks themselves show. In particular, he wanted to address whether the prison tracks provide evidence that this giant sloth could have walked on its hind feet. These tracks, he concluded, do not support this possibility.

The second coda involves the age of the tracks. It would be nice to say that we know how old they are, but we do not. There are no dates for them, no convincing estimates of their age, and no detailed geological study of the sandstone that covered them. The thickness of that sandstone suggests that they may be very old indeed—100,000 years old, or more, would not surprise me. Without a thorough analysis of the Nevada State Prison's geology, however, it is impossible to know.

The third coda involves the fate of the Carson tracks. Only a few of the prints remain visible today. Some have been destroyed and others became inaccessible a few years ago, when access to them was sealed off.[51] Nonetheless, the tracks remain in the public eye.[52] The Nevada State Prison was decommissioned in 2012, but the license plate factory within remains active, the execution chamber on standby status.[53]

In response to concerns that the prison would be torn down, and its architectural and associated history lost, Assembly Bill 356 was introduced in the Nevada State Legislature on March 18, 2013:

> WHEREAS, The Nevada State Prison located on East Fifth Street in Carson City was originally built in 1860 by pioneer Abraham Curry as the Warm Springs Hotel; and

> WHEREAS, The landmark prison was established in 1862 by the Nevada Territorial Legislature and administered by Abraham Curry at the site of the Warm Springs Hotel, and thus represents the first executive agency created in the State of Nevada; and

> WHEREAS, The sandstone quarry on the site contributed to the construction of state, city and private buildings during the early history

of the State of Nevada and Carson City, including the Capitol Building, the United States Mint, the Virginia and Truckee Railroad engine house and many other public and private buildings; and

WHEREAS, The lands and grounds of the prison are known to contain unique specimens of extinct species and the legend of *Homo Nevadensis*; and

WHEREAS, The first lethal gas execution chamber in the world was designed and used at the prison and is currently active; and

WHEREAS, The prison has a well-established and colorful history as a maximum security prison, replete with riots, escapes, gangs, executions, musicians, gambling and Hollywood filming, all of which provide intriguing historical value; and

WHEREAS, The history of the manufacture of Nevada's license plates is integral to the prison; and

WHEREAS, The Nevada State Prison was decommissioned in May of 2012 and is likely to fall into a state of disrepair without continued maintenance and upkeep; and

WHEREAS, Efforts are underway to establish the historic sections of the prison as a site listed on the National Register of Historic Places; now, therefore,

THE PEOPLE OF THE STATE OF NEVADA, REPRESENTED IN SENATE AND ASSEMBLY, DO ENACT AS FOLLOWS:

Section 1. The Nevada Legislature hereby finds and declares:

1. That the Nevada State Prison is an integral part of the history of the State of Nevada, particularly with respect to Carson City and the early development of this State, and should be preserved as a historic place.

2. That Carson City, any nonprofit organization, and any other interested stakeholder are encouraged to work cooperatively with the Department of Corrections, the Office of Historic Preservation of the State Department of Conservation and Natural Resources and the State Land Registrar to:

(a) Develop recommendations for the preservation, development and use of the Nevada State Prison as a historical, educational and scientific resource for the State of Nevada; and

(b) Present the recommendations to the Nevada Legislature along with any recommendations for legislation that may be necessary to fully implement the recommendations.

Sec. 2. This act becomes effective on July 1, 2013.[54]

Harvey Harkness, who had introduced the name *Homo nevadensis* 130 years earlier, would have been pleased at all but the use of the word "legend." The bill was signed into law by Governor Brian Sandoval on May 21, 2013.

2 The Great Basin Now and Then

THE GREAT BASIN NOW

The Rorschach test consists of a standardized set of ten carefully selected ink-blots that mental health experts use to assess the personality characteristics of their patients. A subject can look at one of these things and see something quite normal—a butterfly, for instance, or people. Those are the ones who get to go home. Then there are people who look at a particular inkblot and see something that immediately indicates that they are, in psychiatric terms, wacko. If you are near a computer, you can find these things online. Take a look at the tenth card. If you blurt out that "those are happy members of the Seattle Mariners baseball team celebrating after winning the World Series," then you are seriously delusional and should probably be locked up for the foreseeable future or, what seems to amount to the same thing, be elected to the US Congress.

The Great Basin is something of a Rorschach test for scientists. Different kinds of scientists look at it and see very different kinds of things.[1] There are geologists who are interested in the structure and evolution of the earth's surface and define what is known as the "Physiographic Great Basin," a region of massive mountain ranges that generally trend north–south and that separate broad valleys from one another. There are anthropologists who recognize that part of arid western North America is occupied by Native peoples who share certain distinctive cultural characteristics, often including the languages they speak. These similarities lead them to define a "Cultural Great Basin." Then there are botanists who see a region marked by a fairly distinctive set of plants, leading them to define a "Floristic Great Basin." Finally, there are scientists who define a "Hydrographic Great Basin," a region marked by the fact that none of the precipitation that falls within it drains outside it. Some

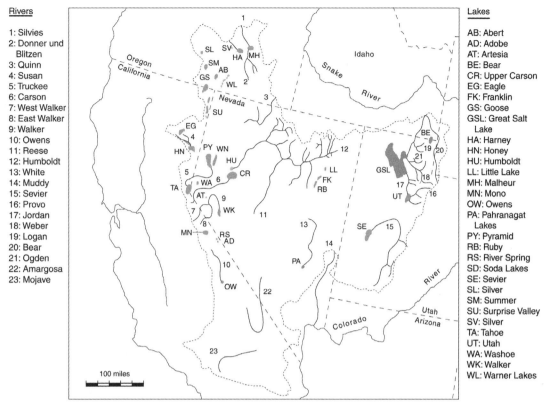

FIGURE 2.1. The hydrographic Great Basin.

of these scientists may be wacko, but if they are, it is not because these are the answers they give to the Great Basin Rorschach test. All of these answers are appropriate.

The Great Basin as a cultural area does not concern us here since it pertains to the contemporary peoples of the region, not to their ancient ancestors. Aspects of the physiography of the Great Basin are important to understanding the history of the Great Basin's extinct Ice Age, but these I will talk about as needed. The two Great Basins that are most important for our purposes are the hydrographic and floristic versions, and these are the ones I discuss in some detail here.

The Hydrographic Great Basin

The hydrographic Great Basin is most easily defined as the largest area of internal drainage in North America, a region of 200,000 square miles marked by the fact that none of the water that falls within it ends up flowing to any ocean. You can see this by picking any river within this Great Basin and following it to its end (fig. 2.1). The Humboldt River ends up in the Humboldt Sink, which

drains nowhere. The Bear River ends up in the Great Salt Lake, which drains
nowhere. The Walker River ends up in Walker Lake, which drains nowhere,
and so on.

The hydrographic Great Basin was the first of the Great Basins to be defined. This accomplishment belongs to John C. Frémont who, in 1843–1844,
led an extraordinary expedition from what is now Kansas City, Missouri, to
Mexican California and back. Along the way, he traversed the eastern and
western edges of the Great Basin—Great Salt Lake, for instance, on the east,
and Summer and Pyramid Lakes (he named both of them) on the west. Combining what he had seen with what he had been told by others, in particular
by the remarkable backwoodsman Joseph Walker (Nevada's Walker Lake and
California's Walker Pass are named after him), Frémont realized the hydrographic truth about the huge region he had skirted. In 1845, in the famous volume he produced about this expedition, he shared that truth with the world
at large:

> The Great Basin—a term which I apply to the intermediate region be-
> tween the Rocky mountains and the next range [the Sierra Nevada],
> containing many lakes, with their own system of rivers and creeks, (of
> which the Great Salt is the principal,) and which have no connexion
> with the ocean, or the great rivers which flow into it.[2]

"The structure of the country," Frémont noted,

> would require this formation of interior lakes; for the waters which
> would collect between the Rocky mountains and the Sierra Nevada,
> not being able to cross this formidable barrier, nor to get to the Co-
> lumbia or the Colorado, must necessarily collect into reservoirs, each
> of which would have its little system of streams and rivers to supply it.[3]

Even those who did not read the whole book could not miss this remarkable geographic discovery. Frémont's 1845 *Report* included a map that is renowned because of its detail, because it illustrated only what had actually
been seen and surveyed, and because it immediately became the correctable
source for many later maps of the American West. The United States clearly
recognized its importance at the time, laying out $9,851.30 to have it, and the
other illustrations in the report, lithographed.[4] In today's dollars, this was an
expenditure of $307,850.[5] It was also an expenditure that was easily justified
by national expansionist interests.

The 1845 Frémont map was prepared by Charles Preuss, the expedition's

dyspeptic but immensely talented cartographer.[6] On this map, the interior of
the Great Basin was left blank except for a long arc of type that again defines

THE GREAT BASIN: diameter 11° of latitude, 10° of longitude: elevation
above the sea between 4 and 5000 feet: surrounded by lofty mountains: contents almost unknown, but believed to be filled with rivers
and lakes that have no communication with the sea, deserts and oases
which have never been explored, and savage tribes, which no traveler
has seen or described.

Three years later, Frémont and Preuss provided another map of the Great
Basin, this time incorporating the work of others along with their own.[7] Many
of the blank spaces that marked the earlier map were now gone. In a very
real sense, this was unfortunate, since the 1848 map shows a "dividing range"
running east–west across south-central Nevada at about 38° north latitude—
that is, at about the latitude of San Francisco. That range, Frémont thought,
marked the southern boundary of the hydrographic Great Basin. A full 20,000
copies of the first edition of this map were printed,[8] and many emigrants to
California carried it with them. A good number of those assumed that they
could follow the edge of the dividing range across Nevada and that this route
would be well watered. The range, however, did not exist, causing significant
misery for those who came this way, including those emigrants who, following the Frémont map, ended up in what they ultimately called Death Valley.[9]
Frémont named many places in the Great Basin, but Death Valley was called
that because of him.

Unlike any of the other Great Basins, Frémont's hydrographic Great Basin
has the advantage of discrete, easily recognizable boundaries. To define the
edges, all you have to do is determine which way the water is draining. If it
heads inward, it is within this version of the Great Basin. If it drains exter-
nally—to the Pacific or the Gulf of California—it is not. So defined, the hydrographic Great Basin extends from the crest of the Rocky Mountains on
the east to the crest of the Sierra Nevada and Cascade Range on the west, and
from the southern edge of the Columbia River drainage system on the north
to the northern edge of the Colorado River drainage system on the south.

Although not the defining characteristic, massive north–south-trending
mountain ranges mark the hydrographic Great Basin (fig. 2.2). Thirty-three
of these have peaks that rise above 10,000 feet in elevation, with the tallest
of them all, White Mountain Peak in the White Mountains on the Nevada-
California border, reaching 14,246 feet. These ranges separate long, broad

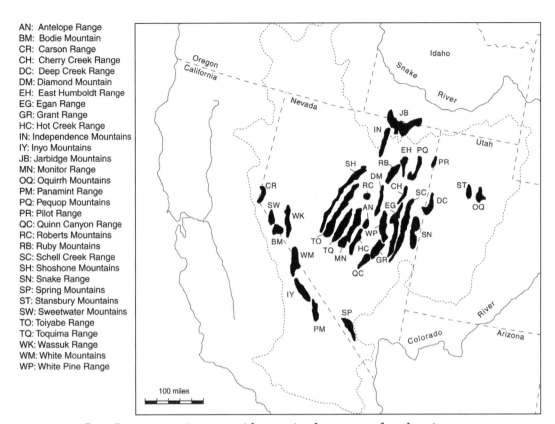

AN: Antelope Range
BM: Bodie Mountain
CR: Carson Range
CH: Cherry Creek Range
DC: Deep Creek Range
DM: Diamond Mountain
EH: East Humboldt Range
EG: Egan Range
GR: Grant Range
HC: Hot Creek Range
IN: Independence Mountains
IY: Inyo Mountains
JB: Jarbidge Mountains
MN: Monitor Range
OQ: Oquirrh Mountains
PM: Panamint Range
PQ: Pequop Mountains
PR: Pilot Range
QC: Quinn Canyon Range
RC: Roberts Mountains
RB: Ruby Mountains
SC: Schell Creek Range
SH: Shoshone Mountains
SN: Snake Range
SP: Spring Mountains
ST: Stansbury Mountains
SW: Sweetwater Mountains
TO: Toiyabe Range
TQ: Toquima Range
WK: Wassuk Range
WM: White Mountains
WP: White Pine Range

100 miles

FIGURE 2.2. Great Basin mountain ranges with summits above 10,000 feet elevation.

valleys. The elevations of the valleys themselves are often high. With the exception of the southernmost hydrographic Great Basin, those elevations generally fall between 4,000 and 6,000 feet. In the southernmost Great Basin, valley-bottom elevations fall beneath 3,000 feet, with Death Valley, at −282 feet, setting the Great Basin's low-elevation record. Even though valley-bottom elevations in the hydrographic Great Basin are high, however, the great heights of the adjacent mountains mean that the elevational differences between valley bottoms and flanking mountaintops can be huge, frequently exceeding a mile.[10]

The precipitation that falls in the mountains flows into streams that then flow into the adjoining valleys. In the wettest parts of the Great Basin, those streams may flow into the lakes that are found in some valley bottoms—Great Salt Lake and Utah Lake in Utah, for instance, or Malheur Lake in south-central Oregon, or Pyramid Lake in northwestern Nevada. The water in those lakes, however, has nowhere to go, unless it seeps into the ground or evaporates. This very distinctive characteristic of the hydrographic Great Basin had a dramatic impact here during the late Ice Age, an impact that may help us

FIGURE 2.3. The floristic Great Basin (after Cronquist et al. 1972).

understand the nature of the extinct mammal communities of the late Ice Age
in this region (chapter 6).

The Floristic Great Basin

Because the hydrographic Great Basin has discrete edges, there is no question
as to where it stops and starts. This is not the case with the floristic Great
Basin. All agree that it is defined as a region that contains a fairly distinctive
set of plants, but not all botanists agree as to what that distinctive set is, or
where it starts and stops. As a result, different versions of the floristic Great
Basin exist.

The version I present here (fig. 2.3) is the one used by the definitive *Intermountain Flora*.[11] The boundary on this map encompasses a Great Basin

FIGURE 2.4. The sagebrush desert, Fort Rock Basin, Oregon. Prior to the introduction of livestock, tall grasses would have been common as well.

whose lower elevations are dominated by saltbush (*Atriplex*), sagebrush (*Artemisia*), and, in some areas, especially in the absence of livestock, grasses. The flanks of the surrounding mountains are in turn dominated by woodland with some combination of pinyon pine and juniper. In much of the Great Basin, the pinyon is singleleaf pinyon (*Pinus monophylla*); in the far eastern parts of the Great Basin, it is twoneedle pinyon (*Pinus edulis*). Through much of the floristic Great Basin, the common species of juniper is Utah juniper (*Juniperus osteosperma*). In the far northern and far western Great Basin, however, western juniper (*Juniperus occidentalis*) takes over. In fact, in the far northern Great Basin, there is no pinyon-juniper woodland at all, since pinyon does not make it north of the Humboldt River.

I was careful to say that it is the flanks of the mountains that are covered with pinyon-juniper woodland in much of the Great Basin, since this woodland generally occurs between 5,000 and 8,000 feet in elevation, though the exact elevations vary from place to place. Below 5,000 feet or so, the pinyon-juniper is gone, the vegetation now dominated by a variety of shrubs and grasses. In places that are better watered, the shrub is routinely big sagebrush (*Artemisia tridentata*); in places that are drier and where the substrate is full of salts, species of saltbush (*Atriplex*), among other things, take over (fig. 2.4).

Get high enough on those mountains and a very different set of conifers appears. Although there are a number of these, the more common ones include white fir (*Abies concolor*), subalpine fir (*Abies lasiocarpa*), limber pine

FIGURE 2.5. Spruce Mountain, northeastern Nevada, showing the classic plant zones of the Great Basin (Billings 1951). The valley bottom is dominated by big sagebrush (*Artemisia tridentata*), and the flanks of the mountains are covered with singleleaf pinyon (*Pinus monophylla*) and Utah juniper (*Juniperus osteosperma*). The uppermost trees are primarily subalpine conifers—subalpine fir (*Abies lasiocarpa*), limber pine (*Pinus flexilis*), and bristlecone pine (*Pinus longaeva*). A small patch of alpine tundra is visible at the top of the peak on the right.

FIGURE 2.6. The alpine tundra zone at the top of the Toquima Range, central Nevada, at an elevation of about 10,000 feet; the trees in the distance are limber pines (*Pinus flexilis*).

(*Pinus flexilis*), whitebark pine (*Pinus albicaulis*), bristlecone pine (*Pinus longaeva*), and Engelmann spruce (*Picea engelmannii*). These, the subalpine conifers, are often found at elevations ranging from 9,000 to 11,000 feet, though they can be found at much lower elevations in certain settings (fig. 2.5). On mountains that are tall enough, the requirements of even these trees are exceeded, and they are replaced by true alpine, and treeless, vegetation (fig. 2.6).

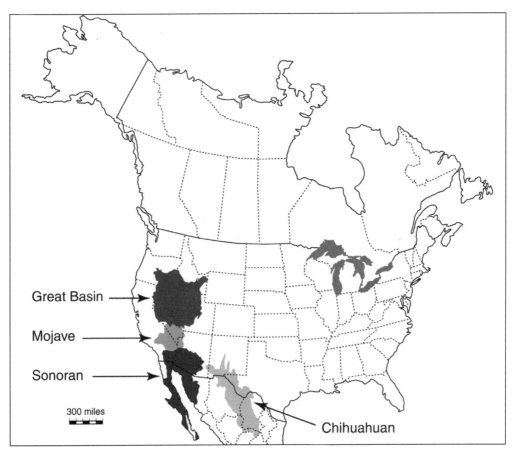

FIGURE 2.7. The four botanically defined North American deserts (after Holmgren, Norris, and Betancourt 2007; Grayson 2011).

When defined in this way, the Great Basin becomes one of North America's four floristic deserts (fig. 2.7). The floristic Great Basin's nearest neighbor in this realm is the Mojave Desert, located just to the south and occupying the southernmost part of the hydrographic Great Basin. The boundary between the two is fairly neatly drawn. The floristic Great Basin stops, and the Mojave Desert starts, where creosote bush (*Larrea tridentata*) appears (fig. 2.8). I will have more to say about this plant in chapter 6, but this is a species that does not like prolonged episodes of temperatures that fall below freezing.

In fact, the floristic Great Basin is a cool desert, with cold winters and relatively cool summers. The deserts to the south—the Mojave, Sonoran, and Chihuahuan Deserts—are warm deserts, with warm winters and relatively hot summers.[12] All the warm deserts are loaded with creosote bush, which the floristic Great Basin lacks, and all have a wide variety of other plants that are also not to be found in their northern neighbor—the famous Joshua tree (*Yucca brevifolia*) in the Mojave Desert (fig. 2.9), for instance, and the equally

FIGURE 2.8. Creosote bush (*Larrea tridentata*) in the Mojave Desert near Barstow, California.

FIGURE 2.9. Joshua tree (*Yucca brevifolia*), Rainbow Basin, Mojave Desert, California.

famous saguaro (*Carnegiea gigantea*) in the Sonoran. Because these plants seem to be telling us something about the Ice Age mammals of the arid West, I return to them in chapter 6.

THE GREAT BASIN THEN

The Ice Age

The term "Ice Age" is an informal replacement for the "Pleistocene Epoch," the official name for the period between 2.6 million and 10,000 years ago, a time that was characterized by the episodic advance and retreat of glaciers over significant parts of the earth's surface. There were over 20 of these major advances and retreats, but only the last of these glacial cycles is of interest to us here since that is when the events at the heart of this book played out. By definition, the Ice Age ended 10,000 years ago and we entered a new epoch, called the Holocene. We remain in the Holocene today, but even given the current concern over global warming, there isn't much reason to doubt that we are actually in the midst of one of the interglacial periods that came and went so many times during the Pleistocene.[13]

In North America, this latest glaciation is known as the Wisconsin, named after the location of the deposits that originally helped define it. Dating to between about 115,000 and 10,000 years ago, the Wisconsin glaciation saw the growth of huge ice masses in northern North America (fig. 2.10).[14] In the west, the Cordilleran Ice Sheet grew primarily from the coalescence of glaciers that formed on mountaintops and then kept growing and growing until they ran into and merged with one another. At its maximum, this process formed a sheet of ice that reached a thickness of at least 6,500 feet and that extended south of where Seattle now sits. Joining it on the east, and extending from the eastern flanks of the Rocky Mountains to the northern Atlantic coast, the Laurentide Ice Sheet reached a thickness of over two miles. In the far north, the Innuitian Ice Sheet formed over the islands of the Canadian Arctic and then merged with the Laurentide Ice Sheet to the south and the Greenland Ice Sheet to the east.[15]

At the same time as this was happening, many isolated mountains to the south grew their own alpine glaciers. In the Great Basin, 25 separate mountains and mountain ranges were glaciated, including Steens Mountain in southeastern Oregon, the Ruby Mountains in northwestern Nevada (fig. 2.11), the Toquima and Toiyabe Ranges in central Nevada, and the White Mountains on the California-Nevada border.[16]

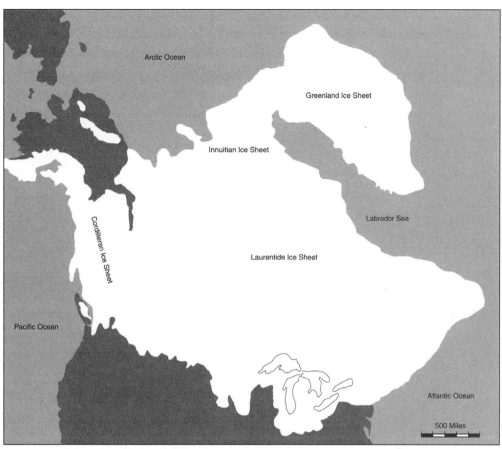

FIGURE 2.10. The Pleistocene glaciation of northern North America (after Booth 1987; Dyke et al. 2002).

The Bering Land Bridge

The water that was locked up in all these glaciers ultimately came from the oceans. When the glaciers melted, it went back to them. Today, there is enough water locked up in the form of ice in Greenland and Antarctica that, were it all to melt, sea level would rise some 210 feet. That is why there is real concern over sea level rise in a time of global warming and why I would think twice about buying shoreline property.[17] Go back to the last glacial maximum, between about 22,000 and 18,000 years ago, and the situation reverses itself. At that time, there was ice over two miles thick sitting on top of Canada and sea levels were dramatically lower than they are now.

They were low enough that, at the time of the last maximum glacial advance, the Atlantic coastline was some 60 miles east of where it is today. Along the Pacific, where the continental shelf lies closer to what is now the mainland,

FIGURE 2.11. Once-glaciated Lamoille Canyon, Ruby Mountains, Nevada. Had this valley not been glaciated, the rocks on the sides would not be so smooth, and the cross section of the valley would be shaped like a *V* rather than a *U*.

the coastline was about 30 miles west of where it is now. That's why you read newspaper reports about the remains of such animals as mammoths and mastodons dredged up from the ocean floor. That was land back then, and those are the remains of animals that were walking around on that land.

Estimates as to how much sea level dropped during the Last Glacial Maximum vary, but they don't vary all that much. Between about 22,000 and 18,000 years ago, sea levels were some 425 feet lower than they are today. Sea level dropped so much that a flat shelf of land emerged to connect far eastern Siberia with far western Alaska. Because this connection occurred in the area of the Bering Strait and Bering Sea, it is known as the Bering Land Bridge (fig. 2.12). The huge region that was connected in this way is known as Beringia, with western Beringia occupying the Siberian side and eastern Beringia, the American one.

It does not take anywhere near a 425-foot drop in sea level for this to happen. Drop sea level 160 feet and the two continents are connected. Drop it 400 feet, and they are connected by an expanse of land that runs 1,000 miles from north to south. Mammoths and horses could, and did, walk from one side to

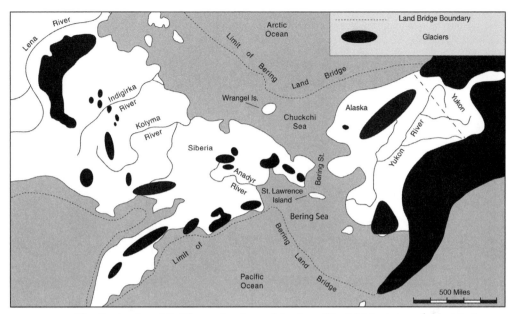

FIGURE 2.12. The Bering Land Bridge at the time of the Last Glacial Maximum (after Morlan 1987; Elias and Crocker 2008).

the other. So did muskoxen, moose, and a wide variety of other mammals, including people (see chapters 3 and 6).

The comings and goings over the Bering Land Bridge through the entire Ice Age are not well understood, but during the late Ice Age, it was open almost continuously from about 50,000 to about 11,000 years ago.[18] It has been closed to land traffic and open to fish ever since.

Whenever it was available, the Bering Land Bridge provided an important corridor for biotic interchange between Eurasia and North America. Had it never existed, horses would not have been able to make their way to Eurasia from their evolutionary homeland in North America. Since, as I discuss later, horses became extinct in the Americas by about 10,000 years ago, we have the Bering Land Bridge to thank for their existence today. Imagine a Great Basin without cowboys on horseback. Without the Bering Land Bridge, that's what would have happened.

Ice Age Lakes in the Great Basin

Just as Frémont said, the defining characteristic of the hydrographic Great Basin is that the waters that fall here must "necessarily collect into reservoirs," that it is a region "filled with rivers and lakes that have no communication with the sea." Since this is the case, the more rain that falls or the less water that evaporates, the larger those lakes must become.

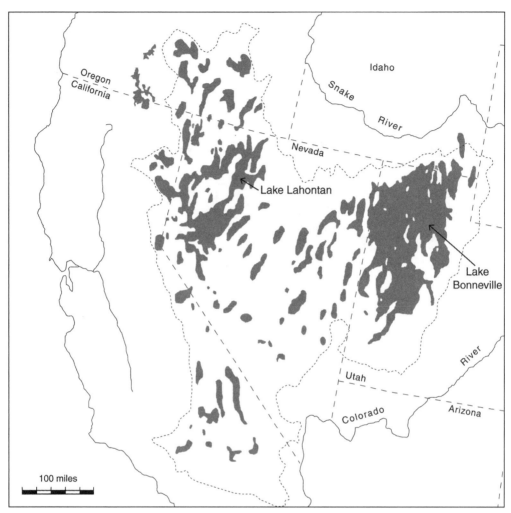

FIGURE 2.13. The Pleistocene lakes of the Great Basin (after Grayson 2011). Lakes Bonneville and Lahontan were the two largest of these lakes.

Pleistocene lakes whose levels were higher because of altered levels of precipitation or evaporation are referred to as "pluvial lakes." Since late Ice Age temperatures in the Great Basin were lower than those of today and precipitation higher, and since those lakes had no outlets, it follows that the Great Basin must have seen the formation of substantial pluvial lakes.

Not only did they form, but there were so many of them, and some of them were so big, that the hydrographic Great Basin is famous for its Pleistocene lake history. Today, there are about 45 valley-bottom lakes in this region. They cover about 2.5 million acres, though many of them have the annoying habit of disappearing entirely during dry times. During the late Ice Age, things were entirely different. Then, there were about 80 pluvial lakes that covered

FIGURE 2.14. Pleistocene terraces carved on the sides of Homestead Knoll, Bonneville Basin, Utah. The terrace marked by the white stripe on the left side of the photograph dates to between about 14,500 and 12,600 years ago (Madsen 2000b).

about 28 million acres of Great Basin lowlands (fig. 2.13). The largest of these, Pleistocene Lake Bonneville, reached its maximum size between 15,000 and 14,500 years ago, flooding 19,800 square miles of western Utah and parts of adjacent Idaho and Nevada to a depth of about 1,200 feet. It also carved distinctive terraces on mountainsides that are visible from great distances (fig. 2.14) and left behind a huge, flat lakebed that, in the form of the Bonneville Salt Flats, has turned out to be perfect for setting land speed records (fig. 2.15). It would have gotten even larger and deeper, except that it got so big that it started to overflow into the Snake River drainage, stabilizing its size. Then, it cut through its overflow point, unleashing one of the most massive floods in earth history, a flood that quickly reached the Columbia River and, through it, the Pacific Ocean.[19]

Lake Bonneville was gone by about 10,000 years ago; land speed records are now set on the western edge of the basin it once occupied. Similar stories played out throughout the hydrographic Great Basin, with substantial lakes in existence during the late Ice Age, only to be greatly diminished as the Pleistocene came to an end. At its maximum, for example, the second largest Great Basin Lake, Lake Lahontan, covered 8,300 square miles of western Nevada and adjacent eastern California. That maximum was reached between about 14,500 and 13,000 years ago, but the lake itself existed, in one form or another, until about 10,000 years ago.[20]

FIGURE 2.15. The Bonneville Salt Flats from the base of Homestead Knoll, Bonneville Basin, Utah. The Bonneville Salt Flats Speedway is in the far distance.

We can be sure that people saw the later stages of at least some of these lakes, since they were in the Great Basin by 12,400 years ago (chapter 6). We can also be certain that such now-extinct animals as mammoths, horses, and camels knew these lakes, because we have radiocarbon dates that place the mammals and lakes in the Great Basin at the same time (chapter 4), and because the bones of these animals are often found associated with deposits laid down by the lakes when they were active.[21]

The Younger Dryas

As the Ice Age was coming to an end, some pluvial lakes in the Great Basin interrupted their general decline by increasing in size and depth. That increase was short-lived—a few centuries—and it does not appear that all Great Basin lakes grew at this time. We do know, though, that it correlated with a brief but very pronounced climatic interval known as the Younger Dryas.

That term refers to a glacial cold snap that occurred between 11,000 and 10,000 years ago[22] and that interrupted the general warming that was occurring at the very end of the Ice Age, only to end abruptly.[23] In northwestern Europe, this cold snap saw forests replaced by Arctic grasses, herbs, and shrubs. One of these herbaceous plants was mountain avens, which belongs to the genus *Dryas* and gives the Younger Dryas half its name (the other half comes from the fact that there is also an Older and an Oldest Dryas). During this interval, temperatures declined dramatically and glaciers expanded in the Northern Hemisphere and elsewhere. The Younger Dryas also saw a significant drop in atmospheric methane levels. Since huge amounts of methane are produced by the microbial decomposition of organic carbon in wetlands,[24] the decrease in methane in the Younger Dryas suggests that the area covered by wetlands decreased substantially or that those wetlands became far less productive, or both. It was during this interval that some Great Basin lakes, including the biggest, increased in size, only to shrink once the Younger Dryas ended.

No one claims that the extinction of North America's Ice Age mammals was caused by this cold snap. However, it has recently been argued that the Younger Dryas and the extinctions were both caused by the same event—a fragmented comet that hit the Laurentide Ice Sheet or exploded in the air above it, causing massive parts of the ice sheet to melt and the animals to die. This explanation, the "Younger Dryas Impact Hypothesis," will be described in detail in chapter 7.

Learning about the Late Ice Age Vegetation of the Great Basin

As the Ice Age came to an end, the Great Basin lost an impressive variety of large herbivores. Giant ground sloths, horses, camels, peccaries, muskoxen, and mammoths, among others, became extinct. With them went the large carnivores that depended on them—the sabertooth cat, for instance, and the American lion (there were lions in the Great Basin during the Ice Age). Before we meet these animals in detail, we need to know something about the vegetation that supported them.

Two prime approaches have been used to learn about the late Ice Age vegetation of the Great Basin, each focused on a different set of plant remains. The first approach takes advantage of the fact that many kinds of plants count on the wind to disperse their pollen. That pollen can be carried far and wide but ultimately falls to the earth. If it lands in the right place, it helps create the next generation of the plants that produced it. Pollen that misses its target, however, can end up in sediments that allow it to be preserved for extraordinary amounts of time. In the Great Basin, the sediments that have been studied for their ancient pollen content have come from caves and rockshelters, from marshes and springs, and from the bottom of lakes. These sediments can be dated and the pollen from them identified. The end result can be a pollen-based vegetation history extending as far back in time as pollen preservation will allow. The Great Basin record for such preservation is set by Utah's Bonneville Basin, where cores extracted for commercial reasons provided a pollen-based vegetational sequence extending back 13.5 million years.[25]

There are some serious complexities to pollen-based reconstructions of vegetation history. First, there are the issues that are always involved in studying change through time, whether we are interested in the history of people or of plants. The deposits from which the pollen comes have to be well layered, or stratified, since the goal of these studies is to analyze changing vegetation through time. These stratified deposits also have to contain material that can be accurately dated so we can know the age of each layer.

Second, there are things that are particular to the study of windblown pollen. Because pollen is carried by the wind, the plant histories that result from this work pertain to large regions, not simply to the plants that happened to be growing around the place where the pollen was found. That Bonneville Basin sequence that extended back some 13.5 million years? It provided pollen that came raining down from the air above as well as pollen that had been introduced by streams flowing into the basin from the surrounding mountains. Even pollen drifting down from the atmosphere can be carried long distances. The pollen of jointfir, or Mormon tea (*Ephedra*), for instance, has been found in sediments in the Great Lakes region, carried there at least 700 miles by the wind.[26]

The bigger problem, though, is that the plants that provided particular pollen grains can often be identified only at a very general level. Want to know the history of Great Basin wildrye (*Leymus cinereus*)? Pollen is not going to help, because most grass pollen can be identified only to the family level. Want to know the history of fourwing saltbush (*Atriplex canescens*)? You're out of luck, because its pollen cannot be distinguished from that of a wide variety of other plants belonging to the same family. Routinely, pollen will not tell us the species, and often not even the genus, of the plant that produced it.

In places like New England and the Great Lakes region that are rich in lakes, the analysis of pollen taken from lake sediments has taught us an enormous amount about the vegetation of the late Ice Age. While the Great Basin has its share of valley-bottom lakes (fig. 2.1), most of them have gone dry from time to time. When that happens, the exposed sediments and the pollen they contain are removed by the wind. The longer and the more often they are exposed, the greater the amount of the ancient pollen record that is destroyed. As a result, while some Great Basin lakes have provided superb pollen-based vegetation histories, including Great Salt Lake, Owens Lake,[27] and Pyramid Lake,[28] many have not and cannot be expected to. That puts a significant crimp on our ability to use lakes in the Great Basin to reconstruct valley-bottom vegetation. To some extent, we can make up for this by using pollen from deposits associated with springs and marshes,[29] but undisturbed deposits from these settings are hard to come by.

The other prime approach to learning about past vegetation in the Great Basin is built on the back of one of the arid West's common rodents, the woodrat (*Neotoma*). These little guys—big ones are about the size of a squir-rel—have the helpful habit of wandering across the landscape, picking up things that interest them, and bringing them home. These things include food items—seeds, twigs, and leaves, or "plant macrofossils"—as well as anything they find interesting that they can carry in their mouths. Because of this habit, they are often called "packrats," though this is not their officially sanctioned name. If, during their meanderings across the landscape, they find something they want more than what they are already carting around, they'll drop the less interesting item for the more interesting one—a seed for a sock or a twig for a toothbrush, to use examples from my own experience. Because of that behavior, they are sometimes called "trade rats."

Whatever we call them, the items woodrats bring back get deposited where they live. There, they often get peed and pooped on. If this accumu-lation of stuff is in a place sheltered from the rain, the pee crystallizes into something with the consistency of rock candy. This crystallized woodrat pee, called amberrat, protects everything it covers, including plant remains. Over time, these accumulations—called packrat middens—can become huge, taller than a person stands, though they are routinely smaller than this. They can also endure for tens of thousands of years—packrat middens are known that are over 50,000 years old.

This means that if you can find an old packrat midden, you can take it apart, remove the plant remains from it, identify them, and date them using the radiocarbon method (chapter 4). Unlike the situation with pollen, the large pieces of plants that woodrats bring home can often be identified to the species level. Because woodrats do not roam far from their homes, you can be certain that these plants came from nearby, again unlike the situation with pollen. Because you can date the plant fragments directly, you can build a time-sensitive, species-level plant history for the area near the midden.

The analysis of ancient packrat middens in the Great Basin comes with its own set of complexities. A lake that doesn't dry up accumulates pollen contin-uously. Woodrats, however, come and go. As a result, their middens provide detailed but discontinuous snapshots of local vegetation. In addition, the only packrat middens that survive for lengthy periods are those that are built in such sheltered places as caves, rockshelters, and the nooks and crannies of cliff faces. Because the little critters don't wander far from home, the vegeta-tion record they accumulate reflects the vegetation of those settings—rock outcrops in valley bottoms, for instance, or caves cut into rock exposures. Fi-nally, while lakes may accumulate thousands upon thousands of pollen grains, the plant samples retrieved by woodrats are far smaller. That makes it harder to build vegetation histories on the basis of changing abundances of plants through time.[30]

As a result, it can be somewhat frustrating to build a detailed, species-level vegetation history for a Great Basin valley bottom, even if that valley has supported a lake for, say, the past 30,000 years. The pollen record from that lake may be continuous, but it will rarely allow the species that contributed pollen to it to be identified. The rock outcrops in that valley may have ancient packrat middens, the contents of which will allow you to identify the species of plants collected by the animals that built them. These plants, however, will primarily reflect the vegetation of the rock outcrop, not of the valley bottom itself, and the record is not likely to be continuous.

There are other sources of information on vegetation history in the Great Basin. Packrat middens also contain pollen, lake sediments also contain plant macrofossils, and some archaeological sites are rich in both pollen and larger plant parts.[31] Our prime scientific access to the late Ice Age vegetation of the Great Basin, however, comes from pollen extracted from lakes, marshes, and springs, and from larger plant pieces retrieved from ancient packrat middens.

The Late Ice Age Vegetation of the Floristic Great Basin

If you were a skilled field botanist who happened to be dropped into the floristic Great Basin 15,000 or 20,000 years ago, you would be struck by three things. First, you would be able to recognize all the species of plants you saw from your knowledge of modern Great Basin plants. This would not have been true if your specialty happened to be birds or mammals. Second, many species wouldn't seem to be where they belonged. Third, one of the Great Basin's most widespread trees—pinyon pine—wouldn't seem to be there at all. You would be right all three times.

I'll start with pinyon pine. Some 17 million acres of the Great Basin are now covered with pinyon-juniper woodland.[32] This is a surprisingly recent development. There were pinyon pines in the Mojave and Sonoran Deserts during the late Ice Age, but none at all in the floristic Great Basin until long after the Pleistocene ended. The earliest record for pinyon in this huge region comes from the northwestern flanks of the White Mountains in eastern California, where a packrat midden provided singleleaf pinyon needles that date to 8,800 years ago. You have to wait far longer than this to find pinyon anywhere else in the floristic Great Basin.[33]

From the perspective of large herbivores, the absence of pinyon pine in the floristic Great Basin during the Ice Age is more likely to have been a good thing than a bad one. While pinyon seeds and foliage are eaten by large herbivores today,[34] they are not a critical food source, and pinyon-juniper woodland reduces the abundance of shrubs and herbaceous plants that make herbivores happy.[35] The fewer the pinyon trees on the landscape, the more forage there is for large mammals that eat plants.

In some parts of the floristic Great Basin, however, the absence of pinyon in the late Ice Age may not have done large herbivores that much good. That is because the mountain flanks in parts of this region—the same flanks that pinyon-juniper occupies today—were then supporting such subalpine conifers as limber pine, bristlecone pine, and Engelmann spruce. Work by paleobotanist Dave Rhode and others has shown that the understory in this woodland included a rich variety of shrubs and herbaceous plants, much as can be found in pinyon-juniper woodland today.[36] While this vegetation could have supported a wide variety of large herbivores, it would not have supported as many as it would have in the absence of the conifers. Along the western edge of the Bonneville Basin, from the Snake Range northward, these trees extended to very low elevations along the edges of mountains and continued to do so until nearly the end of the Pleistocene. They also grew in rocky outcrops on the valley bottoms themselves, surrounded by vegetation dominated by big sagebrush, grasses, and other herbaceous plants—the kind of vegetation often referred to as sagebrush steppe.[37]

Although saltbush-covered bottomlands were present in the late Ice Age floristic Great Basin, sagebrush steppe was far more common than it is today, occupying places that are now often dominated by saltbush. It is not just the pollen and plant macrofossil records that show us this. In a number of places, as the Ice Age ended, mammals that thrive in sagebrush steppe began to be replaced by those that are happier in saltbush habitats, with that replacement generally over by about 8,000 years ago.[38]

From the standpoint of the large, late Ice Age herbivores that we are going to encounter in the next chapter—like horses, camels, and extinct pronghorn—the differences between now and then in the floristic Great Basin are something of a mixed bag. As I noted, the absence of pinyon pine was likely to have been a good thing since, in theory, it would have allowed the more expansive development of vegetation dominated by such things as shrubs and grasses. However, whatever gains might have been had here would have been lost as subalpine conifers extended to very low elevations along mountain flanks. From this perspective, the high times for large mammals in the Great Basin might have occurred between about 10,000 and 8,000 years ago, when pinyon pine had yet to enter the floristic Great Basin, the subalpine conifers were retreating upward, the lakes had retreated to expose vast amounts of

FIGURE 2.16. The distribution of Joshua trees (after US Geological Survey 2013).

new territory, and it was still fairly cool and moist. By then, however, the large mammals that are the focus of this book were gone.[39]

The Late Ice Age Vegetation of the Mojave Desert

The pinyon pines that began to move into the floristic Great Basin after the Ice Age ended were making their way north from what is now the Mojave Desert. Until roughly 10,000 years ago, all but the lowest elevations of the Mojave were covered with pinyon-juniper woodland, as were significant parts of the Sonoran Desert to the south. If that doesn't sound very Mojave Desert–like, that is because it's not.

In fact, the Mojave Desert as a botanical unit simply didn't exist until well after 10,000 years ago. Some of the most characteristic plants that this desert now has to offer were either extremely rare in, or entirely absent from, this area during the late Ice Age. Creosote bush, which today marks the boundary between the Mojave Desert and the floristic Great Basin, has never been found in any of the Ice Age Mojave Desert sites that have provided ancient plant remains. Instead, this species did not begin to appear in this region until early in the Holocene, even though it is extremely widespread in the lower elevations of the Mojave Desert today. Much the same can be said for white bursage (*Ambrosia dumosa*), another widespread Mojave Desert shrub that did not become so until well into the Holocene.[40] In place of plants like creosote bush and white bursage, the shrubby vegetation that was then intermixed with pinyon-juniper woodland was marked by plants that are now far more characteristic of the floristic Great Basin to the north, including big sagebrush.

Just as in the Great Basin, the Mojave Desert's subalpine conifers moved significantly downslope during the late Ice Age and stayed down until as late as 10,000 years ago or so. Limber pine, white fir, and bristlecone pine all did this, in some cases reaching elevations more than 5,000 feet below where they now occur, and living in places where they no longer exist.

The Joshua tree is to the Mojave Desert what the crucifix is to Catholics, the Star of David is to Jews, the maple leaf is to Canadians, and losing is to the Seattle Mariners—a potent and recognizable symbol of the whole.[41] Today, this distinctive tree is discontinuously distributed within the Mojave (figs. 2.9 and 2.16), its future a matter of significant concern. It was far more widespread during the late Ice Age, found from central Arizona to eastern California and from southern Nevada into Mexico. As soon as the Pleistocene ended, Joshua trees began their retreat. By 8,000 years ago, they had reached something close to their current distributional limits. This retreat was probably caused in large part by rising temperatures, but why did they not move farther north, beyond the very southern margins of the floristic Great Basin? Paleobotanist Ken Cole and his colleagues suggest that this might be because the end of the Ice Age also saw the extinction of animals that fed on Joshua trees and by so doing helped disperse their seeds.[42] I explore this possibility in chapter 6.

WHAT IS THE LATE ICE AGE?

There is an official International Commission on Stratigraphy[43] that divides the earth's past into agreed-upon hunks, and assigns dates to those hunks. The most recent of those hunks is the Holocene Epoch of the Quaternary Period, which, as I mentioned earlier, dates from 10,000 years ago to today. The second most recent is the Pleistocene Epoch, also in the Quaternary Period, which dates from 2.6 million to 10,000 years ago. The Pleistocene ("Ice Age") is in turn divided into pieces, with the Late Pleistocene dating from 126,000 to 10,000 years ago. That is one way in which I might mean "late Ice Age."

Another is provided by the division of the deeper past into hunks called Land Mammal Ages. These follow one another in time, each marked by a characteristic set of mammals that allows it to be recognized. The most recent of them in North America is called the Rancholabrean, after the famous site of Rancho La Brea in southern California, a site we will encounter on a number of occasions in what follows. The end of the Rancholabrean is defined by the extinction of the animals that I will be discussing here—mammoths, mastodons, ground sloths, sabertooth cats, and so on. Since those extinctions were over by about 10,000 years ago (chapter 4), that is also when the Rancholabrean ends. Its beginning is defined as the time when bison first managed to cross the Bering Land Bridge from Asia and then make their way down south of glacial ice to the heart of North America. We don't actually know when that was, but it may have been as early as 240,000 years ago.[44]

Either of these would serve as a good definition for the late Ice Age, but I have neither of them in mind in what follows. The late Ice Age I am going to focus on refers to that part of the Pleistocene accessible by radiocarbon dating, the dating method that allows us to place both the extinct animals and ancient archaeological remains accurately in time (see chapter 4). Since this method can reach back to about 50,000 years ago, that is when my late Ice Age begins. Since the extinct mammals were pretty clearly gone by 10,000 years ago, that is when my late Ice Age ends. My late Ice Age begins well prior to the time there were people in the Americas and ends several thousand years after they had arrived.

3

A Zoologically
Impoverished World

With Charles Darwin, Alfred Russel Wallace introduced the concept of evolution by natural selection to the world. Nearly 140 years ago, he also expressed, better than anyone else ever has, the stunning nature of a wave of extinctions that seems to have swept across a significant part of the earth's surface during the recent past. That recent past we now call the Pleistocene.

After surveying what was then known of those extinctions in Europe and the Americas, Wallace told his many readers that

> the first, and perhaps the most startling fact brought out by our sys-
> tematic review, is the very recent and almost universal change that has
> taken place in the character of the fauna, over all the areas we have been
> considering; a change which seems to be altogether unprecedented
> in the past history of the same countries as revealed by the geological
> record. In Europe, in North America, and in South America, we have
> evidence that a very similar change occurred at about the same time. In
> all three we find, in the most recent deposits—cave-earths, peat-bogs,
> and gravels—the remains of a whole series of large animals, which have
> since become wholly extinct or only survive in far-distant lands.[1]

With certain important changes, very much the same could be said today. It is not true that the Ice Age extinctions were unprecedented, as every pre-teen interested in dinosaurs now knows. Nor is it true that these far-flung extinctions occurred all at once. But the general point remains correct, and Wallace was deeply impressed by it. In one of the most striking passages in the vast literature on Ice Age extinctions, he wrote:

> It is clear, therefore, that we are now in an altogether exceptional period
> in earth's history. We live in a zoologically impoverished world, from

which all the hugest, and fiercest, and strangest forms have recently disappeared; and it is, no doubt, a much better world for us now that they have gone. Yet it is surely a marvelous fact, and one that has hardly been sufficiently dwelt upon, this sudden dying out of so many large mammalia, not in one place only but over half the land surface of the globe.[2]

Wallace would have been even more astonished had he known the magnitude of Ice Age extinctions that had impacted Australia and, to a much lesser extent, Africa.[3] Compared to the world of the late Pleistocene, our world really is zoologically impoverished.

Wallace's comment that the extinctions created "a much better world for us now that they have gone" harkens back to the years prior to 1859. That was the year that Darwin's brilliant masterpiece *On the Origin of Species* was published. By a remarkable coincidence, 1859 was also the year when western scientists resolved a debate that had gone on for nearly 60 years.

EXTINCTION GETS REAL

Today, it is hard to believe that there was a time when western natural historians, theologians, philosophers, and other intellectuals believed that extinction could not occur, but it is true and they did.

There were two prime reasons. First, there was the concept of providence, the idea that a kind and caring Creator, having caused a life form to come into existence, would not allow it to pass from that state. Had extinction occurred, it might even imply some flaw in the plan of Creation. Second, there was the notion of plenitude, the idea that all life forms that could exist, did exist, that all of these forms were inextricably linked to one another, and that the loss of one of these links could and would destroy the whole. Together, providence and plenitude implied that extinction was not possible.

No one expressed these ideas more powerfully than Alexander Pope, in his *Essay on Man* (1733–1734):

Vast chain of being, which from God began,
Natures aetheral, human, angel, man,
Beast, bird, fish, insect! What no eye can see,
No glass can reach! From Infinite to thee,
From thee to Nothing!—On superior pow'rs
Were we to press, inferior might on ours:
Or in the full creation leave a void,

Where, one step broken, the great scale's destroyed:
From Nature's chain whatever link you strike,
Ten or ten thousandth, breaks the chain alike.[4]

It was not by accident that Pope began this powerful passage with the "vast chain of being." This was his rendition of the Great Chain of Being, the concept that historian Arthur Lovejoy referred to as a "sacred phrase of the eighteenth century."[5] In this view of life, and as Pope's poetry makes clear, extinction could not occur. Were that to happen, the entire chain of being might be lost.

Thomas Jefferson provided a superb example of these beliefs in action. In 1797, he reported that the skeleton of a remarkable animal had been found in a cave in what is now West Virginia. He referred to this animal as the Great Claw, or Megalonyx; we now call it Jefferson's ground sloth (*Megalonyx jeffersonii*; see later in this chapter). Understandably, he did not recognize it as a huge sloth, comparing it instead to a lion, but he did know that no one living in the area had ever seen it alive. Nonetheless, he concluded, using an argument based entirely on the Great Chain of Being, that it was still out there:

> In fine, the bones exist: therefore the animal existed. The movements of nature are in a never ending circle. The animal species which has once been put into a train of motion, is still probably moving in that train. For if one link in nature's chain might be lost, another and another might be lost, till this whole system of things should evanish by piece-meal.... If this animal then has once existed, it is probable on this general view of the movements of nature that he still exists".[6]

At the same time as Jefferson was reading this paper to the American Philosophical Society, another eminent intellectual, Georges Cuvier, was publishing epochal work in France showing that such huge beasts as mammoths and ground sloths, among others, truly had become extinct. This demonstration ultimately had dramatic impacts on matters ranging from our understanding of life on earth to our understanding of the place of humans in the grand scheme of things. What matters here, however, is that by the time Cuvier presented the results of all his work in his great *Recherches sur les ossemens fossiles de quadrupèdes*, in 1812, he had already made it clear to western intellectuals that extinction was for real.

It proved no difficulty for scholars of intellect to reconcile the theologically based Great Chain of Being, and its denial of the possibility of extinction, with the reality of the extinction of a wide range of beasts in the past. James

Parkinson, whose 1817 *Essay on the Shaking Palsy* led to the naming of Parkinson's disease, helped lead the way to this reconciliation:

> Some very good and learned men have regarded the loss of a single
> link, in the chain of creation, as inadmissible: it implying, they say, such
> a deviation from the first plan of creation, as might be attributed to a
> failure in the original design. But such an inference does by no means
> follow; since that plan, which prevents the failure of a genus, or species,
> from disturbing the general arrangement, and œconomy of the system,
> must manifest as great a display of wisdom and power, as could any
> fancied chain of beings, in which the loss of a single link would prove
> the destruction of the whole.[7]

Shortly after the turn of the nineteenth century, then, western scholars
had become convinced that extinction not only could occur but had occurred
with some regularity. Some of the huge animals—sloths and mammoths, for
instance—that Cuvier used to establish the reality of extinction were, we
know now, of Pleistocene age.

To Cuvier and those who soon followed, it seemed extremely unlikely that
any of these animals had existed alongside people. In this view, our ancestors
had not appeared on earth until after those extinctions had occurred. This
belief followed from the assumption that the history of the earth had been
the history of our planet as a kind and caring Creator was preparing it for our
arrival. Not until those preparations had been completed, not until our planet
became fully modern in form, did our species appear. In that intellectual context, the extinctions became the last significant act in the formation of a fit
habitat for humankind. People, in this view, followed the extinctions.

All this changed in 1859 and 1860, when a series of discoveries in England
and France made it quite clear that people really had coexisted with now-
extinct Ice Age mammals. To western Europeans and those influenced by
them, this discovery removed humankind as the focus of earth history, much
as Copernicus had removed our planet from the center of the universe. This
was the most important lesson ever taught to western scholars by archaeological data.[8]

It also helps explain Wallace's comment that the world became a better
place for us once those beasts were gone. With the extinction of such things as
sabertooth cats and huge bears, the earth had become modern in form. Such
an earth was far more appropriate for our existence.

At its most essential core, Wallace's statement was true. If we use the late
Pleistocene as our baseline, then we do live in a zoologically impoverished

world. It even remains true that the extinctions have "hardly been sufficiently dwelt upon." Although there is now a huge scientific literature on those losses, we still do not know what caused them, as I discuss in chapter 7.

THE AMERICAN CONTEXT

This is a book on Ice Age extinctions in the Great Basin. Those extinctions, however, occurred in the more general context that Wallace was talking about. As a result, understanding the Great Basin requires some knowledge of that broader context. I begin with some basic biology of the sort we all learned in or before high school.

In the Linnaean hierarchy, the basic building block is the species. Very closely related species are placed in the same genus, related genera in the same family, related families in the same order, and related orders in the same class. The class Mammalia contains the order Carnivora (carnivores). The order Carnivora contains the family Canidae (foxes, wolves, jackals, and their allies). The family Canidae contains the genus *Canis* (dogs, wolves, and jackals). In the modern Great Basin, there are three representatives of this genus—the coyote (*Canis latrans*), the wolf (*Canis lupus*), and, thanks to people, the domestic dog (*Canis familiaris*).

Biologists, including paleontologists, prefer to study species, defined as populations of interbreeding, or potentially interbreeding, individuals. I would prefer to do that here, but I cannot, because the species-level status of many extinct North (and South) American mammals is not sufficiently well understood. For some animals, it is clear that there was only one species within a given genus. There is, for instance, only one species accepted for the giant ground sloth genus *Paramylodon* in North America: *Paramylodon harlani*, the same *Paramylodon harlani* we encountered imprisoned in the first chapter. On the other hand, there may have been as many as four, or as few as two, species of tapir (*Tapirus*) in North America.[9] We simply don't know yet, and we probably won't know until enough work has been done on the DNA contained within the remains of these animals.

The horses (*Equus*) are even worse. Twenty or more species of late Pleistocene horses have been described for North America. In an important review of these horses published in 1998, paleontologist Augusto Azzaroli suggested there were about ten such species. On the other hand, analysis of ancient DNA from North American Pleistocene horses suggests there might have been only two.[10] Five species of horses have been reported from the Mexican site of El Cedral alone and, as paleontologist Ismael Ferrusquía-Villafranca

FIGURE 3.1. Paleoecologist Paul Martin (1928–2010) in the Pinaleño Mountains, southeast-ern Arizona. No one has done more to spur interest in the world's extinct Ice Age mammals than he did.

and his colleagues have pointed out, it seems somewhere between highly un-likely and impossible that five species of horses could have coexisted.[11] To make matters worse, paleontologist Eric Scott and his colleagues have shown that the type specimen for the late Pleistocene horse *Equus laurentius*, defined in 1913 from specimens in Kansas, came from a domestic horse only a few hundred years old.[12] In defense of the paleontological enterprise, some pale-ontologists had suggested this not long after the species was first defined, but not until a few years ago was this suggestion confirmed.

Because of such problems, paleoecologist Paul Martin (fig. 3.1), the most influential scientist ever to have worked on the issues raised by the full run of Pleistocene extinctions, argued convincingly that the genus, not the species, was the only rational choice for analyzing these extinctions.[13] In some areas, this is no longer true, but it is certainly true for the Americas. Although I will discuss species when that is appropriate, I focus on the genus level here.

In the pages that follow, I provide a general American context for under-standing the extinctions that occurred in the Great Basin. Because late Ice Age extinctions were so vast and carried away such a remarkable variety of mammals, all I can do is try to provide a general feel for those losses, for what those animals were like. Once we get to the Great Basin, I will provide many more details on the nature of the beasts themselves.

As we will see, most of the mammals that became extinct toward the end of the American Ice Age were large, and some were immense. Throughout,

I provide estimates of their weights, but these estimates should be taken for what they are: estimates. Different scientists calculate the weights of extinct mammals in different ways, and they use different Ice Age specimens to make those estimates. Because of this, their results vary, sometimes dramatically.[14] The weights I provide are simply meant to provide a general indication of how big a particular animal was.

My definition of North America is simple. Following paleontologist Gary Morgan, I define it as the North American continent north of the Tropic of Cancer. South America begins just south of the Isthmus of Panama.[15]

SOUTH AMERICA

The South American late Pleistocene mammal fauna was, to use one of Wallace's words, startling. This region lost some 54 genera of mammals toward the end of the Pleistocene (table 3.1). That is a huge number.

The Cingulates (Armored Xenarthrans)

Not all that long ago, the armadillos, sloths, and anteaters were all placed in one order of mammals called the Edentata, or edentates. Now, however, that order has been divided into two groups. One of these, called the Cingulata, has external body armor (a shell) and today includes only the armadillos. The other, called the Pilosa, lacks external armor and includes the sloths and anteaters as living representatives. The two groups are sufficiently closely related that they are placed in the same superorder of strictly American mammals—the Xenarthra.

The armadillos get their own family, called the Dasypodidae. Only one species makes it into North America—the nine-banded armadillo (*Dasypus novemcinctus*). A big one may weigh 20 pounds. South America has a far richer armadillo fauna, with 21 species distributed across eight genera. The largest of these, the aptly named giant armadillo (*Priodontes maximus*), weighs about 65 pounds.[16]

During the late Pleistocene, South America also supported three genera of huge, now-extinct armadillos, each of which weighed over 400 pounds. There were, in addition, two huge armadillo-like animals that differ enough from armadillos to get their own family, the pampatheres. These are assigned to the genera *Holmesina* and *Pampatherium* and also weighed over 400 pounds.

The size of these animals, however, paled in comparison to that of members of the glyptodont family. South America supported nine genera of these during the late Ice Age.[17] All were marked by being enclosed in a bony,

TABLE 3.1. The extinct late Pleistocene mammals of South America (genera in bold were shared with North America)

Order and family	Genus	Weight (pounds)	Weight references
Cingulata			
Dasypodidae	*Eutatus*	440	1
	Pachyarmatherium	440	1
	Propraopus	440	1
Pampatheriidae	***Holmesina***	440	2
	Pampatherium	440	3
Glyptodontidae	*Doedicurus*	3,230	4, 5
	Glyptodon	4,400	5, 6
	Glyptotherium	2,420	2
	Hoplophorus	620	2
	Lomaphorus	550	2
	Neosclerocalyptus	550	5, 6
	Neuryurus	> 2,200	1
	Panochthus	2,330	5, 6
	Plaxhaplous	2,860	1
Pilosa			
Megalonychidae	*Ahytherium*	660	7
	Australonyx	660 [a]	8, 9
	Diabolotherium	185	10
	Megistonyx	600 [b]	11
Megatheriidae	***Eremotherium***	7,700	2
	Megatherium	13,400	4, 5
Nothrotheriidae	*Nothropus*	220	2
	Nothrotherium	330	2
Mylodontidae	*Catonyx*	3,500	12
	Glossotherium	3,800	5, 8
	Lestodon	7,500	5, 12
	Mylodon	2,200	12
	Mylodonopsis	2,200	7
	Oreomylodon	—[c]	—
	Scelidotherium	2,300	2, 5, 12
	Valgipes	880	7
Carnivora			
Canidae	*Protocyon*	55	13
Felidae	***Smilodon***	660	14
Ursidae	*Arctotherium*	240	2, 15
Rodentia			
Caviidae	***Neochoerus***	330	3
Muridae	*Tafimys*	0.06	—[d]

TABLE 3.1. (cont'd.) The extinct late Pleistocene mammals of South America (genera in bold were shared with North America)

Order and family	Genus	Weight (pounds)	Weight references
Perissodactyla			
Equidae	***Equus***	660	1
	Hippidion	1,065	4
Artiodactyla			
Camelidae	*Eulamaops*	330	1
	Hemiauchenia	650	5
	Palaeolama	650	5
Cervidae	*Agalmaceros*	130	2
	Antifer	175	16
	Charitoceros	130	2
	Morenelaphus	110	1
	Paraceros	90	16
Litopterna			
Macraucheniidae	*Macrauchenia*	2,200	4, 5
Proterotheriidae	*Neolicaphrium*	90	17
Notoungulata			
Toxodontidae	***Mixotoxodon***	8,300	18
	Toxodon	2,600–3,900	4, 18
Proboscidea			
Gomphotheriidae	***Cuvieronius***	11,000	7
	Haplomastodon	13,200	7
	Notiomastodon	13,640	7
	Stegomastodon	16,675	2, 5
Primates			
Atelidae	*Caipora*	45	19
	Cartelles	60	20
	Protopithecus	55	20, 21, 22

References: (1) Prevosti and Vizcaíno 2006; (2) F. Smith et al. 2003; (3) Fariña 1996; (4) Fariña, Vizcaíno, and Bargo 1998; (5) Fariña, Vizcaíno, and De Iuliis 2013; (6) Vizcaíno et al. 2011; (7) Cartelle 2000; (8) Bargo et al. 2000; (9) De Iuliis, Pujos, and Cartelle 2009; (10) Pujos et al. 2007; (11) H. McDonald, Rincon, and Gaudin 2013; (12) De Esteban-Trivigno, Mendoza, and De Renzi 2008; (13) Oliveira, Prevosti, and Pereira 2005; Prevosti, Zurita, and Carlini 2005; (14) Christiansen and Harris 2005; (15) Soibelzon and Tarantini 2009; (16) Ménégaz and Ortiz Jaureguizar 1995; (17) Bond et al. 2001; (18) Elissamburu 2012; (19) Cartelle and Hartwig 1996; (20) Halenar and Rosenberger 2013; (21) Hartwig 1995; (22) Halenar 2011.

[a] From the similarly sized North American *Megalonyx* (F. Smith et al. 2003; H. McDonald, Rincón, and Gaudin 2013).

[b] From the similarly sized *Ahytherium* (De Iuliis, Pujos, and Cartelle 2009).

[c] The body weight of *Oreomylodon* does not seem to have been calculated.

[d] Taken from the body weight of a house mouse.

inflexible carapace akin to that of a tortoise. The largest was about the size of a Volkswagen Beetle (yes, the car) and weighed more than 4,000 pounds.

The Pilosa (Sloths)

As big as the biggest South American glyptodonts were, they were small compared to the biggest of the South American ground sloths. Today, there are two genera of sloths living in South America: the three-toed sloth, placed in the genus *Bradypus* (in the family Bradypodidae), and the two-toed version, placed in the genus *Choloepus* (in the family Megalonychidae).[18] Chubby individuals of these tree-loving animals may weigh as much as 20 pounds.[19] The extinct Pleistocene South American ground-dwelling versions, on the other hand, were an entirely different matter. There were 16 genera of these, arrayed across four different families, of which only the Megalonychidae still exists. What did not carry on at all, however, was the size of the Pleistocene versions. The largest of the South American ground sloths, belonging to the genus *Megatherium*, weighed in at about 13,000 pounds. Even the smallest of them, belonging to the genus *Diabolotherium*, weighed some 200 pounds.[20] If they were not careful, South American Ice Age ground sloths could have trampled two- and three-toed sloths underfoot.

The Carnivores

South America lost only three genera of carnivores toward the end of the Ice Age, compared to the seven lost in North America. Those three include the highly carnivorous canid *Protocyon*, which weighed up to about 50 pounds, as well as the sabertooth cat *Smilodon*, of which I will have more to say later in this chapter. The last of the three was the bear *Arctotherium*, which weighed some 250 pounds. *Arctotherium* is related both to the living South American spectacled bear (*Tremarctos ornatus*), and to the largest land carnivore the Pleistocene had to offer, North America's giant bear *Arctodus*, which I discuss shortly.[21] The scimitar cat *Homotherium*, which I also discuss below, is known from South America, but from deposits that predate the late Pleistocene.[22]

The Rodents

Today, the Central and South American capybara (*Hydrochoerus hydrochaeris*) is the world's largest living rodent, with large males reaching as much as 120 pounds. As we will see, this genus of capybara made it into southern North America during the Pleistocene, only to become extinct here and live on in the far south. A second, closely related genus, *Neochoerus*, was primarily South American but also made its way to southern North America during the Ice

Age. It failed to survive anywhere beyond that time. Estimates of its weight vary, but if it really did weigh some 330 pounds, it would have been the largest late Ice Age rodent the Americas had to offer.[23] For comparison, the largest living American black bears (*Ursus americanus*) usually weigh less than this.[24]

Then, at the other end of the scale, there is *Tafimys*, discovered in a paleontological site in northwestern Argentina's Tafi Valley.[25] This tiny grass-eating rodent was about the size of a house mouse. Although it belongs to the same family as those common animals, it was more closely related to such less familiar Great Basin rodents as the western harvest mouse (*Reithrodontomys megalotis*) and the northern grasshopper mouse (*Onychomys leucogaster*). The deposits that provided the remains of *Tafimys* are at least 10,000 years old, but paleontologist Pablo Ortiz and his colleagues, who brought *Tafimys* to the attention of scientists, wonder whether it might still exist in the higher elevations of northern Argentina and Bolivia. Me, too.

The Perissodactyls (Odd-Toed Ungulates)

Those of us who live or work in arid western North America are used to seeing wild horses. That is not surprising, since there are about 33,000 of them in the western United States. There would be far more had the Bureau of Land Management not removed some 200,000 horses from public lands since the passage of the Wild Free-Roaming Horses and Burros Act in 1971.[26]

Those horses are all descendants, in one way or another, of animals that Europeans brought to North America. Horses evolved in North America, crossed into Eurasia by the Bering Land Bridge (chapter 2), and then became extinct in their homeland by 10,000 years ago. Europeans, in a very real sense, repatriated them.

South America also had its native horses. In fact, it had two genera of them: *Equus*, the same genus found in North America and the genus to which our modern horses belong, and *Hippidion*, which likely evolved in Central or South America, having descended from members of the North American horse lineage.[27] Several species of the latter have been proposed, all of which survived into the late Pleistocene. Compared to modern horses, *Hippidion* was shorter, with relatively shorter legs and a bulkier body; one estimate of its weight puts it at about 1,000 pounds.[28]

The Artiodactyls (Even-Toed Ungulates)

Camels also evolved in North America and then moved across the Bering Land Bridge to Eurasia and south to South America. They became extinct in North America toward the end of the Pleistocene but survive in both of their adopted lands. In South America, they continue in the form of two genera, *Lama* and *Vicugna*.

The genus *Lama* includes the llama, guanaco, and, to some, the alpaca. The genus *Vicugna* includes, obviously, the vicugna. All this can become confusing because there is evidence that the alpaca descended from hybridization between guanacos and vicugnas. This does not matter here, since both of these genera are known from the South American Pleistocene, where they existed alongside three other genera of the lamine group. Two of these, *Palaeolama* and *Hemiauchenia*, weighed about 650 pounds (a very large modern llama can reach 350 pounds) and were also found in North America. Some scientists find the similarities between these two genera to be so great that they recognize only *Palaeolama*.[29] Others keep them emphatically separate,[30] as I do here. The third genus, *Eulamaops*, was smaller, at about 330 pounds, and was confined to South America.

In North America, the family Cervidae is represented by deer (*Odocoileus*), elk (*Cervus*), moose (*Alces*), and caribou (*Rangifer*). They exist here because their ancestors, having evolved in Eurasia, migrated across the Bering Land Bridge, going in the opposite direction taken by camels and horses. Ultimately, they made their way into South America, which today supports six genera of native cervids: marsh deer (*Blastocerus*), guemal (or Andean deer, *Hippocamelus*), brocket deer (*Mazama*), white-tailed deer (*Odocoileus*), pampas deer (*Ozotoceros*), and pudu (*Pudu*). The largest of these is the marsh deer, which can weigh as much as 290 pounds. Pudu fall at the other end of the scale. There are two species of this tiny deer, the smaller of which, the northern pudu (*Pudu mephistophiles*), stands 12 inches tall and weighs about 15 pounds.[31] When we reach the Great Basin, we will encounter a pronghorn antelope of almost the same diminutive size.

During the late Pleistocene, these six South American genera of modern cervids were joined by five others, all listed in table 3.1. Some of these are quite poorly known, having been identified largely on the basis of their distinctive antlers, and all weighed between about 90 and 180 pounds. At times, a sixth extinct genus, *Epieuryceros*, is added to the list,[32] but it is not clear that it actually made it into the late Pleistocene.[33]

The Litopterns and Notoungulates

All North Americans are familiar with camels and deer, but there are two major groups of South American mammals that are pretty much unknown to those of us who live in North America. These are the litopterns and the notoungulates, two separate orders of ungulates whose closest living relatives include the horses and tapirs.[34] Few North Americans have heard of these animals because they evolved in isolation in South America. Even after it became possible to do so, only one of these groups made it to North America, and then just barely.

"Even after it became possible to do so" has a very specific meaning in this context. Today, North and Central America join with South America at the Isthmus of Panama. This, however, has not always been the case. At one time, the area that is now occupied by the Isthmus of Panama was covered by water that connected the Pacific Ocean with the Caribbean Sea. The isthmus then formed as a result of a complex series of geological events that may have begun by 15 million years ago and ended about 12 million years later. That land connection led to what is called the Great American Biotic Interchange, with northern animals allowed access to the south and southern animals access to the north. That is how South America got its squirrels and skunks, and North America its porcupines and opossums.[35]

Prior to this interchange, South American mammals evolved in what paleontologist George Gaylord Simpson called "splendid isolation." The glyptodonts evolved this way, as did the sloths, capybaras, and armadillos.[36]

So did the litopterns and notoungulates. The litopterns contain two genera. *Macrauchenia* weighed about 2,200 pounds and bore a general resemblance to a huge and perhaps long-snouted llama.[37] *Neolicaphrium* weighed about 90 pounds[38] and had a general appearance similar to that of a primitive horse, though with teeth more like those of deer.[39] Neither animal is known to have made its way north through the Isthmus of Panama.

The notoungulates include a single late Pleistocene family, the toxodonts, which in turn include two late Pleistocene genera, *Toxodon* and *Mixotoxodon*. Both were huge, with weight estimates ranging from 2,600 to nearly 4,000 pounds. Both bore a physical resemblance to rhinos and both were apparently capable of eating a very broad set of plants.[40] Unlike the litopterns, one of the toxodonts did make it through the Isthmus of Panama: *Mixotoxodon* had reached southern North America by the late Ice Age, as we will see.[41]

The Proboscidea (Elephants and Their Relatives)

The order of mammals known as the Proboscidea (think "proboscis") still very much exists in the form of the Indian (*Elephas*) and African (*Loxodonta*) elephants, both in the family Elephantidae. Toward the end of the Pleistocene, however, the Americas supported three families of elephant-like animals: the mammoth (*Mammuthus*), in the family Elephantidae; the mastodon (*Mammut*), in the family Mammutidae; and several genera of animals called gomphotheres, in the family Gomphotheriidae. North Americans are generally very aware of both mammoths and mastodons. The same is decidedly not the case for gomphotheres.

The reason is simple. Gomphotheres were found only at the very southern edge of North America during the late Pleistocene, apparently in small numbers. Mammoths and mastodons, on the other hand, were common late Pleistocene North American mammals. Mastodons made it into Honduras during this period, and mammoths to Costa Rica, but neither got farther south.[42] The late Pleistocene elephant-like animal of South America was the gomphothere.

From a distance, gomphotheres would have looked like elephants, with straight to curved tusks and a massive body. If you were an elephant dentist, however, you would quickly realize that you were dealing with a very different animal. In elephants and mammoths, the business surface of the molar teeth consists of a series of enamel plates arranged at right angles to the main axis of the tooth. Gomphothere teeth are completely different, with surfaces consisting of a series of large, cone-like cusps. They are more mastodon-like in having this form, but they are easily distinguished from mastodon molars because the cusp pattern is more complex.

Earlier in this chapter, I pointed out that most analyses of the Pleistocene mammals of the Americas are done at the genus level, even though all of us recognize that the truly appropriate unit to use would be the species. We tend to use the genus instead because we do not know how many species there were within some genera, with horses being the most problematic example. For North America, using the genus solves the problem because there is little disagreement about the genera of mammals that were present here during the late Pleistocene.

This is not the case for South America. For instance, in table 3.1, I have listed the ground sloth genera *Catonyx* and *Scelidotherium*, following the work of sloth expert Greg McDonald and his colleagues.[43] Many accept a third genus, *Scelidodon*, as valid,[44] but McDonald and his coworkers argue that it is no different from *Scelidotherium*. Still others think that the material assigned to all three of these genera should actually be assigned to *Scelidodon*.[45]

The situation with the gomphotheres is similar. I have listed four genera in table 3.1. While some agree with that assessment,[46] many do not. Some accept only three;[47] even more accept only two. To make matters even worse, different paleontologists choose those two differently. Some accept *Cuvieronius* (named after Georges Cuvier), and *Stegomastodon*.[48] Others accept *Cuvieronius* and *Notiomastodon*.[49] Someday, ancient DNA will probably tell us the answer.

No matter what decisions are finally made here, gomphotheres were elephant-sized animals. *Cuvieronius, Haplomastodon,* and *Notiomastodon* weighed about 13,000 pounds, the same size as African elephants. *Stegomastodon* would have required an even more substantial scale. With an estimated weight of 16,700 pounds, it was more massive than any living elephant. Gomphotheres were widespread in South America, found nearly everywhere that sufficient paleontology has been done to detect them.[50] In fact, the oldest securely dated archaeological site in the Americas, Monte Verde, located near the coast of far southern Chile, provided *Stegomastodon* remains alongside a wide variety of artifacts, all dated to 12,500 years ago.[51]

The Primates

Then there are the primates. Three genera of extinct late Pleistocene primates are known from South America, all from caves in Brazil. The first of these to be reported was found in 1836 by the Danish natural historian Peter Lund in the remarkable Lagoa Santa cave system of southeastern Brazil. Lund recognized it as a fossil monkey and assigned it to the genus *Protopithecus.* In 1995, anatomist Walter Hartwig carefully analyzed the specimens and showed that this animal was closely related to the spider monkeys (*Brachyteles*). He also showed that, at about 55 pounds, it was more than twice the size of its living relatives. More recent weight assessments have confirmed that original estimate.[52] *Protopithecus* was, for an American monkey, huge.

Another Brazilian cave, Toca da Boa Vista, about 750 miles north of Lagoa Santa, has provided the remains of two other kinds of extinct monkeys. One of these, assigned to the genus *Caipora,* is also closely related to the living spider monkeys and weighed about 45 pounds. The other, named *Cartelles,* is related to the South and Central American howler monkeys and, with a weight of about 60 pounds, was, as far as we know, the largest of the extinct late Ice Age monkeys of South America.[53]

Counting South American Genera

How many genera of mammals were lost during the late Pleistocene in South America? It is unfortunate that this question must be asked, since genera are artificial constructs, composed of species felt to be sufficiently closely related to one another to deserve to be in the same genus. But since we can't analyze the American losses at the species level, count them we must if we want to compare South American to North American losses. This is not a problem for North America, where, as we shall see, the number is 37. In South America, on the other hand, there is no accepted number. Were there four genera of gomphotheres or two? Do the closely related sloths *Catonyx*, *Scelidodon*, and *Scelidotherium* deserve to be in three genera or one? These are not the only debates that paleontologists have about the genus-level assignments of the late Pleistocene mammals of South America.[54] We could choose 56, which is what you get if you count all the genera in table 3.1, or go with two gomphotheres and only *Scelidodon*, which would give us 52, or make some other decision. In what follows, I will split the difference and use 54.

NORTH AMERICA

While South America was losing some 54 genera of mammals as the Pleistocene came to an end, North America was busy losing 37 (table 3.2). Six of these live on elsewhere: the dhole (*Cuon*), the spectacled bear (*Tremarctos*), the capybara (*Hydrochoerus*), the horse (*Equus*), the tapir (*Tapirus*), and the saiga (*Saiga*). Of the 37, 20 are known from the Great Basin and a few more may be found here as we learn more about what this region was like during the Ice Age. Here, I review all 37 of those genera, but my prime focus is on those that lived in the Great Basin. I provide the names of the species involved only when there is real agreement among paleontologists as to what those species are.

I also provide maps of the distribution of all but one of these animals.[55] The exception is known from only one specimen (*Mixotoxodon*, discussed below) and so hardly needs a map. Each black dot on those maps represents a site where the animal has been reported. In many instances, we have only a general idea as to how old these sites are, and this is especially true for the Great Basin. It is important to realize this, because looking at all the dots on some of these maps—for the mammoth, for instance, or for horses—might give the impression that the animal involved must have been very common. In some cases, this might have been true. In others, however, this impression might be very misleading because the sites depicted on the maps may represent a very long time span and so give an incorrect idea of how many of the animals might have been wandering around the landscape at any one time. Chapter 6 takes a close look at this issue for the Great Basin.

The Cingulates (Armored Xenarthrans)

Cingulates are those armored relatives of sloths that I discussed in the South American context. Three now-extinct genera of these mammals made it as far north as what is now the southern United States. These included two kinds of huge, armadillo-like pampatheres (figs. 3.2 and 3.3), both of which weighed

TABLE 3.2. The extinct late Pleistocene mammals of North America (genera marked with an asterisk live on elsewhere; genera in bold are known from the hydrographic Great Basin)

Order and family	Genus	Weight (pounds)	Common name
Cingulata			
Pampatheriidae	*Pampatherium*	550	Southern pampathere
	Holmesina	550	Northern pampathere
Glyptodontidae	*Glyptotherium*	2,400	Simpson's glyptodont
Pilosa			
Megalonychidae	**Megalonyx**	1,300	**Jefferson's ground sloth**
Megatheriidae	*Eremotherium*	7,700	Laurillard's ground sloth
Nothrotheriidae	**Nothrotheriops**	660	**Shasta ground sloth**
Mylodontidae	**Paramylodon**	2,000	**Harlan's ground sloth**
Carnivora			
Mustelidae	**Brachyprotoma**	5	**Short-faced skunk**
Canidae	*Cuon**	35	Dhole
Ursidae	*Tremarctos**	615	Florida cave bear
	Arctodus	1,500	**Giant bear**
Felidae	**Smilodon**	500	**Sabertooth**
	Homotherium	400	**Scimitar cat**
	Miracinonyx	200	**American cheetah**
Rodentia			
Castoridae	*Castoroides*	150	Giant beaver
Caviidae	*Hydrochoerus**	120	Holmes's capybara
	Neochoerus	220	Pinckney's capybara
Lagomorpha			
Leporidae	**Aztlanolagus**	4	**Aztlán rabbit**
Perissodactyla			
Equidae	**Equus**	660	**Horses**
Tapiridae	*Tapirus**	660	Tapirs
Artiodactyla			
Tayassuidae	*Mylohyus*	190	Long-nosed peccary
	Platygonus	240	**Flat-headed peccary**
Camelidae	**Camelops**	2,400	**Yesterday's camel**
	Hemiauchenia	240	**Large-headed llama**
	Palaeolama	180	Stout-legged llama
Cervidae	**Navahoceros**	490	**Mountain deer**
	Cervalces	1,070	Stag-moose
Antilocapridae	**Capromeryx**	25	**Diminutive pronghorn**
	Tetrameryx	130	**Shuler's pronghorn**
	Stockoceros	115	Pronghorns

TABLE 3.2. (cont'd.) The extinct late Pleistocene mammals of North America (genera marked with an asterisk live on elsewhere; genera in bold are known from the hydrographic Great Basin)

Order and family	Genus	Weight (pounds)	Common name
Bovidae	*Saiga**	80	Saiga
	Euceratherium	1,000	**Shrub ox**
	Bootherium	1,300	**Helmeted muskox**
Notoungulata			
Toxodontidae	*Mixotoxodon*	8,300	Toxodont
Proboscidea			
Gomphotheriidae	*Cuvieronius*	11,000	Cuvier's gomphothere
Mammutidae	***Mammut***	10,000	**American mastodon**
Elephantidae	***Mammuthus***	1,600–17,600	**Mammoths**

References: Unless otherwise indicated in the text, weights are from C. Johnson (2002), F. Smith et al. (2003), and, where relevant, table 3.1.

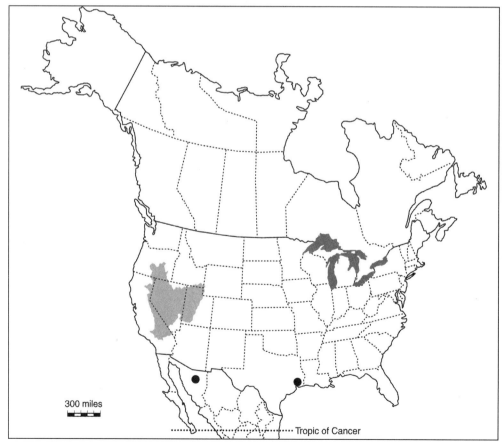

FIGURE 3.2. The late Ice Age distribution of the armadillo-like southern pampathere (*Pampatherium*) in North America.

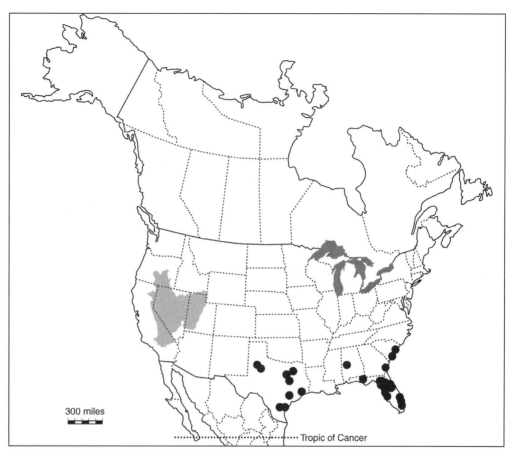

FIGURE 3.3. The late Ice Age distribution of the armadillo-like northern pampathere (*Holmesina*) in North America.

about 550 pounds,[56] and the even huger, 2,400-pound Simpson's glyptodont (*Glyptotherium*) (fig. 3.4). Simpson's glyptodont was some 6.5 feet long and 5 feet tall, with that tortoise-like carapace that I mentioned earlier, along with an armored tail and skull. Because the remains of Simpson's glyptodont have often been found near lakes, marshes, and streams, some paleontologists have reasonably inferred that it may have been semiaquatic.[57] None of these three genera appear to have made it close to the Great Basin, and it is unlikely that future work will show that they lived here during the late Ice Age.

The Pilosa (Sloths)

The sloths, however, were a different issue. Of the four genera of ground sloths in North America during the late Ice Age, only one is unknown from the Great Basin. The one with the unfortunate taste in geography is Laurillard's ground sloth (*Eremotherium laurillardi*).[58] This astonishing animal combined the

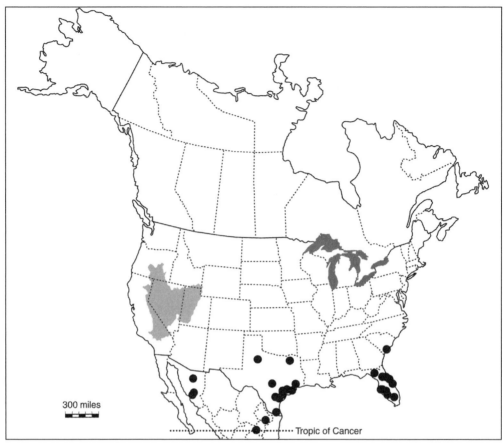

FIGURE 3.4. The late Ice Age distribution of the glyptodont (*Glyptotherium*) in North America.

height of a giraffe with the bulk of an elephant (it weighed an estimated 7,700 pounds) and made it up from South America to the southern United States, although there is also an outlying record from coastal New Jersey (fig. 3.5).[59] Although the other three sloths did make it to the Great Basin, they seem to have lived only along the edges of it.

Jefferson's Ground Sloth (*Megalonyx jeffersonii*)
The 1,300-pound Jefferson's ground sloth (*Megalonyx jeffersonii*; fig. 3.6)—the animal that Jefferson assumed could not be extinct—belongs to the same family as the two-toed tree sloth (*Choloepus*), which may weigh 20 pounds. This was the most widespread of the North American ground sloths, known from Florida to Alaska, although the far northern examples predate the latest Pleistocene expansion of glacial ice that occurred in this region (fig. 3.7).

As part of that widespread distribution, Jefferson's ground sloth is also known from a few sites along the very edge of the Great Basin. The material from the deposits of Pleistocene Lake Manix, which formed along

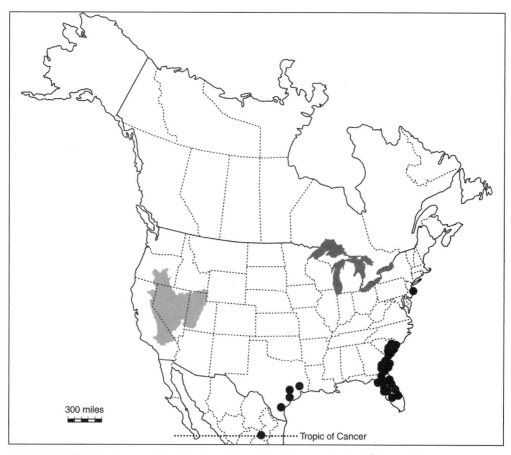

FIGURE 3.5. The late Ice Age distribution of Laurillard's ground sloth (*Eremotherium laurillardi*) in North America.

FIGURE 3.6. Jefferson's ground sloth (*Megalonyx jeffersonii*). Drawing by Wally Woolfenden.

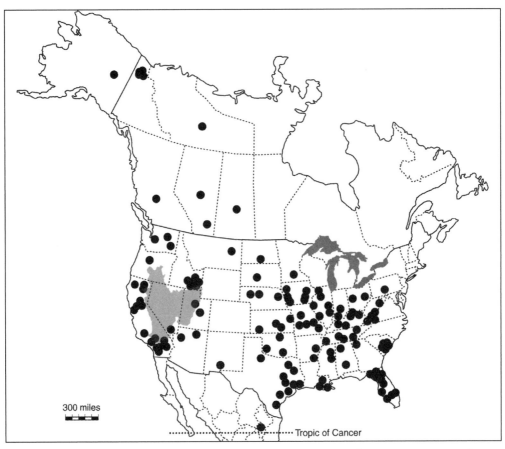

FIGURE 3.7. The late Ice Age distribution of Jefferson's ground sloth (*Megalonyx jeffersonii*) in North America.

southeastern California's Mojave River drainage, has not been fully described,[60] but that from the famous Tule Springs site, in Las Vegas Valley about ten miles north of Las Vegas, has. This site provided a single *Megalonyx* tooth, along with four toe bones that might or might not have come from this animal.[61] The most complete Jefferson's ground sloth material from the Great Basin, however, comes from the Orem site, located just east of Utah Lake along the eastern edge of Pleistocene Lake Bonneville. This site provided a partial *Megalonyx* skeleton embedded in deposits laid down shortly after the lake had declined from its late Pleistocene high stand, about 14,500 years ago (chapter 2). Jefferson's ground sloth remains are fairly common in sites in the Snake River Plain just to the north of this portion of the Great Basin (fig. 3.7), and it is very possible that it was from this area that these sloths made their way southward, moving along the eastern edge of Pleistocene Lake Bonneville.[62]

FIGURE 3.8. The Shasta ground sloth (*Nothrotheriops shastensis*). Drawing by Wally Woolfenden.

The Shasta Ground Sloth (*Nothrotheriops shastensis*)

The Shasta ground sloth (*Nothrotheriops shastensis*) was, at about 660 pounds, the smallest of the North American ground sloths (fig. 3.8). It was confined to western North America, with most, but not all, of its known sites in the Southwest (fig. 3.9). Although clearly happy in the arid Southwest, it appears to have been less so in the Great Basin, with nearly all of its sites in what is today the Mojave Desert of southern Nevada and adjacent California.

The superb preservation of Shasta ground sloth remains in caves in the arid West has provided us with hairs of that animal, along with skin and other soft tissue. In fact, the hairs of a Shasta ground sloth from southern New Mexico's Aden Crater were so well preserved that, in 1930, paleontologist Richard Lull was able to report that they may have had algae growing on them. If this is correct, Shasta ground sloths were similar to modern sloths in this way, since the hairs of the latter often support greenish algae that might help camouflage them.[63]

We know an impressive amount about the diet of these animals, since not only did they leave their skeletons and soft tissue in dry caves, but they had the

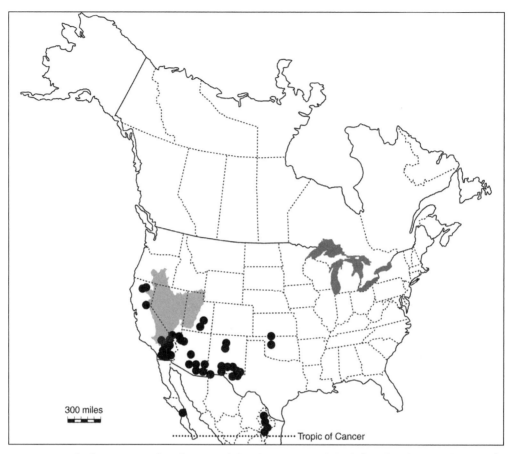

FIGURE 3.9. The late Ice Age distribution of the Shasta ground sloth (*Nothrotheriops shastensis*) in North America.

fortunate habit of pooping in them (fig. 3.10). A series of sites rich in the fossil dung ("coprolites") of these animals is known from this region, and the dung of many of these sites has been analyzed for what it might tell us about the animals' lifestyle. These sites include Muav and Rampart Caves in Arizona's Grand Canyon,[64] Gypsum Cave near Las Vegas (see chapter 5),[65] southern New Mexico's Aden Crater,[66] Shelter Cave in south-central New Mexico,[67] and Cave 08 and Williams Cave in the Guadalupe Mountains of northwestern Texas.[68] The scientists who have done this work have identified leaves, stems, seedpods, and seeds from these coprolites; studied the pollen they contain; analyzed their chemical constituents; and begun to study the plant DNA preserved within them. As if this were not enough, they have extracted and identified parasites from Shasta ground sloth dung. They have even studied the scratches on sloth teeth to see what the scratches can tell us about Shasta ground sloth diet.[69]

FIGURE 3.10. Shasta ground sloth dung balls from Rampart Cave, Grand Canyon, Arizona.

The analysis of ancient plant DNA from Shasta ground sloth dung has so far been able to identify plants only to the family level. The analysis of the plant macrofossils from those coprolites, however, has provided plant identifications down to the genus, and at times the species, level. As a result, we know that these animals dined on a diverse variety of plants. Among many others, these included agave (*Agave*), yucca (*Yucca*), Nevada jointfir (*Ephedra nevadensis*), saltbush (*Atriplex*), catclaw acacia (*Acacia greggii*), common reed (*Phragmites australis*), desert globemallow (*Sphaeralcea ambigua*), Fremont cottonwood (*Populus fremontii*), cactus (*Opuntia*), and even Utah juniper, along with a wide variety of grasses. An obvious oddity, and one that has been pointed out since the very first Shasta ground sloth coprolite was analyzed,[70] is that although the animal is gone, its diet lives on without it.

Scattered specimens of Shasta ground sloth are known from the deposits of Pleistocene Lake Manix and from the Fort Irwin National Training Center, both in the Mojave River drainage of southeastern California. Not far away, however, Newberry Cave, in the Newberry Mountains just south of the Mojave River, provided 156 specimens from a juvenile Shasta ground sloth along with what may have been a coprolite from this animal. A second individual was found in Devil Peak Cave, at the south end of the Spring Mountains just north of the Nevada-California border,[71] and a single tooth at Las Vegas Valley's Tule Springs.[72] None of these sites can come close to Gypsum Cave, east of Las Vegas. This site provided Shasta ground sloth bones and dung, the

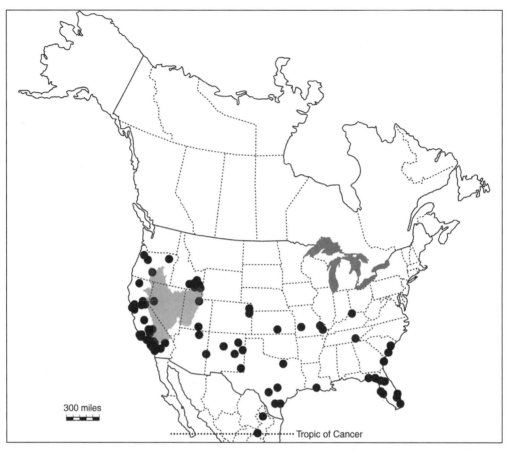

FIGURE 3.11. The late Ice Age distribution of Harlan's ground sloth (*Paramylodon harlani*) in North America.

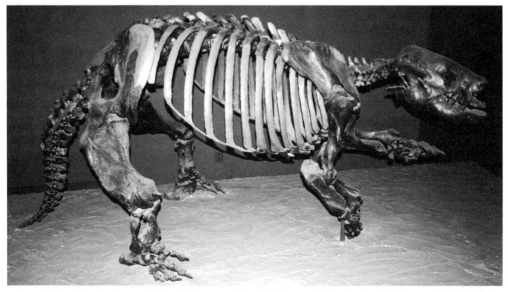

FIGURE 3.12. The skeleton of Harlan's ground sloth (*Paramylodon harlani*), National Museum of Natural History specimen V15164, Los Angeles County, California.

latter in profusion. This site is so important that I talk about it in detail in the next chapter.

Harlan's Ground Sloth (*Paramylodon harlani*)

Weighing an estimated 2,000 pounds, Harlan's ground sloth was widespread south of glacial ice in North America, known from Florida to Oregon (figs. 3.11 and 3.12). As part of that broad distribution, it was also found in the Great Basin, though again just from the very edges of it. This was the sloth that left the Nevada State Prison footprints that I discussed in chapter 1.

In general, *Paramylodon* remains have been found as single individuals, suggesting that this was a solitary animal.[73] On the other hand, the Boney Spring site in southwestern Missouri yielded the remains of four animals that might have died at the same time. If so, it is possible that this animal did, at times, lumber across the landscape in small groups.[74]

Harlan's ground sloth belongs to the family of sloths known as the Mylodontidae. These sloths were abundant and diverse in Ice Age South America (table 3.1), but Harlan's ground sloth was the only late Ice Age representative of this group in North America. Among other things, the mylodont ground sloths are characterized by having small, pebble-like bones—called dermal ossicles—embedded in their skin.

That skin, and those ossicles, led to one of the most famous episodes in ground sloth lore. As naturalist Francisco P. Moreno told the story, in November 1897, he was leading a team that was charting the southern border between Chile and Argentina when he saw a piece of dried skin hanging from a tree. That skin, he said, "attracted my attention most strangely, as I could not determine to what class of Mammalia it could belong, more especially because of the resemblance of the small incrusted bones it contained to those of the Pampean *Mylodon*."[75] The specimen he had in hand, he soon learned, was part of a larger piece that had been found two years earlier in a nearby cave.

That cave is now known as Mylodon Cave and is the heart of Patagonia's Mylodon Cave National Monument. We now know that *Mylodon* is extinct, but the discovery of a fresh-looking piece of skin suggested otherwise to many people.

Moreno himself noted that "ancient chroniclers inform us that the indigenous inhabitants recorded the existence of a strange, ugly, huge hairy animal which had its abode in the Cordillera to the south of latitude 37°."[76] He also reported that Native peoples "have mentioned similar animals to me, of whose existence their ancestors had transmitted the remembrance; and in the neighborhood of the Rio Negro [Patagonia], the aged cacique Sinchel, in 1875, pointed out the supposed lair of one of these animals."[77]

Moreno did not think the animal still existed. Others, however, did, including the famed naturalist Florentino Ameghino. In 1898, Ameghino observed that he, too, had been told of the existence of a "mysterious quadruped" that lived in dens it hollowed out in the ground—just as, it now appears, some South American ground sloths actually did.[78] This mysterious animal was usually active at night, and, "according to the reports of the Indians, it is a strange creature, with long claws and a terrifying appearance, impossible to kill because it has a body impenetrable to firearms and missiles."[79] The Mylodon Cave skin, he argued, documented that this quadruped still lived, just as Patagonian Natives had so often reported. The title of his 1898 paper summed it all up well: "An Existing Ground Sloth in Patagonia."

Although numerous scientists doubted that ground sloths were to be found alive,[80] the belief that they might still be out there was credible. It was credible enough that the explorer, author, and war hero H. Hesketh Prichard traveled to Patagonia to determine whether "the prehistoric Mylodon might possibly still exist hidden in the depths of the forests of the southern Andes."[81]

Hesketh Prichard spent nearly a year in Patagonia between 1900 and 1901. In the end, he found "no single scrap of evidence"[82] to support belief in the continued existence of the animal. That, fortunately, did not stop him from producing a deeply interesting account of his Patagonian travels, one that remains very much worth the read.

This, though, was not the end of the story. In 2001, sloth expert David Oren reported stories he had been told of a creature that the Native peoples of the Brazilian Amazon called the *mapinguari*. They described that beast as powerful, hairy, scary, and as having an overpowering stench. When he first heard these stories, Oren immediately thought "ground sloth." Later, he encountered seven different hunters who claimed to have killed these animals. They described them as six feet tall and repulsively smelly, with strong claws, hair that was long and coarse, a powerful build, and the ability to walk either on all fours or on their hind feet alone. Even more intriguing, Oren reported that *mapinguari* footprints were said to be either "roundish," produced when the animal walked quadrupedally, or like those of people, when it walked on its hind legs.[83] It is hard not to be reminded of the supposed human footprints of the Nevada State Prison, which turned out to be footprints of *Paramylodon*. Thomas Jefferson might have been the first to suggest in print that ground sloths must still exist, but many have followed him in the enticing thought that they might still be around. The occasional sightings of individuals in certain parts of Las Vegas that bear a striking resemblance to them do not count.

Three Great Basin sites have provided the remains of Harlan's ground sloth, all from the very edges of this region. The Nevada State Prison in Carson City provided the *Paramylodon* footprints that raised such a ruckus in the late nineteenth century, which I discussed in the first chapter. On the eastern edge of the Great Basin, the Silver Creek site in Utah's Wasatch Range, about 20 miles east of Salt Lake City, provided two teeth and a vertebra of this animal.[84] Oregon's Fossil Lake has yielded a much richer assemblage of Harlan's ground sloth remains, ranging from toe bones to a mandible, but all represent perhaps only two individuals.[85]

The Carnivores

North America lost seven genera of carnivores toward the end of the Ice Age: three cats, two bears, a member of the dog family, and a skunk. All but two of these are known from the Great Basin.

Of the two that are not known from the Great Basin, it is very unlikely that one will ever be found here—the extinct Florida cave bear (*Tremarctos floridanus*). The genus lives on today, in diminishing numbers,[86] as the spectacled or Andean bear (*Tremarctos ornatus*), found in the mountains of northwestern South America. During the North American late Pleistocene, *Tremarctos* was found from northeastern Mexico to the southeastern United States (fig. 3.13). This was not a bear we can expect to find in what is now arid western North America.

The pack-hunting dhole (*Cuon*), however, is a different issue. Dholes (*Cuon alpinus*) also exist today, this time in southeastern Asia but again in declining numbers.[87] During the Pleistocene, the genus was remarkably widespread, occurring from southwestern Europe's Iberian Peninsula[88] across Eurasia and into far northwestern North America. It is also known from a single site in northeastern Mexico, San Josecito Cave (see fig. 3.14).[89] Obviously, it had to occur someplace in between, and the Great Basin is one possibility.

The Short-Faced Skunk (*Brachyprotoma obtusata*)

In life, the short-faced skunk (*Brachyprotoma obtusata*) would have looked much like a spotted skunk (*Spilogale putorius*) with short and powerful jaws.[90] Although we have no way of knowing what its coat might have looked like, *Brachyprotoma* likely had warning coloration just like that of other skunks. This animal is not particularly well known, with only seven late Ice Age North American sites having provided its remains (fig. 3.15). Two of these are in the Great Basin: Nevada's Mineral Hill Cave[91] and Utah's Crystal Ball Cave.[92] Both of these sites are discussed in the next chapter.

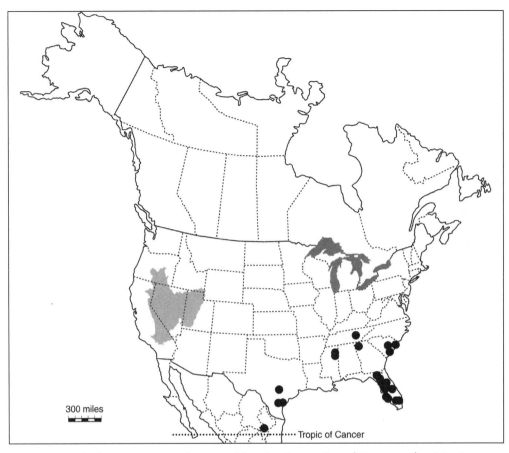

FIGURE 3.13. The late Ice Age distribution of the Florida cave bear (*Tremarctos*) in North America.

The Giant Bear (*Arctodus simus*)

There are some Ice Age mammals whose prior existence I find hard to fathom even after decades of experience with them. A Volkswagen Beetle–sized glyptodont is one; a ground sloth as tall as a giraffe and as bulky as an elephant is another. For me, the giant bear (*Arctodus simus*) falls in the same category (fig. 3.16).

Arctodus shares a number of distinctive physical characteristics with the living spectacled bear of South America, with the extinct *Tremarctos floridanus* of North America, and with the extinct South American *Arctotherium* that I discussed earlier in this chapter. As a result, they are all placed in the same subgroup of the bear family—the tremarctines.

There is a reason *Arctodus simus* is called the giant bear. Estimates of its weight vary, but all agree that this was the largest land carnivore the Americas had to offer during the late Ice Age. Just like living bears, *Arctodus* was sexu-

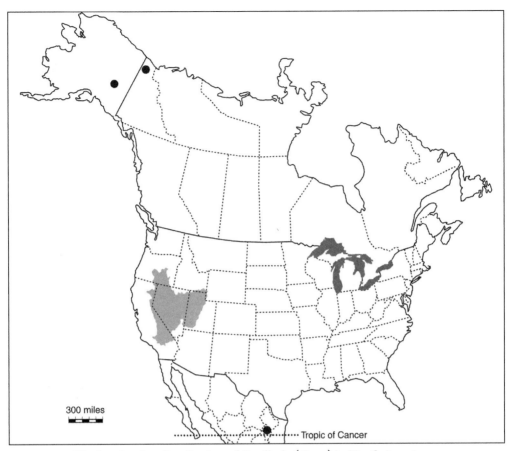

FIGURE 3.14. The late Ice Age distribution of the dhole (*Cuon*) in North America.

ally dimorphic, with males larger than females. Large individuals, presumably males, weighed between 1,500 and 1,800 pounds; the largest may have exceeded 2,000 pounds, giving us a ton of bear. Even the smallest adults, presumably females, weighed over 600 pounds.[93]

Not only were these bears large, they were also widespread, known from Florida to Alaska, including the Great Basin (fig. 3.17). One of the Great Basin locations, the Huntington Mammoth Site on Utah's Wasatch Plateau,[94] has provided the youngest trustworthy radiocarbon date available for the giant bear; this site is discussed in chapter 5.

A second location, Labor-of-Love Cave, is on the eastern edge of the Schell Creek Range, in eastern Nevada. Initially reported by scientifically concerned cavers (many of them are), this site was carefully collected by paleontologists Steven Emslie and Nicholas Czaplewski.[95] Their work showed the presence of two partial skeletons of giant bears, both lying in pools of water after having been eroded out of the cave's sediments by a stream flowing through the

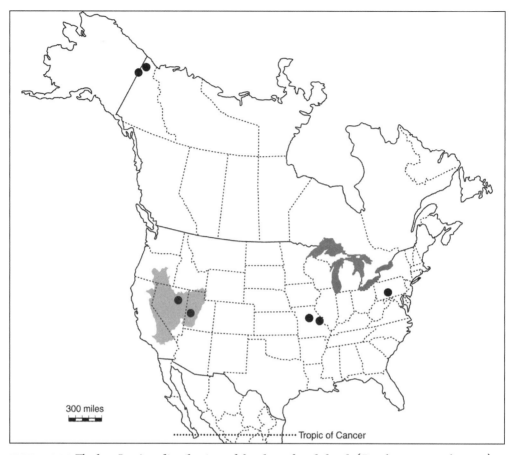

FIGURE 3.15. The late Ice Age distribution of the short-faced skunk (*Brachyprotoma obtusata*) in North America.

FIGURE 3.16. The giant bear (*Arctodus simus*). Drawing by Wally Woolfenden.

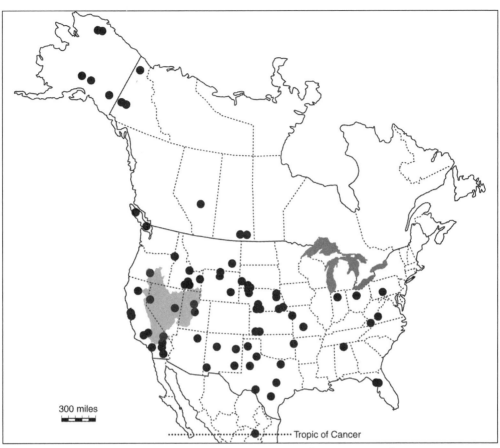

FIGURE 3.17. The late Ice Age distribution of the giant bear (*Arctodus simus*) in North America.

site. One of these two bears was a juvenile, about three or four years old; the second was a full adult. Both had apparently died while denning in the winter, something we know these bears did.[96] The bones recovered by this work now reside at the Natural History Museum of Los Angeles County, but a reconstruction of an adult giant bear skeleton can be seen in its full glory at the White Pine County Public Museum in Ely, Nevada.

Until very recently, *Arctodus* was referred to as the "giant short-faced bear." I haven't called it that because biologist Borja Figueirido and his colleagues have shown that, for a bear of its size, the length of *Arctodus*'s face was just right.[97] "Giant," however, remains as appropriate as ever.

What the giant bear did for a living has long been a topic of debate among paleobiologists. In the past, this bear has been seen as a highly mobile predator, as a specialized scavenger, and as a true omnivore, like so many living bears.[98] Now, as a result of detailed analyses of the bear's skeleton, it is appearing more and more likely that giant bears were, in fact, omnivores. They killed

FIGURE 3.18. The sabertooth cat (*Smilodon fatalis*). Drawing by Wally Woolfenden.

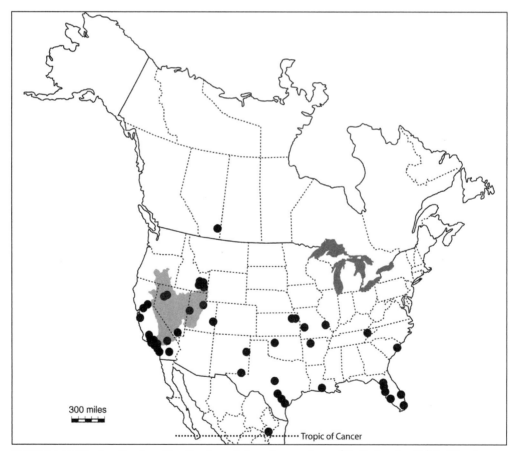

300 miles

Tropic of Cancer

FIGURE 3.19. The late Ice Age distribution of the sabertooth cat (*Smilodon fatalis*) in North America.

when they could kill, scavenged when they could scavenge, and ate whatever else their large hearts desired, including plant material.[99]

Although the giant bear was the largest land carnivore that the American late Ice Age had to offer, it was not the largest terrestrial carnivore of the American Pleistocene as a whole. That honor goes to the South American early and middle Pleistocene bear *Arctotherium angustidens*, the South American giant (short-faced) bear. This animal weighed an astonishing 3,600 pounds or so.[100] As far as we know, it was the largest bear to have ever existed. To balance it on a set of scales, you would need about ten spectacled bears on the other side.

In addition to the Huntington Mammoth site and Labor-of-Love Cave, the remains of giant bears have been found at the Monroc Gravel Pit south of Salt Lake City, at Fossil Lake in south-central Oregon, at Duck Flat in western Nevada, in the deposits of Pleistocene Lake Manix in the Mojave River drainage system, and from an otherwise undescribed site in Lincoln County, Nevada.[101]

The Sabertooth Cat (*Smilodon fatalis*)

Like the mammoth, the sabertooth cat (*Smilodon fatalis*) is one of the iconic mammals of the North American Ice Age (fig. 3.18). Most of us learned what it looked like while we were growing up, even if this is all we know about it.

Widespread south of glacial ice in North America (fig. 3.19), sabertooth cats are best known from southern California's Rancho La Brea. For at least the last 50,000 years, animals have been trapped by the viscous asphalt deposits found here.[102] Herbivores, perhaps coming to drink from water pooled on the surface, became stuck. Once stuck, they attracted carnivores and scavengers, who in turn became stuck. This was, as paleontologist Loye Miller put it long ago, a "trap that was practically automatic, its demands insatiable, and its patience unwearied."[103]

The remains of over 2,000 individual sabertooth cats have been exhumed from these deposits, providing an unparalleled opportunity to learn what these animals were like. The Great Basin record is far scantier than this, doing little more than showing us that these animals were here.

Although they are sometimes referred to as sabertooth tigers by professionals and the general public alike, these animals were not tigers at all. Along with lions, jaguars, and leopards, tigers belong to the genus *Panthera*; they are morphologically and genetically very distinct from the cats that are placed in the genus *Smilodon*.[104] With an estimated body weight of 350 to 600 pounds, however, sabertooth cats were about the same size as the Siberian tiger (*Panthera tigris altaica*), which weighs between 480 and 570 pounds.[105] These were massive animals.

Sabertooth cats not only were massive in general but had particularly massive forelimbs, with the upper bone of the forelimb—the humerus—having very thick walls and a very large diameter. Those forelimbs would have been resistant to bending and, coupled with the huge muscles attached to them and broad paws, would have enabled *Smilodon* to safely bring down and hold its prey.[106] Along with its powerful forelimbs, *Smilodon* had relatively short legs, producing, as paleontologist Mauricio Antón has so appropriately described it, a "hyper-robust, wrestling physique."[107] It was also a physique appropriate to an ambush hunter, one that emerged from concealing vegetation to attack its victims.[108]

Once the prey was down, it was time for the sabers to get to work. Those sabers are greatly elongated upper canines: 8 to 10 inches long, with about three-fifths of the length protruding from the gum line. They are also flattened, with the inner side somewhat flatter than the outer, and about twice as long from front to back as from side to side. Both the front and back of the tooth have fine serrations, though these are more obvious in younger individuals than in older ones since they wore down with use.[109] The exposed portion of these teeth grew quickly, about ¼ inch per month, and continued to grow at that rate until the tooth reached its final form. For a saber to have 6 inches exposed above the gum line would require 24 months. That is about the same amount of time it takes the much slower-growing, and far shorter, canines of a lion (*Panthera leo*) to reach their final dimensions.[110]

How sabertooth cats used those upper canines to dispatch their prey has long been a matter of debate. It has been argued that they were used to stab prey animals, to slash them, to shear through their flesh using the lower front teeth (the incisors) as anchor points, or to slice the prey without the lower incisors playing a major role. It now seems likely that the sabertooth used this last approach, slicing its sabers with an arc-like cut through its unfortunate prey.[111] To avoid breaking the slender and fragile sabers, those teeth would have had to avoid contact with bone. The best guess is that a sabertooth cat used its powerful forelimbs to hold its prey down and then applied the finishing blow to the prey's neck.[112]

We don't know exactly what that prey was, but we can make some pretty good guesses. *Smilodon* seems to have been capable of taking down animals more than twice its own body weight.[113] Chemical analyses of the bones of *Smilodon* and of the cat's potential meals from Rancho La Brea suggest that they were feeding on such animals as camels, horses, and Harlan's ground sloth (*Paramylodon harlani*).[114] Adult mammoths and mastodons were probably beyond their reach, but juvenile members of these species would not have been.

Although more than 2,000 individual sabertooth cats have been unearthed at Rancho La Brea, it is not the most abundant carnivore known from there. That honor belongs to the dire wolf (*Canis dirus*),[115] which I discuss later in this chapter. Dire wolves seem to have been social animals, much like contemporary wolves (*Canis lupus*), and the great abundance of dire wolves at Rancho La Brea is thought to reflect the fact that packs of them were attracted to animals that had become mired in the tar-like deposits of the site. Once drawn in, the dire wolves themselves became mired.

It could be that *Smilodon* was also a social animal and that its great abundance at Rancho La Brea is to be accounted for in the same way. In addition, the teeth of the Rancho La Brea sabertooth cats show heavy wear, and significant numbers of them are broken. Paleobiologists Wendy Binder and Blaire Van Valkenburgh have argued that this sort of tooth damage resulted from members of a social species competing with one another for food, eating quickly and consuming large carcasses fairly thoroughly.[116] It has long been known that all age classes of sabertooth cats are represented at Rancho La Brea, from juveniles to old adults, and this, too, is consistent with groups of social animals becoming trapped together.[117]

Not all agree that *Smilodon* was a social animal, suggesting instead that it was a solitary hunter.[118] If it were solitary, or if it hunted in pairs with cubs, an animal like a bison would have provided more dinner than a single sabertooth or even a pair with cubs could eat. The result might have been a boon for scavenging mammals and birds. On the other hand, if sabertooth cats were social, an added advantage would have been the ability to protect the carcass from such other predators as dire wolves. That would have allowed a significant number of cats to be fed with a single victim and would thus have reduced the number of potentially dangerous kills that had to be made.[119]

None of the five Great Basin sites that have yielded *Smilodon* material have yielded it in any abundance. Utah's Crystal Ball Cave provided a wrist bone, a claw, and a vertebra.[120] Silver Creek, in northern Utah's Wasatch Range, yielded a fragment of a saber, a humerus, and a vertebra. Each of two sites in northwestern Nevada's Black Rock Desert provided a saber fragment, not far from the more complete remains of mammoths.[121] The precise nature of the *Smilodon* remains from the Ridgecrest site, in the China Lake Basin of southeastern California, has not been described in detail, but there is no reason to think that this location was any richer in such material.[122] The most recently

FIGURE 3.20. The scimitar cat (*Homotherium serum*). Drawing by Wally Woolfenden.

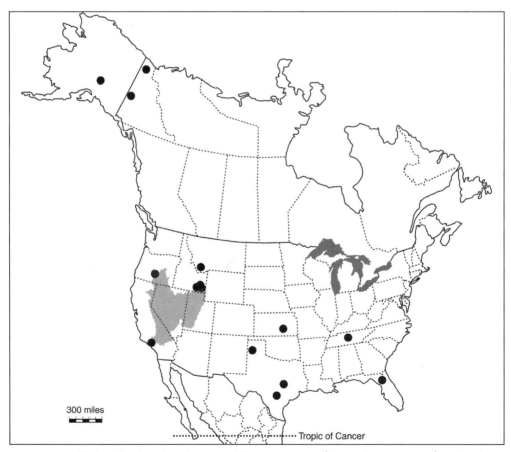

FIGURE 3.21. The late Ice Age distribution of the scimitar cat (*Homotherium serum*) in North America.

reported Great Basin sabertooth remains—two limb bones—come from Las Vegas Wash, near Las Vegas.[123] Someday, perhaps, those of us who work in the Great Basin will get lucky, and a complete, or at least more complete, skeleton of this animal will be found.

The Scimitar Cat (*Homotherium serum*)

During the late Ice Age, North America supported three different kinds of cats that weighed 300 pounds or more. One of these is the sabertooth, which I have already discussed. The second is the American lion (*Panthera leo atrox*), which I discuss later in this chapter. The third is the scimitar cat (*Homotherium serum*; fig. 3.20). The three are distinct in many ways, including the shape of their upper canines. The American lion had conical upper canines, with the fore-and-aft dimensions roughly equal to the side-to-side ones. The saber-tooth cat had enormously long and flattened upper canines. Scimitar cats had long, flattened canines as well, but these were shorter and flatter than those shown by *Smilodon*. *Homotherium* was North America's late Pleistocene scimitar cat.

Widespread in North America but known from only a fairly small number of sites (fig. 3.21), *Homotherium* was longer legged and far more agile than the stockier *Smilodon* and was likely to have been as nimble as a modern lion.[124] With claws less fully retractable than those of either the sabertooth or lion, the scimitar cat was capable of running down its prey with short bursts of speed.[125] Weighing some 350 to 400 pounds—about the size of an African lion—it was also capable of taking down animals weighing at least 1,200 pounds.[126] We know this because it has been found directly associated with such prey.

Friesenhahn Cave, in south-central Texas, provided the remains of at least 33 scimitar cats, including the skeletons of kittens between two and four months old and the remains of old adults.[127] Everything about this site suggests that it served as a *Homotherium* maternal den and shelter across an unknown, but substantial, period during the late Ice Age. It also served as a dining room for these predators. Something on the order of 1,000 mammoth bones have been found here, nearly all of which came from juveniles. Using the development of modern elephants as his guide, paleontologist Russell Graham has shown that these juveniles averaged about two years old when they died and that they would have weighed between about 1,100 and 1,300 pounds. From the work of paleobiologists Curtis Marean and Celeste Erhardt, we also know that some of the bones of these mammoths carry marks left by *Homotherium* teeth.[128] These young mammoths were scimitar cat prey.

It is not clear whether the Friesenhahn scimitar cats were dragging back entire carcasses of young mammoths or were bringing back only the meatiest parts. It also is not clear whether adult mammoths were too big for *Homotherium* to kill. It is possible, Graham has pointed out, that they could kill both young and old but could haul back only the young ones.[129] Nonetheless, given the size of a scimitar cat and the size of a full-grown mammoth, it seems most likely that juvenile mammoths were at the upper end of the size range these predators could handle.

Like those of sabertooth cats, the sabers of the scimitar version were serrated on both the front and back edges, though these serrations were coarser than those sported by *Smilodon*. Scimitar cat sabers also grew much more slowly than those of *Smilodon*. At 0.1 inch per month, their growth rate was virtually identical to that of modern lion teeth.[130] As I mentioned, the sabers were shorter than those of *Smilodon*, with crowns much less than half the length of the crowns of their sabertooth relative. They were also more compressed from side to side. Given this, their diets were likely far broader than *Smilodon* diets.[131]

Homotherium teeth also differed from those of the sabertooth in that they were all serrated.[132] *Homotherium* incisors were not only serrated, they were also larger and more robust than those of sabertooth cats and, unlike those in *Smilodon*, slanted forward from their bases. The Friesenhahn Cave scimitar cat incisors had broken during life with surprising frequency. About 22 percent of the upper incisors, and 60 percent of the lower, were damaged while their owners were still alive, well above the 5–10 percent frequency with which modern lions have been reported to break these teeth.[133] We know the animals were still alive when this happened because the broken edges are rounded and polished from further use. This, along with the heavy wear on these incisors, broken or not, shows that these were important scimitar cat tools, likely used to subdue prey, drag meals back home, and remove soft tissue from bone.[134]

In keeping with its generally scanty North American record, the scimitar cat is known from only one site in the Great Basin: Fossil Lake, Oregon (chapter 5). It is very possible that it was far more widespread in the Great Basin than our current knowledge would suggest.

The American Cheetah (*Miracinonyx trumani*)

The pronghorn antelope (*Antilocapra americana*), or, more properly, the pronghorn, is impressively fast on its feet. It can run long distances at 40 miles an hour and reach bursts of 60 miles an hour.[135] Given that it is far speedier

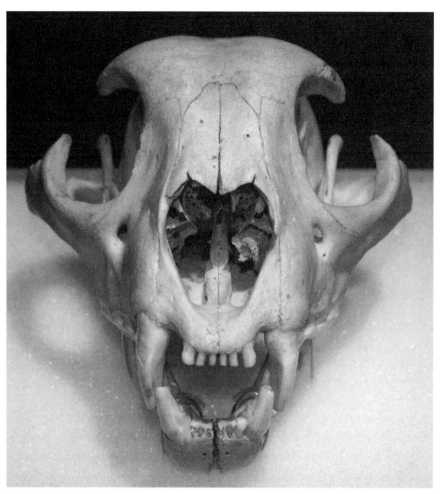

FIGURE 3.22. The skull of the American cheetah (*Miracinonyx trumani*) from Crypt Cave, Nevada, on display at the Nevada State Museum.

than any of its potential predators—"ridiculously too fast" is the way zoologist John Byers puts it [136]—we may well wonder why such an apparently over-engineered animal evolved.

To some, the answer to this question is provided by the American cheetah (*Miracinonyx trumani*; fig. 3.22).[137] Picture the form, but not necessarily the coloration, of an African cheetah (*Acinonyx jubatus*), and you will have an excellent idea of what this animal looked like: long legged, sleek, and powerful (fig. 3.23). There were also significant differences between the two. The late Ice Age *Miracinonyx* weighed an estimated 200 pounds compared to the African cheetah's 90 pounds.[138] In the African cheetah, the long bones of the lower part of the hind limb—the tibia (the inner bone) and the fibula (the outer)—are so tightly bound to one another that the lower limbs cannot rotate very

FIGURE 3.23. The African cheetah (*Acinonyx jubatus*), Nairobi National Park, Kenya. The skeleton of the American cheetah (*Miracinonyx trumani*) is so similar to that of the African cheetah that, until recently, the two were thought to be closely related. Photograph by James F. O'Connell.

much. In the American cheetah, this was not the case, allowing the lower limbs to be much more mobile.[139]

To get a feel for the difference this makes, find the two bumps at the very top of your ankle. The outer bump, called the lateral malleolus, marks the bottom end of the fibula. The inner bump, the medial malleolus, marks the bottom end of the tibia (your shin is part of the body of the tibia). Rotate your foot to the left and right. Then, tightly grab your leg just above the lateral and medial malleoli and try to rotate your foot again. You will immediately recognize that your ability to do this is greatly restricted when the tibia and fibula are not able to move relative to one another. The condition in the African cheetah is akin to the latter situation, of restricted mobility. In the American cheetah, it is akin to the former situation, of increased mobility. There is a real trade-off here. The American cheetah had greater mobility in the lower part of its hind limb; the African cheetah has traded mobility for structural strength and stability.

In addition, and unlike those of other cats, the claws of the African cheetah cannot be fully retracted. Because of this, the claws, along with ridges on the foot pads, provide greater traction and stability in running.[140] The American cheetah, on the other hand, had fully retractile claws, suggesting de-creased traction and stability when running compared to that of the African cheetah.[141]

There are other differences,[142] but in general these are morphologically very similar animals. They are so similar that when paleontologist Daniel Adams described the American cheetah in detail in 1979, he assigned it to a subgenus, *Miracinonyx*, of the genus to which the African version belongs, *Acinonyx*. The "mira" part of the name comes from the Latin word *mirum*, meaning "surprising" or "amazing." It was, in all ways, an appropriate choice.

About a decade later, enough new material of this cat had been found that a more detailed analysis of its skeleton became possible. That analysis showed the full range of differences between the skeletons of the American and Afri-can cheetahs[143] and made it clear that the American version deserved to be in its own genus. Adams's subgenus *Miracinonyx* was raised to the genus level and has stayed there ever since.

Adams's demonstration of the similarities between these versions of cheetahs led him to wonder whether the ancestry of the African cheetah might lie in North America. Others have wondered the same thing.[144] Thanks to the analysis of ancient DNA from the bones of American cheetahs, we now have the answer. The closest relative of the American cheetah is the cougar (*Puma concolor*). The ancestor of both the cougar and the American cheetah entered North America via the Bering Land Bridge, perhaps some eight mil-lion years ago. Both the cougar and the American cheetah evolved from this ancestor.[145] This is what many, though not all, paleontologists had thought all along but, until the genetic evidence came along, could not prove.[146]

If that is the case, why are African and American cheetahs so similar? The answer to this also seems clear. They look similar because they developed a similar lifestyle, involving the high-speed pursuit of their prey. We do not know how fast American cheetahs could run, but African cheetahs can reach speeds of about 65 miles an hour for short distances.[147] Since it seems likely that pronghorn formed part of the American cheetah's diet, the speedy nature of this potential prey, and its seemingly overengineered nature, may have been the result of millions of years of trying to escape *Miracinonyx*. The faster ones escaped; the slower ones became dinner.

There are not many late Ice Age records for *Miracinonyx* in North America (fig. 3.24), but two of these are from the Great Basin. In the 1950s, paleontol-ogist Phil Orr excavated a complete skull of what he argued was a new spe-cies of cat from Crypt Cave, in the Winnemucca Lake Basin of northwestern

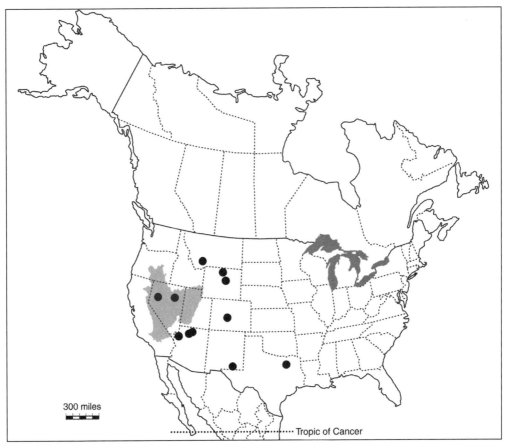

FIGURE 3.24. The late Ice Age distribution of the American cheetah (*Miracinonyx trumani*) in North America.

Nevada (fig. 3.22).[148] He called that cat *Felis trumani*, thereby providing the species name for what we now recognize as the late Ice Age American cheetah, *Miracinonyx trumani*.

The second Great Basin record for the American cheetah comes from Mineral Hill Cave, in northwestern Nevada (chapter 5). This site provided a fourth metatarsal that paleobiologists Bryan Hockett and Eric Dillingham identified as "cf. *Miracinonyx trumani*."

This identification needs a bit of explanation. First, the metatarsals are the long bones of your feet, the bones to which the bones of your toes are attached. Your fourth metatarsal is one stop in from the outside of your foot. Second, "cf." comes from the Latin word *confer*, which means to "bring together" or "join." Paleontologists use this word to mean "compares best with." By "cf. *Miracinonyx trumani*," Hockett and Dillingham meant that the fourth metatarsal of their cat compared best with the fourth metatarsal of *Miraci-*

FIGURE 3.25. The capybara (*Hydrochoerus hydrochaeris*). Photograph by E. R. Congdon, courtesy of the Mammal Image Library (MIL image 1898).

nonyx trumani. Given their detailed anatomical description of this specimen, there is little reason to doubt that this is exactly what they found.[149]

The Rodents

The three genera of rodents that were lost toward the end of the North American Ice Age were remarkable for their size. I have already discussed two of these, because they were also found in South America: the capybaras *Hydrochoerus* and *Neochoerus*. As I mentioned, *Hydrochoerus*, the world's largest living rodent (fig. 3.25), lives on in Central and South America. During the late Pleistocene, it made it into far southeastern and far southwestern North America. In addition to the sites shown in figure 3.26, this capybara also made its way to coastal California, roughly midway between Los Angeles and the Mexican border, but the age of this specimen is unknown.[150] As far as we know, the much larger *Neochoerus* was confined to the southeastern United States (fig. 3.27).

The giant beaver *Castoroides* was 8 feet long and weighed about 150 pounds, some eight times more than modern beavers.[151] Although a few sites are known from Alaska and the Yukon, this was primarily an animal of the

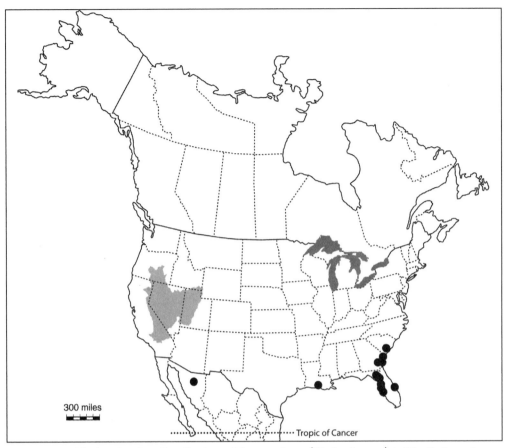

FIGURE 3.26. The late Ice Age distribution of the capybara (*Hydrochoerus*) in North America.

eastern half of North America (fig. 3.28). Compared to the chisel-like incisors
of today's beavers, those of the giant beaver were blunt and rounded; this, and
other aspects of its skeleton, tell us that giant beavers were not tree-felling dam
builders. Like modern beavers, they lived near water, but everything we know
about them suggests that their habits were far more like those of muskrats
(*Ondatra zibethicus*) than those of modern beavers.[152]

The Lagomorphs (Rabbits, Hares, and Pikas)

The lagomorph order includes the rabbits, hares (jackrabbits), and pikas. The
first two of these will be familiar to anyone who has been bold enough to go
outside in the arid West. The third, the American pika (*Ochotona princeps*),
will be familiar to those who have spent time hiking at higher elevations in
the Sierra Nevada, the Rockies, the Cascades, or toward the top of a number
of isolated mountain ranges in the Great Basin. The deeper histories of all
of these animals are reasonably well known. Special attention has been paid
to pikas, whose fate under warmer climatic conditions is of great concern

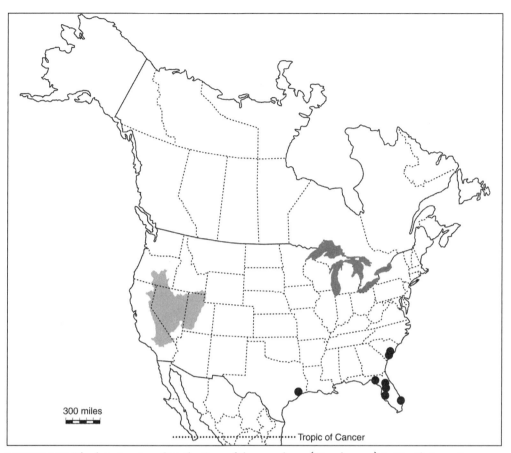

FIGURE 3.27. The late Ice Age distribution of the capybara (*Neochoerus*) in North America.

to many and whose history since the late Pleistocene has been one of fairly steady geographic retreat.[153]

The Aztlán Rabbit (*Aztlanolagus agilis*)

To these familiar animals, the late Pleistocene of western North America adds a third: the Aztlán rabbit (*Aztlanolagus agilis*). Smaller than a cottontail, this rabbit has been securely identified only from its distinctive teeth. These teeth, however, have been found associated with skeletal material that could belong to no other known animal. These bones show that the Aztlán rabbit had slender limbs compared to those of both jackrabbits (*Lepus*) and cottontails (*Sylvilagus*), and that its running habits were probably far more similar to those of the former than to those of the latter. This, in turn, suggests that it was also an animal of fairly open landscapes.[154]

The Aztlán rabbit is not particularly well known, but we do know that it was found along what is now the US-Mexico border in the Southwest, south into Mexico (fig. 3.29). There are also two records from the Great Basin. The

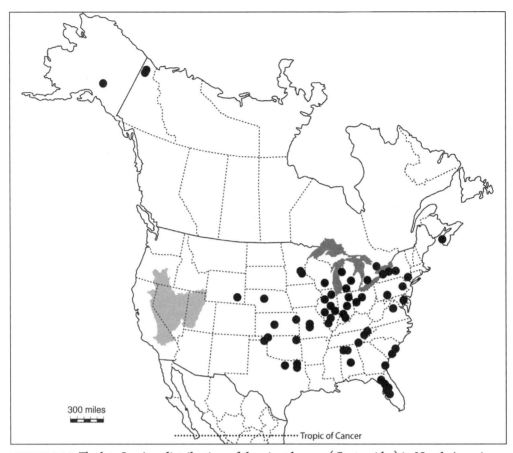

FIGURE 3.28. The late Ice Age distribution of the giant beaver (*Castoroides*) in North America.

first of these is from Cathedral Cave, on the east flank of eastern Nevada's
Snake Range. Here, paleontologist Chris Jass reported the presence of the
Aztlán rabbit in sediments that may be as much as 150,000 years old.[155] Jass
also discovered that previously unanalyzed late Pleistocene material from
Smith Creek Cave (chapter 5), on the other side of the narrow valley in which
Cathedral Cave sits, contains the remains of this animal. It is a good guess that
as more excavations are done in the Great Basin, more remains of the Aztlán
rabbit will be found, as long as the work is done with the painstaking care that
Jass brought to his work at Cathedral Cave.

The Perissodactyls (Odd-Toed Ungulates)

Not surprisingly, the odd-toed ungulates—the horses, tapirs, and rhinos—are
characterized by an odd number of weight-supporting toes on their feet, with
the middle toe carrying the main burden. The horses have one per foot, the
tapirs and rhinos three.

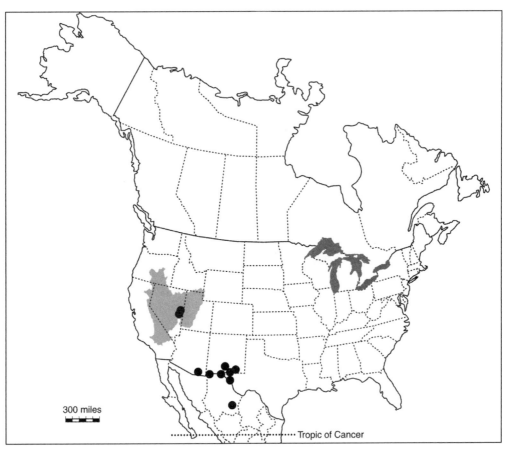

FIGURE 3.29. The late Ice Age distribution of the Aztlán rabbit (*Aztlanolagus agilis*) in North America.

There were no Pleistocene rhinos in the Americas. Although the Eurasian woolly rhinoceros, *Coelodonta antiquitatis*, made it to the Siberian side of the Bering Land Bridge during the late Pleistocene, it never made it across, perhaps because the Land Bridge did not provide suitable habitat for it.[156]

Tapirs, however, did make it across (fig. 3.30). The ancestors of modern tapirs originated in Eurasia and crossed the Bering Land Bridge long before the Ice Age began.[157] Today, they come in five varieties—four species in southern Mexico and South America and one in southeastern Asia.[158] During the late Pleistocene, large and small versions were found from coast to coast in North America (fig. 3.31), though how many species these different-sized tapirs represent remains unknown.[159] They are not known from the Great Basin, but they came so close that that it would not be astonishing were one to be discovered here.

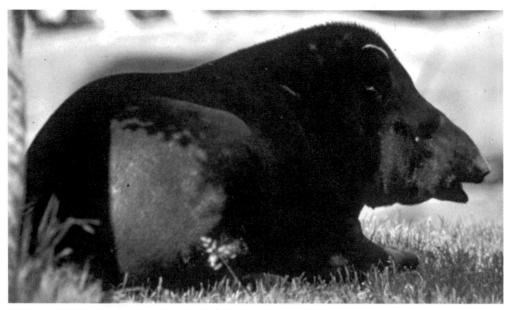

FIGURE 3.30. The South American tapir (*Tapirus terrestris*). Photograph by L. L. Master, courtesy of the Mammal Image Library (MIL image 602).

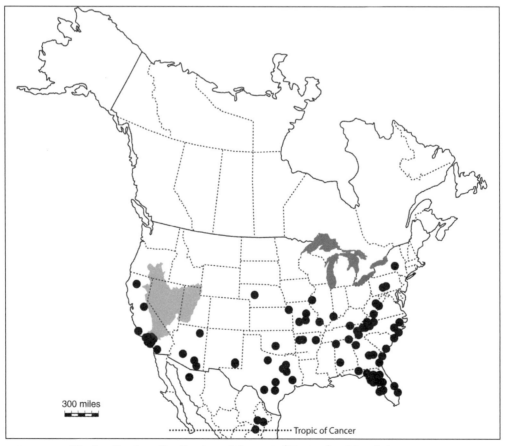

FIGURE 3.31. The late Ice Age distribution of the tapir (*Tapirus*) in North America.

The American Horse (*Equus*)

Horses and their relatives, all in the genus *Equus*, evolved in North America. They spread across the Bering Land Bridge into Eurasia, giving us the wild horse (*Equus caballus*) and the Asian asses (the onager, *Equus hemionus*, and the kiang, *Equus kiang*). From Eurasia, they spread into Africa, giving us the African wild ass (*Equus asinus*), three species of zebras (Burchell's zebra, *Equus burchellii*; Grevy's zebra, *Equus grevyi*; and the mountain zebra, *Equus zebra*), and the recently extinct South African quagga (*Equus quagga*). They also spread into South America, providing the two genera of South American horses that I discussed earlier in this chapter—*Equus* and *Hippidion*.

By 10,000 years ago, all of the American horses were extinct. As I have discussed, the horses that are found here today were repatriated by Europeans, who brought them over as part of their great colonizing enterprise and who continue to bring them over today. Those wild horses you see in the American West owe their existence to Europeans, but Europeans had them only because of the Bering Land Bridge.

The bones and teeth of horses are immediately recognizable. See one, and you know it came from an animal belonging to the genus *Equus*. Identifying the species of horse that might have been involved is a different matter. In fact, if you want to confuse yourself, try reading the literature on American Ice Age horses. About 60 species of them have been defined over the years. In 1989, paleontologist Melissa Winans did a massive statistical analysis and reduced them all to five major groups, but for all the work she did, she still could not say how many species were in each group. As I mentioned earlier, an important review published some ten years after that reduced all this confusing clutter to ten species, but even the species in that review did not necessarily match the species other people used afterward.[160]

We do know that North American horses fall into two broad groups. There are horses with long and slender bones (metacarpals and metatarsals) in their lower limbs. These are called stilt-legged horses. In general form, they are quite similar to the Asian wild asses. In fact, they are often referred to as "hemiones," the informal subset of the genus *Equus* that includes the onager and kiang. Then, there are horses that are far more similar to the domestic horse and its wild relatives of the same species, *Equus caballus*. These are often referred to as "caballine" horses. This is the animal that Europeans began bringing over almost as soon as they began to settle in the Americas.

Beyond that, there is not much agreement as to how many kinds of horses there were during the North American Ice Age. Fortunately, the scientists who have made it their business to study ancient DNA have begun to figure this out for us. Paleogeneticist Jaco Weinstock and his colleagues took a close look at the DNA from a wide variety of American Ice Age horses, and the results began to make sense out of the confusing variety of species that have been named. Although their sample of fossil material was not large enough to answer all the questions paleontologists have had about American Pleistocene horses, their findings and conclusions were delightfully straightforward. Their work confirmed that there really were three major groups of genetically distinct horses in the American Pleistocene: the South American *Hippidion*, the stilt-legged horses, and the caballine horses. They showed that the stilt-legged horses were distinct from the Asian asses and represent a genetic lineage that evolved in North America. That was a big deal for those interested in horses, whether ancient or modern. And, a bigger deal yet, they suggested that the North American Pleistocene may have supported only one very variable species of stilt-legged horse and one very variable species of caballine horse. Later work has supported these conclusions.[161]

Because this work is still in its relative infancy, we still do not know how many species of horses there were in the North American Ice Age. We won't know that until a far larger sample of Pleistocene horse material has been analyzed. At least, though, we have learned that all those paleontologists who divided North American Ice Age horses into two large groups—the stilt-legged ass-like ones, and the stouter-legged domestic horse–like ones—were right to do so. We also know that there may have been only two species of horses in North America as the Pleistocene came to an end. To paleontologists who have struggled with this material for so long, that is refreshing. We can look forward to being increasingly refreshed, and far less confused, as more work of this sort is done.

Horses were both widespread and common in North America during the late Pleistocene, both south and north of glacial ice (fig. 3.32; the sites in central and western Canada likely represent animals that were moving northward as the western and eastern ice masses parted in response to warming at the end of the Ice Age). The Great Basin is no exception to this abundance. In fact, as we will see in chapter 4, it is the most commonly reported extinct Ice Age mammal in the Great Basin.

The Artiodactyls (Even-Toed Ungulates)

If it wasn't surprising to learn that the odd-toed ungulates support their weight on an odd number of toes, it won't be surprising to learn that the even-toed ungulates, the cloven-hoofed beasts, support their weight on just two, the third and fourth toes. Pigs, cattle, goats, sheep, and giraffes are obvious examples.

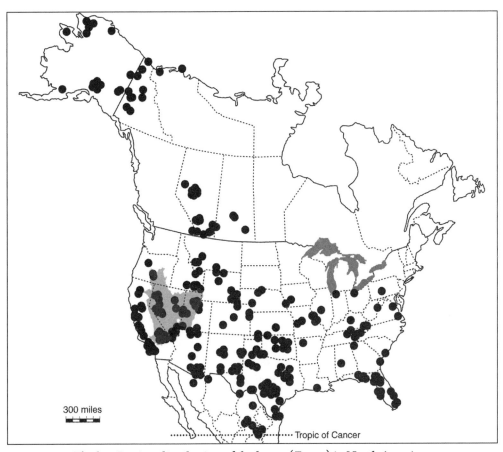

FIGURE 3.32. The late Ice Age distribution of the horse (*Equus*) in North America.

Today, North America supports ten genera of artiodactyls, all of which are familiar to us (table 3.3). Deer, bison, caribou, elk, moose, pronghorn antelope, mountain goats, bighorn sheep, and collared peccaries (javelinas) are all artiodactyls—ten genera in four families.

All these genera are known from the North American late Ice Age, though, as I discuss below, elk and moose are very late arrivals and collared peccaries seem to have been very rare. Alongside them, there were 13 other genera, including one family that no longer occurs here—the Camelidae, or camel family. During the late Pleistocene, North America supported 23 genera of artiodactyls compared to the 10 it supports today. These came in an almost bewildering variety of forms.

The Tayassuids (Peccaries)

Although collared peccaries are known from the late Pleistocene of Florida, they seem to have been exceedingly rare in North America until very recent times.[162] To make up for that rarity, we were graced by the presence of two

TABLE 3.3. The contemporary artiodactyls of North America

Family	Scientific name	Common name
Tayassuidae	*Pecari tajacu*	Collared peccary
Cervidae	*Alces americanus*	Moose
	Cervus elaphus	Elk
	Odocoileus hemionus	Mule deer
	Odocoileus virginianus	White-tailed deer
	Rangifer tarandus	Caribou
Antilocapridae	*Antilocapra americana*	Pronghorn
Bovidae	*Bison bison*	Bison
	Oreamnos americanus	Mountain goat
	Ovibos moschatus	Muskox
	Ovis canadensis	Bighorn sheep
	Ovis dalli	Dall's sheep

other genera of peccaries. The long-nosed peccary (*Mylohyus nasutus*) was primarily an eastern animal, though it has been discovered as far west as western Texas (fig. 3.33).[163] Marked by a long and narrow snout, by long and slender legs built for running, and by eye placement suggesting it had long-distance vision better than that of most modern peccaries, it appears to have been a solitary animal that preferred forested habitats.[164]

The Flat-Headed Peccary (*Platygonus compressus*)

The flat-headed peccary (*Platygonus compressus*) was distinctly different from the long-nosed version (fig. 3.34). With the latter, it shared long and slender legs suggesting good running abilities and eye placement suggesting strong vision. The differences, however, are substantial.

Most obviously, *Platygonus* was far more widespread, found from coast to coast and even known from the Yukon (fig. 3.35).[165] It lacked the long snout of *Mylohyus*, and the part of its skull just behind the nose was flatter than in the long-nosed peccary (hence the name). Not only did the flat-headed peccary have better long-distance vision than most modern peccaries, it also had a better sense of smell than both its modern relatives and *Mylohyus*.

This we know because those parts of the facial skeleton that are associated with the sense of smell are greatly enlarged compared to those of most of its modern relatives and to those of *Mylohyus* (though the long snout of the long-nosed peccary would have helped here). The bones in this part of the skull, called turbinates, are draped with soft tissue that has three prime functions, depending on which kind of turbinate is involved. The first function is to humidify and warm air that is about to enter the lungs. The second

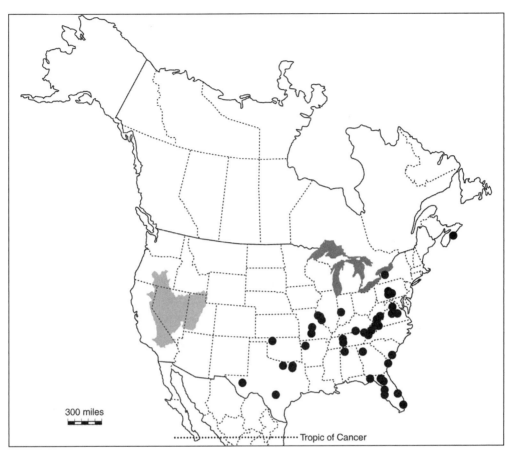

FIGURE 3.33. The late Ice Age distribution of the long-nosed peccary (*Mylohyus*) in North America.

is to retain moisture that would otherwise be lost when the animal exhales. Air that passes over the turbinates deposits that moisture on them, which is then picked up by the next round of incoming air. The third prime function is to help detect whatever scents are carried by the air, since the soft tissue that covers some of these turbinates contains the nerve endings that put the sense of smell into play.[166] The expanded portions of the flat-headed peccary's skull suggest that while all three of these features may have been enhanced, the evolutionary target was likely to have been the animal's sense of smell.

When peccaries fight one another, they use their elongated lower canine teeth as slashing tools, jabbing upward and outward. We can be certain that *Platygonus* did this because it had slender versions of those elongated lower canines, and because those portions of the skull that were involved in peccary fights were strengthened. To help support the canines and strengthen

FIGURE 3.34. The skeleton of the flat-headed peccary (*Platygonus compressus*), National Museum of Natural History specimen V23706, Bradford County, Pennsylvania; see Ray, Denny, and Rubin (1970).

the mandible (the lower jaw), the bottom of the mandible has a bony buttress where the two halves meet. In addition, the bar of bone that helps form the lower portion of the eye socket, called the zygomatic arch, is enlarged. That enlargement would have protected the eye, and a diverse set of muscles, from being damaged by an enemy's jabbing lower canine.[167]

We think that *Mylohyus* was a solitary animal because it is generally found only one at a time. This is unlike most modern peccaries, which tend to organize themselves into herds.[168] We know that *Platygonus* was organized into at least family groups, because true fossil herds have been found. Denver, Colorado, provided an assemblage of five flat-headed peccaries, consisting of two adult males, an adult female, a juvenile, and a fifth individual that was either a fetus or a newborn.[169] Western Kentucky yielded five specimens that ranged in age from a young adult to fully mature individuals, all of which died while

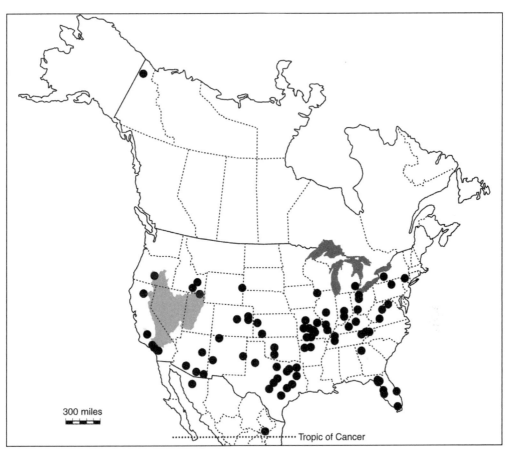

FIGURE 3.35. The late Ice Age distribution of the flat-headed peccary (*Platygonus compressus*) in North America.

they were walking in a line, head to tail.[170] Central Indiana also provided five animals that died together: a juvenile, one verging on adulthood, and three adults.[171] Bigger groupings are known as well, but the cumulative implication is clear: these were gregarious animals.

In 1975, the scientific world was introduced to a new genus of living peccary, found in the Gran Chaco of Argentina, Paraguay, and Bolivia. Named the Chacoan peccary (*Catagonus wagneri*), this animal had been known to science before that since it had been described from paleontological and archaeological sites, but scientists did not know that it still existed. The Chacoan peccary differs from other living peccaries in a substantial number of ways. It has longer legs, more slender canine teeth, eyes located for enhanced long-distance vision, and an expanded version of those parts of the skull associated with the sense of smell. If that sounds familiar, it is because these are all traits that characterize the flat-headed peccary.[172]

FIGURE 3.36. The living Chacoan peccary (*Catagonus wagneri*). The skeleton of the Chacoan peccary is so similar to that of the flat-headed peccary that when it first became known to science in the mid-1970s, it was initially thought that the animals might be one and the same. Photograph by D. G. Huckaby, courtesy of the Mammal Image Library (MIL image 1812).

In fact, the two forms are similar enough that some of the South American fossil material that we now know came from *Catagonus* was, at the time it was discovered, assigned to *Platygonus*. These are very similar animals, at one time making people wonder whether *Platygonus* had actually been discovered alive and well in South America. There are enough differences between the two to make it clear that this is not the case, but the similarities are impressive. Looking at *Catagonus* (fig. 3.36), you can get a good impression of what it might have been like to see a flat-headed peccary walk by.

The discovery of the *Platygonus*-like *Catagonus* brings me back to Hesketh Prichard's search for a living ground sloth in Patagonia and to the more recent report by David Oren that these animals might still exist. Ground sloths are known only from fossil material. *Catagonus* was first described from fossil material, but in this case, the animal lives on.

Some years ago, I wrote to biologist Jim Patton, an expert on the mammals of Amazonia, among many other places. I asked him whether he thought ground sloths might still be walking around in scientifically unexplored regions of South America. Here is what he said in response:

> Phil Myers, the "co-discoverer" to the *Catagonus* in Paraguay, wrote me from the field that the Guarani recognized three different kinds of peccaries out in the Chaco, and what did we (Oliver Pearson and I, his thesis advisors) think about that? We both answered, "well, everyone knows that there are only two living peccaries," a response that turned to the proverbial egg on our faces quickly. I long ago became more than just a little impressed by the knowledge of local peoples regarding the biological diversity they encountered, so I wouldn't dismiss Oren's tale—especially given the amount of western Amazonia that is unexplored and unknown.[173]

This turned out to be true not only for *Catagonus*, but also for Amazonian tapirs, a new species of which was described in 2013. The same is true for the peccary genus *Pecari*, a new species of which was described from the Amazon in 2007. All of these recently described species had long been known to local people.[174] Perhaps, one hopes, Hesketh Prichard's only mistake was to have looked for ground sloths in the wrong place.

The long-nosed peccary did not come close to the Great Basin, but the flat-headed peccary is known from two sites on the far northern edge of this region: Fossil Lake, Oregon (chapter 5), and a site in Idaho's Cache Valley, in the very northern edge of the area occupied by Pleistocene Lake Bonneville.[175]

The Camelidae (Llamas, Guanacos, Alpacas, Vicugnas, and Camels)

Everyone knows what camels look like: the two-humped Bactrian camel (*Camelus bactrianus*) of Asia, and the one-humped dromedary (*Camelus dromedarius*) of the Arabian Peninsula and northern Africa. Many also know that South America has its own members of the camel family—the llama, guanaco, and alpaca, now placed in a single species (*Lama glama*), and the vicugna (*Vicugna vicugna*). What is generally less well known is that, like horses, camels are native North American mammals. They evolved here and then, some seven million years ago or so, made their way to Eurasia through the Bering Land Bridge area.[176] Much later, early in the Ice Age, they also made their way into South America. Then, by 10,000 years ago, the North American versions became extinct, leaving two widely separated groups of camels with no relatives in between. These two groups are routinely placed in their own subsections of the camel family, the Camelini for the Old World forms, and the Aucheniini for their South American relatives.[177]

Toward the end of the Pleistocene, North America sported three distinctly different kinds of camels. Two of these—the stout-legged and large-headed llamas—would have looked very much as we would expect llamas to look and are most closely related to their South American cousins. The third, yesterday's camel, is far more closely related to the modern camels of Africa and Asia.[178]

The stout-legged llama (*Palaeolama mirifica*) had, for a llama relative, relatively robust and stocky limbs, as well as very low-crowned cheek teeth—the lowest, in fact, of any member of the lamine group.[179] As we have seen, this genus was also found in South America.

The fact that *Palaeolama* had such low-crowned cheek teeth for a llama is surely telling us something. High-crowned, or hypsodont, cheek teeth are often thought to have evolved in response to a diet heavy in grasses. Grasses are loaded with abrasive material: they contain silica, absorbed from ground water, and often have a healthy supply of grit on their surface. The higher the crowns on its teeth, the better equipped a mammalian herbivore is to deal with having its teeth worn down as it eats. As a result, low-crowned cheek teeth suggest an animal that was eating other things.

Detailed studies of *Palaeolama* teeth suggest that it was either a browser—a herbivore that eats plants other than grasses and sedges—or a mixed feeder, browsing while at the same time feeding on grassy plants low to the ground.[180] In fact, paleontologist Russell Graham has described a stout-legged llama specimen with odd wear on the tips of its incisors, as if it had been using them to strip leaves or other material from tall plants, and the same kind of wear has been reported elsewhere.[181]

In North America, *Palaeolama* was primarily an animal of the east, with sites known only as far west as central Texas (fig. 3.37). In these areas, it seems to have hung around in woodlands. Its body form evolved, Russell Graham has suggested, to evade predators in just such an environment.[182]

The Large-Headed Llama (*Hemiauchenia macrocephala*)

The large-headed llama (*Hemiauchenia macrocephala*) had longer and more slender limbs than *Palaeolama* and was clearly a better runner (fig. 3.38). Its teeth were also distinctly different, higher crowned and covered with cementum, the soft tissue that helps teeth adhere to the bone in which they are embedded and that helps bolster their strength when they are under pressure.[183] Sometimes thought to be an animal of open woodland,[184] *Hemiauchenia* was found from coast to coast in North America, though there are far more records for it from the western than the eastern half of this region (fig. 3.39).

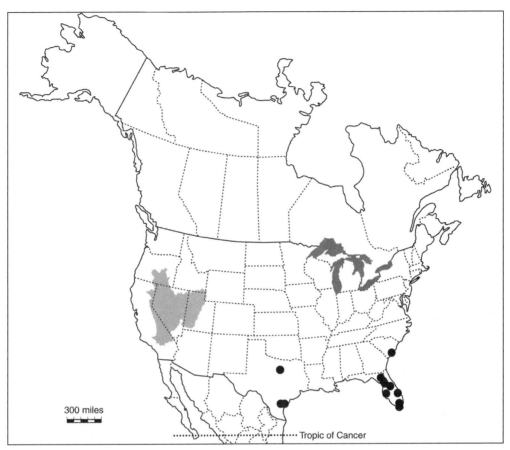

FIGURE 3.37. The late Ice Age distribution of the stout-legged llama (*Palaeolama*) in North America.

That in itself suggests that this was an animal comfortable in many different kinds of environmental settings. Matching this broad distribution, studies ranging from the shape of the animal's skull to the chemical content of its skeleton have shown that it was a mixed feeder, able to both browse and graze.[185]

It is likely that large-headed llamas organized themselves into herds. Paleontologists David Webb and Frank Stehli have described a site from coastal Florida that contained the remains of at least nine *Hemiauchenia* individuals that seem to have died at the same time.[186] To judge from tooth eruption, only one of these individuals was over two years old, and some had been born only about three months before they were killed by whatever catastrophe overtook them. Webb and Stehli suggest that these llamas lost their lives in a summer storm, and that only older, stronger adults were able to escape. Because this site dates to the middle Pleistocene, and the species of *Hemiauchenia* is not

FIGURE 3.38. The large-headed llama (*Hemiauchenia macrocephala*).
Drawing by Wally Woolfenden.

the same as the late Ice Age one, we cannot be sure that the large-headed llama
behaved in the same way, but it probably did.

Hemiauchenia was widespread in the Great Basin during the late Ice Age,
with sites known from the Mojave Desert of southeastern California and Gypsum Cave, Nevada, in the south, to Smith Creek Cave, Mineral Hill Cave, and
Fossil Lake in the north. We will meet most of these sites in the next chapter.

Yesterday's Camel (*Camelops hesternus*)
Excluding such animals as the capybara and horse, which live on elsewhere,
yesterday's camel is one of the easiest of North America's extinct mammals
to visualize (fig. 3.40). Begin with the one-humped dromedary. Make its legs
about 20 percent longer, move the hump slightly farther toward the animal's
head, make that head longer and narrower, bend the face downward, and
make the split upper lip heavier and stronger, and you've got something very
close to *Camelops hesternus*, even if not a spitting image. There were other dif-

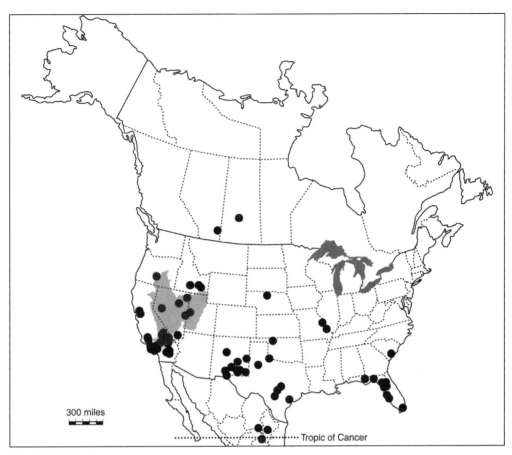

FIGURE 3.39. The late Ice Age distribution of the large-headed llama (*Hemiauchenia macrocephala*) in North America.

ferences, but they were more subtle. For instance, the two toes on each foot that supported the animal's weight were not as flat as they are in the existing camels. From a distance, the extinct yesterday's camel might well be mistaken for a dromedary.[187]

Armed with a long neck, a powerful but not stiff upper lip, long legs, and high-crowned teeth, yesterday's camels were mixed feeders, just as modern camels are. If there is any debate about this, it is over the proportions of grass *Camelops* included in its diet, with some studies suggesting a fairly small amount and others suggesting that they partook roughly equally of leaves, fruit, grasses, and other plant parts. These differences are not at all surprising since the diets of mixed feeders can be expected to vary depending on what is available to be eaten.[188]

Yesterday's camels were widespread in the more westerly parts of North America (fig. 3.41). Although they are known in small numbers from Alaska

FIGURE 3.40. Yesterday's camel (*Camelops hesternus*). Drawing by Wally Woolfenden.

and the Yukon, they, like the giant bear *Arctodus*, failed to cross the Bering
Land Bridge and so are unknown from Siberia.[189]

They are, however, known from throughout the Great Basin. We even
know that while they were here, they were eating saltbush (*Atriplex*), one
of the most common of the modern Great Basin's low-elevation shrubs (see
chapter 2). With modern camels, they seem to have shared a fondness for salty
plants.[190] To judge from the number of sites that have provided their remains,
camels seem to have been one of the most abundant of now-extinct late Ice
Age mammals in this area.

The Cervidae (Deer, Moose, Elk, Caribou, and Their Allies)
The deer family is represented in North America by four genera and five species: two species of deer and one each of caribou, elk, and moose (table 3.3).
Only the first three of these species seem to have been present much before
the tail end of the late Ice Age. Moose and elk, it appears, did not wander
across the Bering Land Bridge from Siberia until shortly after 13,000 years ago,
not long before the land bridge itself was flooded for the last time.[191] It is also
possible that elk did not arrive in North America south of glacial ice—that is,
south of about the Canadian border—until after about 10,000 years ago.[192]

On the other hand, caribou, which we now associate with the cold winters

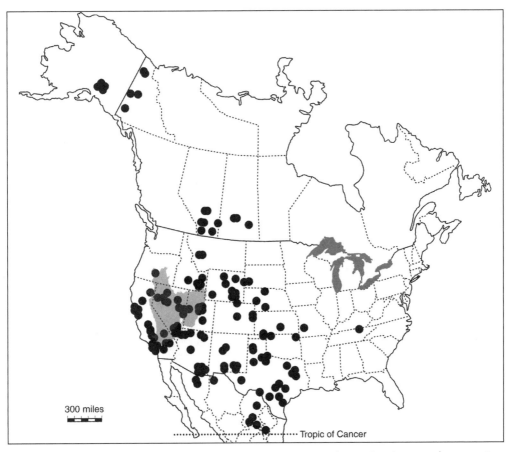

FIGURE 3.41. The late Ice Age distribution of yesterday's camel (*Camelops hesternus*) in North America.

of the far north, made it as far south as northern Mississippi during the late Pleistocene.[193] Much the same thing happened in Europe, where animals of the same species as our caribou are known as reindeer. Today, reindeer are also animals of the far north, but during the Pleistocene they were found in such places as western Italy and northern Spain.[194] Their European retreat northward seems to have been caused by ever-warmer summer temperatures, and the same may be true for their North American counterparts.[195]

To these animals, the North American Pleistocene added two other genera, only one of which, the stag-moose (*Cervalces*), is particularly well known. The stag-moose was found in two far-flung parts of North America: the eastern United States, especially the upper Midwest, and Alaska and the Yukon (fig. 3.42). In life, *Cervalces* would have looked very much like a moose, except for its longer legs and shorter hump, less massive neck, and distinctively different antlers.

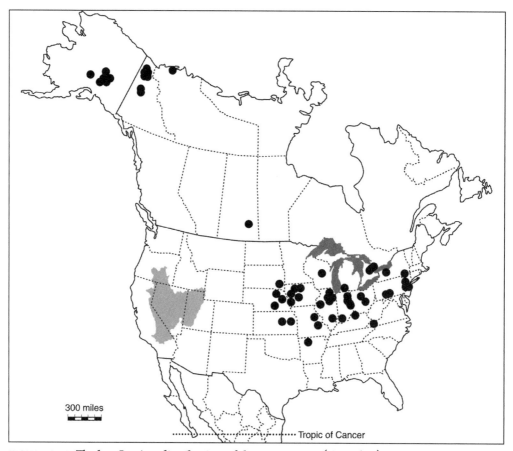

FIGURE 3.42. The late Ice Age distribution of the stag-moose (*Cervalces*) in North America.

In fact, most specimens of the stag-moose have been identified on the basis of those antlers, since the antler "beam"—that part of the antler that attaches directly to the skull—shot out horizontally from the side of the head. For the animals in the eastern United States, the beam was up to about 14 inches long; for those in eastern Beringia, up to about 20 inches long. The rest of the antler was palmated, as in moose, but the forms of those palmations differed. It is, however, the beam that is particularly distinctive about this animal, leading to antlers that are broad and low slung. Because of the less massive neck, long antler beam, and the low-slung nature of the antlers as a whole, it seems very unlikely that *Cervalces* used these antlers for male-male dominance bouts. Instead, they were likely used as ornaments, to convince girl *Cervalces* that they would be appropriate mates.[196]

The antlers are routinely used to identify this animal because, except for certain aspects of the skull, the rest of the stag-moose skeleton is so very moose-like that the two are very difficult, and sometimes impossible, to tell apart with fragmentary material.[197] The similarities in their skeletons suggest that the stag-moose lived in habitats similar to those of the moose. In line with this, its remains are frequently found in bogs and in marsh-side settings, often in habitats marked by open spruce woodland. Water, zoologist Valerius Geist has suggested, may have provided this animal with its escape route from predators.[198]

Earlier in this chapter, I mentioned that there is healthy debate among paleontologists over such things as the number of genera of South American ground sloths and gomphotheres, but that there is not much debate over similar issues for North America. *Cervalces* is an exception to this general rule. A number of paleontologists think that it should be placed in the moose genus *Alces*.[199] There is good reason for this, given the similarities in the postcranial skeletons. On the other hand, there are significant differences in the anatomy of the skull and antlers, and most paleontologists retain *Cervalces* as a separate genus,[200] as I do here.

Then there is that huge gap between the eastern Beringian and eastern North American populations. The animals from eastern North America are placed in the well-defined species *Cervalces scotti*. The problem lies with the material from Alaska and the Yukon. *Cervalces* existed not only in North America but also in Eurasia, where it is assigned to the species *Cervalces latifrons*. Some have suggested that the eastern Beringian animals belonged to that species,[201] but it is going to take better specimens from far northwestern North America, and perhaps an analysis of ancient DNA, to settle this particular matter.

The Mountain Deer (*Navahoceros fricki*)

The genus *Cervalces* was first recognized and described in the nineteenth century. Our knowledge of the mountain deer (*Navahoceros fricki*) is much more recent (fig. 3.43). In 1936, three fragmentary bones that differed from those of any known North American deer were described from Burnet Cave in New Mexico.[202] Because they looked more like caribou bones than anything else, they were tentatively assigned to the caribou genus, *Rangifer*. Because they clearly were not caribou bones, they were placed in a new species, *Rangifer fricki*, named after the eminent paleontologist Childs Frick. Not until 1975 had enough material of this animal been collected to allow the insightful paleontologist Bjorn Kurtén to make the case that it was so different from anything else that it deserved to be placed in its own genus, *Navahoceros*. He also recommended that, because of its intriguing adaptations, it be called the mountain deer.[203] It has been called this ever since.

FIGURE 3.43. The mountain deer (*Navahoceros fricki*).
Drawing by Wally Woolfenden.

Navahoceros fricki is known from east of the Cascade Range and Sierra
Nevada to the central United States and south into southern Mexico (fig.
3.44).[204] Between mule deer and elk in height, the mountain deer differed
from both of these animals in having simple three-tined antlers, thick limb
bones, and very short, robust metapodials (the long bones of the feet). Kurtén
pointed out that these characteristics of the limbs reflected adaptations akin to
those of such mountain-loving animals as the European chamois (*Rupicapra*)
and ibex (*Capra*), suggesting that they were adept rock climbers. Deer expert
Valerius Geist has compared them to mountain goats (*Oreamnos*), which
seems appropriate as well. Kurtén also suggested that they might have had
the same body mass as a female elk, but mountain-deer expert Lisa Blackford
thinks they might have been even heavier than that species.[205]

Until someone analyzes mountain deer genetic material, we will not have
a good understanding of its closest kin.[206] To judge from details of the skull,
its closest North American relative may be the caribou,[207] reminiscent of
the initial, tentative assignment of the Burnet Cave specimens to the caribou
genus. On the other hand, paleontologists have also pointed out that the skeleton as a whole seems most similar to that of the South American guemal

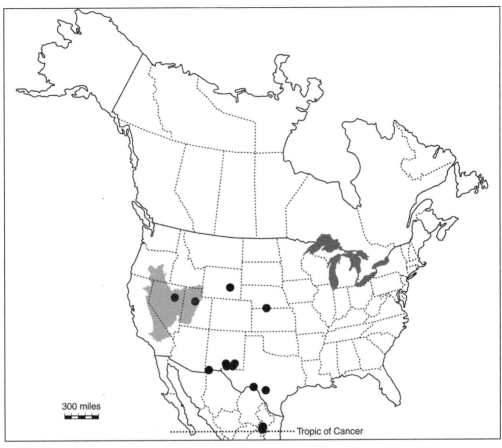

FIGURE 3.44. The late Ice Age distribution of the mountain deer (*Navahoceros fricki*) in North America.

(*Hippocamelus*).[208] Given that caribou and guemal are genetically distinct at the same time as they are fairly closely related,[209] the analysis of *Navahoceros* DNA should be able to resolve this issue.

Of the relatively few North American late Ice Age sites that have provided the remains of mountain deer, two are found in the Great Basin and provide the westernmost records for the genus. The first of these was found during highway construction at the northern end of the Oquirrh Mountains in north-central Utah. Those who take Interstate 80 west from Salt Lake City past the southern end of the Great Salt Lake drive right by the spot, Black Rock Canyon, where this specimen, the upper end of a femur (thigh bone), was found.[210] Mineral Hill Cave in northeastern Nevada (chapter 5) provided the other Great Basin specimens—two toe bones—of this intriguing deer. Paleobiologists Bryan Hockett and Eric Dillingham were cautious in identifying these specimens as mountain deer, but they provided excellent

FIGURE 3.45. The American pronghorn (*Antilocapra americana*). Photograph by T. L Best, courtesy of the Mammal Image Library (MIL image 1310).

FIGURE 3.46. The skull of the American pronghorn.

illustrations of them and there seems no reason to doubt that this is what they are.[211] One of these toe bones provided the youngest radiocarbon date there is for *Navahoceros* (chapter 4). Three mountain deer specimens from two sites—not much, but enough to tell us they were here.

The Antilocaprids (Pronghorn)

The pronghorn—or, as pretty much all of us say even though it is incorrect, antelope—is a common sight in the Great Basin. Marked by a white rump, a distinctly striped neck, and a single pair of narrow, flattened horns, the pronghorn is the speediest land mammal in the Americas (fig. 3.45). As I discussed earlier in this chapter, some have suggested that the great speeds pronghorn can attain reflect the fact that they were a favorite prey of the extinct American cheetah (*Miracinonyx trumani*).[212]

Both male and female pronghorn have those narrow, flattened horns, but the female version is far smaller and may be barely, if at all, visible from a distance. There are actually two parts to those horns, an inner bony core, and an outer keratinous sheath (keratin is the stuff of which our fingernails are made). The horn cores of males are covered by a sheath that is distinctly pronged, with the shorter, lower prong pointing forward, the other upward and often back (fig. 3.46). During the late Pleistocene, the two-horned pronghorn was joined by three other kinds of pronghorn, all of which had four horns.

To call the extinct North American Pleistocene pronghorn "four-horned" is actually something of a misnomer. All four of the horn cores—two in front, and two in back—emerge from a very short base or shaft, so each of the four is actually a "tine," much as the separate branches of deer antlers are tines. Antlers, though, are shed every year, unlike horn cores, which are permanent. In pronghorn, including the four-horned kind, the horn cores are permanent; it is the horn sheath that is shed on an annual basis.[213]

The horns of modern pronghorn suggest that their evolutionary past included a four-horned version. On occasion, individuals of the modern, two-horned version sometimes have four horn cores,[214] but other things matter more than this. First, the horn sheath, with that anterior prong, has long been thought to reflect the prior existence of a second tine that no longer exists. Second, each horn has a flattened projection pointing toward the front of the skull that seems like it might be the remnant of what was once, in the deep past, a second horn core (fig. 3.46). Third, there are two narrow grooves on the outer face of the horn core, one heading up the main shaft of the bone, the other heading toward the front, as if each once contained a blood vessel meant to serve separate horns.[215]

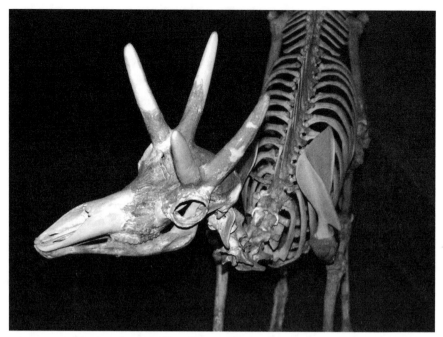

FIGURE 3.47. The four horns of *Stockoceros*, National Museum of Natural History specimen V16802, Papago Springs Cave, Arizona; see Skinner (1942).

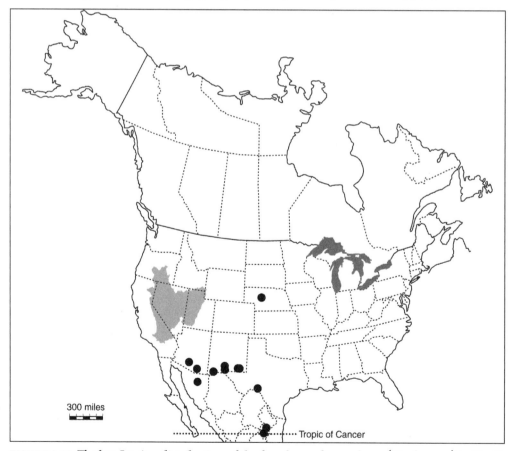

FIGURE 3.48. The late Ice Age distribution of the four-horned pronghorn (*Stockoceros*) in North America.

The four-horned pronghorn *Stockoceros* (fig. 3.47) came in what might or might not have been two separate species: Burden's pronghorn (*Stockoceros onusrosagris*), and Conkling's pronghorn (*Stockoceros conklingi*). If these two animals actually could interbreed and produce fertile offspring and so did belong to the same species, that species would be *Stockoceros conklingi* since it was the first to be named.

That would get us out of having to deal with "*onusrosagris*." Paleontologist Richard White has figured out that the two young men who discovered this animal concocted *onusrosagris* by translating their own last names into Latin. Those men were Quentin Roosevelt (Teddy Roosevelt's grandson) and his friend Joseph Burden.[216] *Onus*, in Latin, means "work" or "burden." In Dutch, *roosevelt* means "red field," or *rosagris* in Latin. The two jammed together provided *onusrosagris*, which paleontologist Walter Granger called "the most unpronounceable specific name I have ever seen."[217] It would be nice to be rid of it.

Burden's pronghorn approached the modern pronghorn in size and was larger than Conkling's pronghorn; both were somewhat stockier than *Antilocapra* and were probably slower on their feet. Most characteristic, though, are those four horn cores, apparently possessed only by the males. Those horn cores are rounded, unlike the flattened versions in the existing species. In adult animals, the front and back pairs diverged from one another at about a 45° angle (fig. 3.47) and tended to flare outward from the top of the skull; each was also covered by a sheath. Unlike those of the other four-horned pronghorns we will encounter, each of the four horns reached about the same height.[218]

Stockoceros was found primarily in the southwestern United States and northeastern Mexico (fig. 3.48). Although not a lot of sites have provided its remains, the skeleton of the animal is quite well known because it tends to be found in bunches—over 60 individuals in Arizona's Papago Springs Cave, for instance, and over 50 in northeastern Mexico's San Josecito Cave. These kinds of discoveries make it clear that *Stockoceros* was a herd animal.[219]

Shuler's Pronghorn (*Tetrameryx shuleri*)

In the late 1880s, commercial quarrying in the sand and gravel pits of the Trinity River near Dallas, Texas, revealed the presence of a rich Ice Age fauna. Many of these specimens were simply discarded by the workers, but, thanks to geologist Ellis Shuler of Southern Methodist University,[220] some were saved. Shuler provided them to Yale paleontologist Richard S. Lull—the same Lull who showed that the hairs of at least some Shasta ground sloths might have supported a crop of algae.

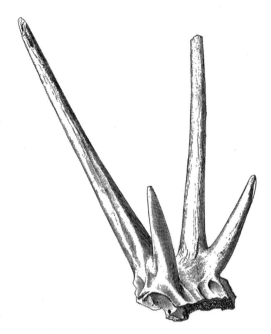

FIGURE 3.49. The four horn cores of Shuler's pronghorn
(*Tetrameryx shuleri*). The rear horn cores are about 12 inches
long; the front ones, about 4 inches long (from Lull 1921).

Lull described the Dallas Sand Pits fauna in 1921. Many of the animals
he described we are already familiar with—sabertooth cat (*Smilodon*), camel
(*Camelops*), horse (*Equus*), and, to be discussed shortly, mammoth (*Mammuthus*). In addition to these, though, Lull also described what he called "an
aberrant genus of the family Antilocapridae."[221] Aberrant, he said, not because of its teeth or size, both of which resembled those of the modern pronghorn, but because of its remarkable horn cores (fig. 3.49).

There are four of these, all more rounded than the narrower horn cores
of today's pronghorn. In adults, the hind pair tilts backward and slightly outward; the front pair, forward and slightly outward, and all four were covered
by a sheath. All this is very much as in *Stockoceros*, but, unlike in *Stockoceros*,
the posterior pair is far longer than the anterior one.[222]

Lull named this new and distinctly different animal *Tetrameryx shuleri*, or
Shuler's pronghorn, in honor of Ellis Shuler. This is not the only thing named
after Shuler: Southern Methodist University's Shuler Museum of Paleontology is named after him as well.[223]

Shuler's pronghorn is known from only a small number of sites (fig. 3.50):
three in Texas and, perhaps, two in the Great Basin. I say "perhaps" two in
the Great Basin, because neither of the Great Basin sites has provided the
horn cores that are so critical to identifying this animal. The possible record

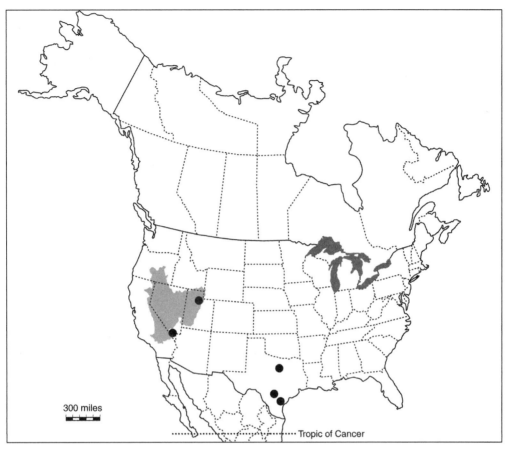

FIGURE 3.50. The late Ice Age distribution of Shuler's pronghorn (*Tetrameryx shuleri*) in North America.

for Shuler's pronghorn in northern Utah comes from a mandible only tentatively identified as belonging to *Tetrameryx*. Since the horn cores provide the most secure way to identify this animal, this mandible might, in fact, come from a different pronghorn.[224] The record from southern Nevada comes from Tule Springs in the Las Vegas Valley. This, too, is based on a single mandible. While paleontologist John Mawby felt it more likely that it belongs to *Tetrameryx* than to the modern pronghorn, he could not be certain.[225] As a result, there might be two sites in the Great Basin from which Shuler's pronghorn is known, or there might not be any.

The Diminutive Pronghorn (*Capromeryx furcifer*)

The diminutive pronghorn (*Capromeryx furcifer*) is known from far more sites than is *Stockoceros*, ranging from eastern Texas across the American Southwest and adjacent Mexico to coastal California. It also occurred in the Great Basin, although our record from here is not rich (fig. 3.51).

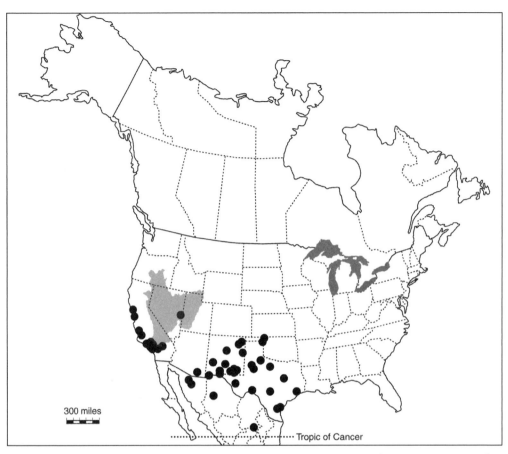

FIGURE 3.51. The late Ice Age distribution of the diminutive pronghorn (*Capromeryx furcifer*) in North America.

The horns of all three extinct Ice Age North American pronghorns allow them to be told apart from one another. In *Stockoceros*, the four horns are of roughly equal height and angle away from one another. In Shuler's pronghorn (*Tetrameryx*), the horns also angle away from one another, but the front set is smaller than the hind pair. In the diminutive pronghorn, the four horns project vertically from the skull, parallel to one another, with the front horns much smaller than the back ones and sometimes reduced to small nubbins. As with the other extinct late Ice Age pronghorns, the horns of the diminutive pronghorn were sheathed and the full versions were carried only by the males.[226]

However, the outstanding characteristic of the diminutive pronghorn is not its four horns, but its very small size. The diminutive pronghorn weighed an estimated 25 pounds and stood only some 24 inches tall at the shoulder, with both males and females about the same size.[227]

Were it still alive, *Capromeryx furcifer* would not be the world's smallest artiodactyl. As I mentioned earlier, the world's smallest deer, the northern pudu (*Pudu mephistophiles*) of South America, weighs in at no more than 15 pounds or so.[228] The living record, though, belongs to the royal antelope of western Africa (*Neotragus pygmaeus*). This animal is 12 inches tall at the shoulder; one that weighed seven pounds might be considered overweight by other members of its species.[229] Nonetheless, for a pronghorn, the diminutive pronghorn was tiny. In fact, the earliest known member of the genus, found from eastern Washington into central Mexico and dating to as early as about 3.5 million years ago, weighed twice as much. Why members of the genus became smaller as time went on is unknown, but it happened.[230]

Paleontologists Richard White and Gary Morgan have pointed out that there are modern African antelope that share important physical characteristics with the diminutive pronghorn, combining small body size with relatively simple horns, with males and females being about the same size.[231] These similarities, they suggest, might be telling us about other ways in which they were alike.

African antelope are very different animals from American pronghorn, even though the latter are often called by the same common name. African antelope belong to the cattle family, the Bovidae, while pronghorn are placed in their own family, the Antilocapridae. Pronghorn aren't particularly closely related to any bovid. As hard is it might be to believe, the genetics of today's pronghorn show that they are more closely, if extremely distantly, related to giraffes.[232] White and Morgan fully recognize that they are comparing very different animals, but they also recognize that physical similarities in artiodactyls often imply behavioral ones, and that there are ways that the implications they draw can, with sufficient work, be explored in the paleontological record.

The comparisons they make suggest that the diminutive pronghorn lived singly or in very small groups. This is consistent with the fact that, unlike the situation for *Stockoceros* (or the flat-headed peccary), large numbers of these animals are not found together. Given their small size, they probably hid to avoid predators, since they could hardly have outrun them or stared them down. They probably also had protective coloration of the sort that young deer have. All these things together—being solitary, being protectively colored, and hiding from predators—make full sense for both modern small antelope and for the diminutive pronghorn, since a predator could have easily eliminated large numbers of them at once if they had wandered around the landscape in herds. In fact, this description also catches many relevant aspects

FIGURE 3.52. The mysterious model of the diminutive pronghorn hidden on the Dugway Proving Ground, central Utah. The size is correct, but the legs are too short relative to the size of the body, and the front and rear horns are placed at too great an angle to one another. The aging field worker in this photograph has requested that he not be identified. Photograph by David B. Madsen.

of the behavior of the only living mammal that has four horns, the small four-horned antelope *Tetracerus quadricornis* of India and Nepal.[233]

I wish it were otherwise, but there are only three records for the diminutive pronghorn from the Great Basin. They have been reported from the Twenty-nine Palms area of the far southwestern Great Basin, though this material does not appear to have been described in detail.[234] Schuiling Cave, in southeastern California's Mojave Desert, provided two fragments of lower limb bones identified as diminutive pronghorn. That identification is undoubtedly correct because these specimens are clearly from a pronghorn and no other adult pronghorn is that small.[235] Smith Creek Cave, in eastern Nevada's Snake Range (chapter 5), yielded nine specimens of this animal, nearly all of them from the feet, again identified on the basis of their small size.[236]

In a certain way, there is another record of the diminutive pronghorn from the Great Basin. While doing archaeological fieldwork in the Dugway Proving Ground, a highly secure military base in central Utah, David Madsen and his colleagues came across a remarkable wicker model of what was clearly meant to be *Capromeryx* (fig. 3.52). This had been placed amid low-growing shrubs, invisible from the dirt road that ran nearby. Who made this and why they put it here is a complete mystery, but the model is accurate enough that it is worth mentioning the ways in which it has gone wrong: the legs are too short relative to the size of the body, and the front and rear horns are placed at too great an angle to one another. If anyone knows who did this, I would love to know as well. I would congratulate the artist, find out why he or she put it there, and see whether I could have one of my own. As far as I know, the wicker *Capromeryx* is still there, though at last sighting it was in weather-beaten shape.

The Bovids (Cattle, Antelope, Sheep, and Goats)

Today, the native bovids of North America are represented by bison, bighorn sheep, Dall's sheep, mountain goats, and muskoxen (table 3.3). These four genera were joined by three others during the late Ice Age.

Although the largest males may reach 110 pounds, saiga (*Saiga tatarica*) generally weigh about 80 pounds and are no more than about 30 inches tall at the shoulder. Today, they are found in central Asia, from Mongolia to the Caspian Sea, though in declining numbers in parts of this area.[237] They were far more widespread during the Ice Age, found as far southwest as northern Spain, across central and northern Eurasia, and into far northwestern North America (fig. 3.53).[238] Had they been able to get south of glacial ice in North America, we would likely know it by now.

Today, we think of muskoxen (*Ovibos moschatus*) as those great, shaggy beasts of far northern North America and Greenland whose prime means of defense against predators is to stand in a circle staring outward, horns on alert, doing their best to look mean. During the Pleistocene, they glowered over a far more expansive landscape; they have been found in North America at least as far south as Virginia and across Eurasia through France and into Spain.[239] In much of their Eurasian territory, they were gone by the end of the Ice Age, but in northern Siberia, they survived until just a few thousand years ago.[240]

Late Ice Age North America held two other genera of muskoxen, neither of which survived the end of the Pleistocene: the shrub ox (*Euceratherium*) and the helmeted muskox (*Bootherium*). Both of these animals are closely related to the living muskox,[241] and both are known from the Great Basin.

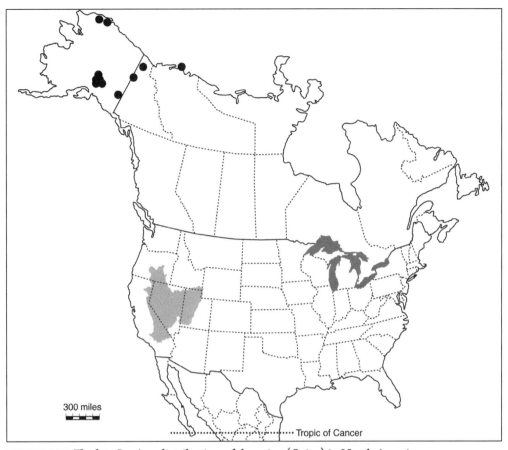

FIGURE 3.53. The late Ice Age distribution of the saiga (*Saiga*) in North America.

The Shrub Ox (*Euceratherium collinum*)

Between a muskox and a bison in size, the shrub ox (*Euceratherium collinum*) is known from a variety of sites in western North America (fig. 3.54). From the time it was first recognized over a century ago, from material discovered in caves in Shasta County, California, similarities to the modern muskox seemed clear,[242] but the horn cores are so very different that it took the analysis of ancient DNA to show that the two are, in fact, closely related.[243]

Related or not, shrub oxen horn cores are distinctly different from those of contemporary muskoxen (fig. 3.55). The bases of shrub oxen horn cores leave the skull from well behind the eyes, pointing up and back and then curling around toward the front, ending with an upward flourish. In 1905, the skull of a shrub ox–like animal was discovered with horns that differed somewhat from those of *Euceratherium*, so it was assigned to an entirely different genus (*Preptoceras*). But it did not take long to recognize that these differences likely reflected differences between the horns of male (*Preptoceras*) and female

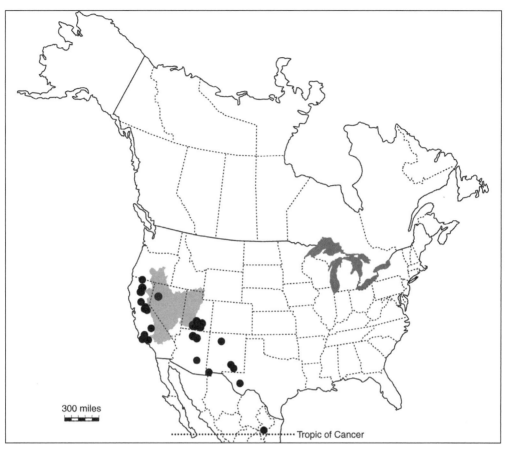

FIGURE 3.54. The late Ice Age distribution of the shrub ox (*Euceratherium collinum*) in North America.

(*Euceratherium*) shrub oxen, rather than the existence of two very similar animals on the late Ice Age landscape of western North America.[244] Today, the remains of this animal are placed in a single species: *Euceratherium collinum*.

Thanks to the work of paleobiologists Manny Knopf, Jim Mead, and Scott Anderson, we know quite a bit about the dietary preferences of the shrub ox. Knopf and his colleagues showed that the muscle attachments, shape, and teeth of *Euceratherium*'s skull imply that it could not have relied heavily on such plants as grasses for its diet. Instead, it was built to either browse or to eat a mixture of both grasses and browse. Just as important, they also showed that caves and rockshelters in the Escalante River Basin of southern Utah contained the well-preserved and distinctively shaped dung pellets of this animal—a shape they referred to as the "Hershey's Kiss" morphology, since that is what some of them looked like. By analyzing the content of a large series of these small pellets, they showed that the shrub oxen of this area were

FIGURE 3.55. The shrub ox (*Eucheratherium collinum*). Drawing by Wally Woolfenden.

browsers, with at most 5 percent of the pellets composed of grasses. They were
even able to identify the plants the animals were browsing—acacia (*Acacia*),
evening primrose (*Oenothera*), oak (*Quercus*), rabbitbrush (*Chrysothamnus*),
big sagebrush, and sumac (*Rhus*).[245]

Perhaps, the skeptic might point out, shrub oxen were eating these plants
in such abundance because there was nothing to graze. Knopf and his coworkers recognized this possibility and dismissed it for a very good reason.
One of the sites they analyzed, southern Utah's Bechan Cave, also provided
the dung of mammoth (*Mammuthus*) and bison from the same stratigraphic
unit as provided some of the shrub oxen Hershey's Kisses. The dung of those
animals was full of grasses,[246] making it clear that the shrub oxen of at least
the southern Colorado Plateau were browsers by choice.

The shrub ox has been reported from two sites in the Great Basin, but if
you look closely at figure 3.54, you will see only one. That is because the initial
report of this animal, from Mineral Hill Cave, turned out to be based on a mis-
identified bison bone.[247] The site that is on the map is one of the caves on Falcon Hill, on the northwestern edge of the Winnemucca Lake Basin in northwestern Nevada. This site yielded a shrub ox mandible that has now provided
one of the youngest radiocarbon dates we have for this animal (chapter 4).[248]

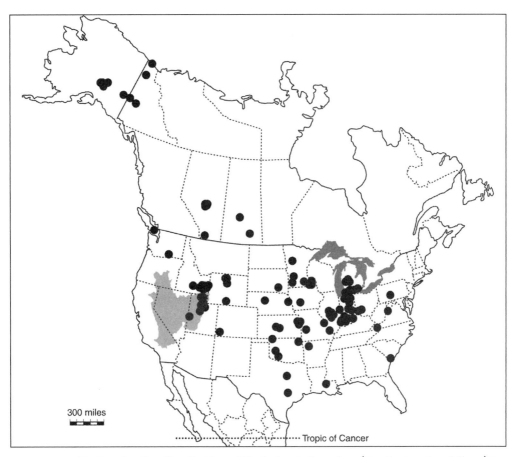

FIGURE 3.56. The late Ice Age distribution of the helmeted muskox (*Bootherium bombifrons*) in North America.

The Helmeted Muskox (*Bootherium bombifrons*)

The helmeted muskox (*Bootherium bombifrons*) was far more widespread than the shrub ox, known from all parts of unglaciated North America except for the southwest and the far southeast (fig. 3.56). Even where its distribution overlapped that of the shrub ox, these two closely related animals were clearly doing different things. As paleontologists Gary Morgan and Spencer Lucas have pointed out, the remains of shrub oxen tend to come from caves, while those of helmeted muskoxen tend to be found in the open.[249]

We know a lot more about *Bootherium* than we do about *Euceratherium*. Not only have a lot more of its skeletal remains been found, but soft tissues, including a nearly complete carcass, have been preserved in the frozen sediments of far northwestern North America.

Compared to the modern muskox, the helmeted muskox stood taller on longer legs but was shorter head to toe. It had a longer head than today's muskox, with bony eye sockets that protruded less from its skull. From that nearly complete carcass, we know that it had a darker but shorter coat. The shorter coat matches the less-protruding eye sockets, since the more protruding versions in the modern muskox help keep its longer coat out of its eyes. It is the horns of the helmeted muskox, though, that have gained the most attention.

Bootherium horn cores emerged horizontally from the skull, then curled downward, forward, and to some degree outward. How far outward they curled depended on whether they were possessed by a male or female, since the horn cores, and associated horn sheaths, differed dramatically between the sexes.

In fact, they differed so dramatically that for quite some time, the animals we now place in the genus *Bootherium* were placed in two separate genera, *Bootherium* and *Symbos*, distinguished primarily by the very different forms of the horn cores of each. Today, we know that the individuals that led to the definition of *Symbos* were males, and to the definition of *Bootherium*, females. It is now accepted that all the late Ice Age material that was once assigned to these two genera actually belongs not just to the same genus, *Bootherium*, but to the same species, *Bootherium bombifrons*, the helmeted muskox.

The horn cores of male helmeted muskoxen were much more massive than those of the female. They spread farther onto the top of the skull, with additional bony growth between their bases. Each horn core had its own sheath running from tip to base, with the skull ends of the sheaths joining in the middle, covering the bone at the top of the skull between the two horns. It is this form that gives *Bootherium* its name—the helmeted muskox.

The female horn cores were more simply constructed, not extending as far across the top of the skull, with sheaths that did not meet in the middle. In addition, they pointed more forward than did those of the males, the latter moving away from the front of the skull as they reached their ends.[250]

Zoologist Dale Guthrie has used these male-female differences in *Bootherium* horn cores to probe the behavior of this extinct animal. His analysis, directed toward Alaskan helmeted muskoxen, is insightfully speculative and very likely correct.

Among modern muskoxen, males establish dominance by lowering their heads, running at one another, and crashing into each other at full tilt. Guthrie argues that this behavior, only recently outlawed by the National Football League, characterized male *Bootherium* as well. There is no other way to account for the massive bases of the horn cores of those animals. The females, with more lightly built horn cores, are not likely to have engaged in this behavior. This also explains why the ends of male *Bootherium* horns angle away from one another. It was head butting, not the clashing of horns, that was the point of it all.

Muskoxen, modern or not, have to protect themselves and their young from predators. Guthrie observes that the more-forward-pointing horn tips in female *Bootherium* are actually better suited for this than are the more angled horn tips of the males. In the males, he argues, the form of the horns has evolved in response to two needs, one involving male-male dominance bouts and the other involving defense against predators. For the females, on the other hand, defense played the bigger role.

Guthrie also notes that the forward-pointing tips of *Bootherium* horns provided defense only against predators that came at the animals from the front. Your basic Ice Age predators were probably too clever to have tried this, or they probably wouldn't have lasted into the Pleistocene. Given that *Bootherium* horns were best suited for defense against frontal attacks, the animals must have defended themselves by gathering into lines or clusters, just as modern muskoxen do.[251]

Today's muskoxen are confined to the cold north, as were those that lasted until a few thousand years ago in Siberia. This was clearly not true of *Bootherium*, given its very broad North American distribution during the late Pleistocene. Its longer legs made it more energetically efficient than our muskoxen, meaning that it was likely far more mobile than its modern relative. As Guthrie points out, its shorter coat, which would weigh less and be less liable to get wet, probably also helped make it more mobile. Even though today's differently constructed muskox made it south of glacial ice into what is now the United States, it did not come close to matching the broad distribution that *Bootherium* managed to attain.

The helmeted muskox is known only from the eastern edge of the Great Basin, but here it has been found in abundance on the late Pleistocene shorelines along the eastern edge of Lake Bonneville.[252] They are abundant enough that the Utah Division of State History once used them, quite literally, as a poster child to increase interest in preserving the state's paleontological resources (fig. 3.57).[253]

The Notoungulates

It would take a dedicated reader to recall that, in my discussion of the extinct Ice Age mammals of South America, I mentioned the family of mammals known as toxodonts—huge animals that somewhat resembled rhinos. I also mentioned that these animals made it north through the Isthmus of Panama,

FIGURE 3.57. The Utah Division of State History request that people keep their eyes open for additional specimens of the helmeted muskox eroding from the deposits of Pleistocene Lake Bonneville.

and that one of the two genera, *Mixotoxodon*, is known from as far north as North America. We know that this animal made it into North America because paleontologist Ernie Lundelius and his colleagues recently reported a single *Mixotoxodon* tooth from late Pleistocene deposits on the Gulf Coastal Plain of east Texas.[254] Radiocarbon dates from the same deposit suggest that this specimen may have been deposited around 20,000 years ago. Since sea levels were far lower then than they are now, it is very possible that significant numbers of other toxodonts now lie in the sediments deep beneath the water in the Gulf of Mexico. No other toxodont is known from North America. In fact, until this specimen was reported, they were not known from farther north than Michoacán and Veracruz, in southern Mexico.[255]

The Proboscideans (Gomphotheres, Mastodon, and Mammoths)

Earlier in this chapter, I pointed out that neither mastodons nor mammoths made it to South America. Instead, the elephant-like animal of late Ice Age South America was the gomphothere, an animal North Americans generally haven't heard of because gomphotheres were rare on the late Ice Age North American landscape. Although there were between two and four genera of gomphotheres in South America, depending on how one decides to classify the remains of these animals, all of the North American records are for Cuvier's gomphothere (*Cuvieronius*), a genus no one argues about.

The northernmost record for this animal comes from South Carolina and is something like 125,000 years old.[256] The three remaining records are from the southern edge of North America—one from southeastern Texas[257] and two from Sonora, Mexico (fig. 3.58). The Sonoran sites are fairly well dated. The more eastern of the two Sonoran dots on the map represents the site of Térapa, which dates to between 43,000 and 40,000 years ago.[258]

The latest North American record for this huge animal comes from the recently excavated site of El Fin del Mundo, in northwestern Sonora. Here, archaeologist Guadalupe Sanchez and geoarchaeologist Vance Holliday and their colleagues found four of the distinctive artifacts known as Clovis points (chapter 6) near the remains of at least two individuals of Cuvier's gomphothere (fig. 3.59), all associated with radiocarbon dates of about 11,550 years ago. The surfaces of the gomphothere bones were too weathered to be able to tell whether they had been butchered by people, but the archaeological association is certainly intriguing, even though the archaeological phenomenon known as Clovis generally dates to a few hundred years younger than the El Fin del Mundo site.[259]

The gomphothere bones from this site do not contain enough protein to allow them to be dated directly (chapter 4), so we can't be certain of their exact age. Nonetheless, the careful work at El Fin del Mundo has provided our first and only record for very late Ice Age gomphotheres in North America, gomphotheres that seem to have had what was for them an unfortunate encounter with people (chapters 6 and 7).

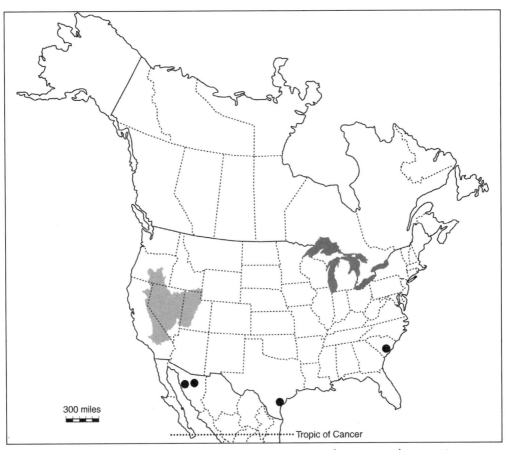

FIGURE 3.58. The late Ice Age distribution of the gomphothere (*Cuvieronius*) in North America.

The American Mastodon (*Mammut americanum*)

The deep ancestors of the American mastodon evolved in Africa. Much later, descendants of those animals crossed the Bering Land Bridge, and their descendants ultimately gave rise to the American mastodon. By the late Pleistocene, *Mammut americanum* was widespread in North America, from Florida to Alaska, and from Nova Scotia south through Mexico into Honduras (fig. 3.60). It never made it through the Isthmus of Panama and so is unknown from South America, nor did it cross the Bering Land Bridge into Siberia.[260]

With shoulder heights ranging from about 6 to 9 feet and a weight of 6,500 pounds or more, the mastodon was a "long, low, and stocky" animal,[261] with a broad, relatively flat-topped skull that gave rise to upper tusks that projected more horizontally than did those of mammoths and that tended to turn upward and outward before curving back toward one another (fig. 3.61). These upper tusks may be paired with much shorter lower tusks, though these are routinely lost in adults. Both sets of tusks are often worn, as if the animal

FIGURE 3.59. The mandible of the gomphothere (*Cuvieronius*) from the late Pleistocene site of El Fin del Mundo in northeastern Sonora, Mexico. Photograph by
Iván Alarcón Durán, courtesy of Joaquin Arroyo-Cabrales and Vance Holliday.

had used them to help gather dinner. That gathering, we know from detailed
studies of mastodon skulls, would also have been done with the help of a
powerful trunk and a tongue that was up to about 3 feet long. Once in the
mouth, whatever they were eating would have been processed with a very
distinctive set of cheek teeth. Those teeth are marked by pairs of pointed cusps
that sit on between two and five ridges that are parallel to one another and at
right angles to the main axis of the tooth. These teeth belonged to an animal
that had evolved to crush, not grind, its food.[262]

Late in the Pleistocene, mastodons were clearly most abundant in the
Great Lakes region. At that time, much of this region was marked by spruce
forest and woodlands. Because mastodon teeth had evolved to fit a browser,
it was long assumed that the diet of mastodons in this region included significant amounts of spruce (*Picea*). Not until the actual remains of mastodon
meals were discovered and analyzed from a wide range of sites, however, was
it truly clear what they did for a living. Now, thanks to work by many different
scientists, we know that mastodons did, in fact, browse on spruce, just as had
been surmised from their teeth and their environmental context.

However, we also know that they dined on far more than that. They ate
grass and a broad variety of water plants; they also chewed on willows (*Salix*)

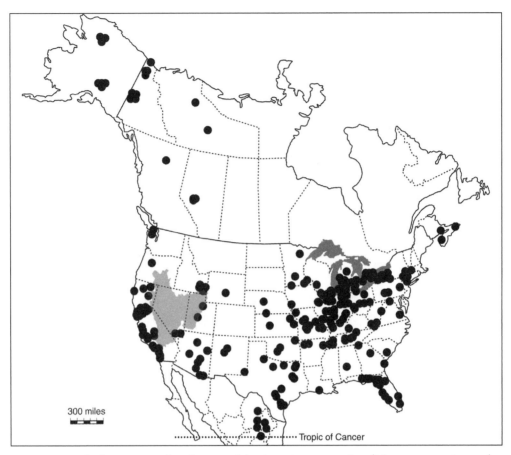

FIGURE 3.60. The late Ice Age distribution of the American mastodon (*Mammut americanum*) in North America.

and, in Florida, on cypress (*Taxodium*), often incorporating a significant amount of bark into their diet. Although a large part of that diet came from browse, and they clearly browsed a wide variety of trees, mastodons were to some extent mixed feeders, eating both high and low.[263] Their diet was also markedly diverse. For instance, the mastodons who left behind a thick deposit of roughly 12,400-year-old dung at northern Florida's Page-Ladson site had dined on 27 different genera of plants. The meals of those animals ranged from cypress twigs and hickory (*Carya*) nuts to gourds (*Cucurbita*) and the fruits of blackberry (*Rubus*) and wild persimmon (*Diospyros*).[264]

Opinions about the social organization of mastodons have varied widely over the years. Some have thought that they were primarily loners, which would account for the fact that when you have excavated one mastodon at a site, you have usually excavated all the mastodons at that site. Others, though, suggest that they lived in small herds of perhaps 10–15 individuals, with those

FIGURE 3.61. The American mastodon (*Mammut americanum*). Drawing by Wally Woolfenden.

groups made up largely of adult females and their young, perhaps with adult males attached to them to some degree. That would account for some finds of mastodons consisting of more than just solitary individuals: the Boney Spring site, in southwestern Missouri, yielded at least 31 of them.[265]

Assuming, as now seems most likely, that mastodons formed social groups, those groups could have communicated with one another, and with solitary individuals, at a distance. Basing their arguments on the structure of mastodon ear bones, and of the bones to which the tongue attaches, elephant expert (and, sadly, terrorist attack victim) Jeheskel Shoshani and his colleague G. H. Marchant showed that mastodons had the ability to make trumpeting calls that might have traveled a mile or more.[266]

Every seven years, *Star Trek's* Mr. Spock had to endure the aggressive, sexually driven agonies of what is known in the Vulcan language as "pon farr" (I looked up the name on *Wikipedia*). Modern adult male elephants undergo something quite similar, an episode of greatly heightened male-male aggression known as the "musth."

While these battles might end without great harm to either male, this is not always the case.[267] Paleontologist Daniel Fisher, who has done very important work on the American mastodon, provides a description of such a battle between two male African elephants (*Loxodonta africana*), originally written by zoologist R. W. Carroll:

The noise made by the two contestants was terrific. The battleground was about an acre in extent. They continued charging each other, parrying with their tusks, growling, and screaming. The brute force of the charges was amazing...as their tusks engaged one gave a mighty twist with his head, catching his opponent's tusk about midlength and broke it off.... Sparring continued but was uneven and difficult with just the one tusk, but suddenly he dropped his head, turning it at the same time, thus aiming his tusk at his opponent's throat. Dropping his head allowed his opponent's tusks to go high on his head, tearing a terrible gash above his eye and ripping a great hole in his ear. But that one tusk thrust home and caught his opponent in the throat, going deep. With a tremendous heave One Tusk raised his head, lifted the impaled bull off his front feet, and ripped a great hole in his neck. As he caught his balance his guard dropped, and again the one tusk went home, this time through the trunk and deep into the head. Both elephants went down to their knees. The one tusker immediately arose, tossed his head, and again thrust his tusk deep into the head of his opponent. With this blow the stricken bull went over on his side, feet flaying the air. One Tusk quickly stepped around and repeatedly drove his tusk into the fallen bull's back, all the while trumpeting and screaming.[268]

Mastodons, Fisher has shown, did exactly the same thing: a series of late Pleistocene sites from New York State and elsewhere have provided mastodon skeletons that show bone breakage consistent only with those animals having been on the unfortunate end of a musth battle. Broken tusks, circular tusk wounds, shattered skull bones, splintered ribs, fractured vertebrae—all can be explained only if male mastodons engaged in the kind of aggressive behavior known from their modern relatives.[269]

As if mastodons did not have enough to contend with—males pounding one another into submission and carnivores large enough to take down their young—there is also evidence that people preyed on, or at least scavenged, these animals.[270] The Kimmswick site in eastern Missouri provided paleontologist Russell Graham and his colleagues with two distinctive Clovis fluted points tightly associated with mastodon remains (see fig. 3.62 and chapter 6).[271] Although Michigan's Pleasant Lake site provided no stone tools, or artifacts of any sort, it did provide mastodon remains that appear to have been butchered and burned by people.[272] Kimmswick alone is enough to establish that people and mastodons interacted in a way that was not cheery for the latter.

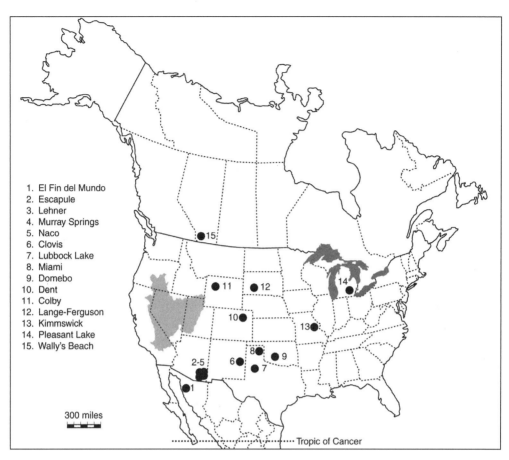

1. El Fin del Mundo
2. Escapule
3. Lehner
4. Murray Springs
5. Naco
6. Clovis
7. Lubbock Lake
8. Miami
9. Domebo
10. Dent
11. Colby
12. Lange-Ferguson
13. Kimmswick
14. Pleasant Lake
15. Wally's Beach

300 miles

Tropic of Cancer

FIGURE 3.62. Archaeological sites with evidence of human hunting or scavenging of extinct Ice Age mammals. Site 1 contains the gomphothere *Cuvieronius*; sites 13 and 14, mastodon; site 15, horse and camel. The others contain mammoths. (After Grayson and Meltzer 2015.)

The Manis Mastodon site, on the northern edge of Washington State's Olympic Peninsula, is perhaps the best known of all the sites that have been claimed to provide evidence for human predation on these animals. Emanuel Manis discovered the Manis mastodon in 1977 as he was using his backhoe to build a pond on his property. When he realized what he had, he and his wife, Claire, contacted Washington State University to ensure that the material was removed properly. Thanks to the insight and generosity of the Manis family, professional excavations were conducted at what came to be known as the Manis Mastodon Site, the site has been donated to the national Archaeological Conservancy, and the mastodon skeleton is on display at the Museum and Arts Center in nearby Sequim. In 1978, the site was placed on the National Register of Historic Places.[273]

FIGURE 3.63. The mysterious bony object protruding from the Manis mastodon rib. This photograph was taken by the author in 1983; see Waters et al. (2011) for a photograph of the same specimen in its current condition.

None of this would have happened without the concern shown by Emanuel and Claire Manis. In the end, it turned out that this was not just any old mastodon. Instead, this male mastodon had what appeared to be a bone projectile point embedded in a rib, just behind the spot where the rib met the animal's vertebral column (fig. 3.63). Since the end of this object was embedded in bone, X-rays were taken to determine whether or not the thing really was pointed. It was, so some then interpreted the mastodon as having been killed by people, even though it was still hard to explain how someone was able to stick a spear into the top of its back. It was also argued that the bones showed signs of having been butchered, but that claim has never been adequately supported.[274] When, thanks to the kindness of Sequim's Museum and Arts Center, I was able to examine the skeleton, I saw no suggestion of butchery.

Recently, Michael Waters and his colleagues reanalyzed the rib that makes the Manis mastodon so potentially important. They obtained a large series of precise radiocarbon dates (chapter 4) showing that the animal had died around 11,700 years ago. They took a very high-resolution X-ray computed tomographic (CT) scan of the rib that clearly shows the pointed nature of the intrusive object, just as the earlier, conventional X-rays had suggested. Waters's team also analyzed DNA from the intrusive bony object and showed that it had come from a mastodon.[275]

In short, the Manis Mastodon site has provided us with a pointy thing intruding into a mastodon rib. What we still don't know is whether the pointy thing is a projectile point made by human hands, or something else. The other obvious possibility is that it is a piece of the mastodon's own skeleton. Recall the skeletal damage that Daniel Fisher has shown that male mastodons could do to one another—shattered skull bones, broken ribs, and splintered vertebrae, for instance. If we use African elephants as our modern guide,[276] the Manis mastodon, estimated to have been 45 years old when it died,[277] was certainly not too old to be in musth. Could it be that the intrusive object came from the Manis mastodon's own skeleton, forced there by the tusk of its opponent?

There are two obvious ways to see whether this is the case. Theoretically, DNA from the rib could be compared to DNA from the intrusive bony object to see whether they came from the same individual, but that's only theory. In practice, according to the geneticist who did the work,[278] not enough DNA was recovered to allow this. And, again theoretically, what was recovered of the mastodon's skeleton could be examined to see whether there is a missing piece that would fill the pointed-object bill—if the right part had, in fact, been found when the site was excavated.

No matter how the animal died, it was not healthy when it succumbed to whatever killed it. Its skeleton is so shot through with pathologies that it would have been easy pickings for a sex-crazed male mastodon.[279] And, if the pointy thing protruding from the rib really is a bone projectile point, the unhealthy state of the Manis mastodon might suggest that it was able to be speared in the top of the back because when that happened—if it happened— it was not on its feet.

Mastodon remains have been reported from only three sites in the Great Basin. Over a century ago, geologist J.E. Spurr reported the discovery of "some mastodon teeth and bones" from the Tule Springs area of southern Nevada, but the location of these specimens is unknown and no one who has worked in this area since then has found mastodon remains here. It seems extremely unlikely that the report is correct.[280] The remaining two records both come from Utah—a single tooth from near Salina,[281] and other remains from what I will call the Mastodon Sinkhole site (it does not have a formal name), on the Wasatch Plateau not far from Mount Pleasant.

The Mastodon Sinkhole site is actually located a very short mastodon-walk east of the eastern edge of the Great Basin drainage divide, so short that it might as well be in the Great Basin. What makes the site remarkable is not that it provided the remains of two mastodons, or that at least one of those individuals may date to 10,800 years ago, at the very end of the Pleistocene (chapter 4). Instead, the site is remarkable for its elevation. At 9,780 feet, this is the highest elevation at which mastodons have ever been found. To judge from the roughly contemporaneous vegetation at the nearby Huntington Mammoth site (chapter 5), the vegetation surrounding the Mastodon Sinkhole site at the time these high-elevation mastodons died included healthy amounts of Engelmann spruce. As we saw earlier, mastodons routinely ate spruce when it was available. About 10,800 years ago, there were mastodons on the Wasatch Plateau, literally getting high—9,780 feet high—and probably doing it on spruce.[282]

The Columbian Mammoth (*Mammuthus columbi*)

There is a reason—above and beyond the great story, animation, and voicing—that the movie *Ice Age* (2002) connected so well with an American audience. Diego, the giant cat, is a sabertooth (*Smilodon*), the iconic carnivore of the North American late Pleistocene. Manny is a mammoth (*Mammuthus*), Diego's herbivorous counterpart. We were familiar with the animals long before we saw the movie. We just didn't know they could talk. Sid, by the way, is a ground sloth, Lenny a scimitar cat (*Homotherium*), and there are glyptodonts and notoungulates as well. Whoever came up with all this knew about American Ice Age mammals. Unfortunately, animals like Scrat, the sabertooth squirrel, never existed. If they did, they probably wouldn't have eaten acorns. Unless you count chemical defenses, acorns don't fight back.

Given that tens of thousands of mammoth specimens—Manny bones and teeth—have been found in virtually all parts of unglaciated North America (fig. 3.64) and as far south as Costa Rica,[283] it may come as a surprise to learn that we are still not sure how many species there were. The most common interpretation, which I follow here, sees three species of American mammoths: the woolly mammoth (*Mammuthus primigenius*), the Columbian mammoth (*Mammuthus columbi*), and the pygmy mammoth (*Mammuthus exilis*).[284] Some add a fourth species to this list, a midwestern form known as Jefferson's mammoth (*Mammuthus jeffersonii*).[285] Still others accept only the woolly and Columbian mammoths as distinct species, treating the others as subspecies of *Mammuthus columbi*.[286] We will see, as we go along, why this confusion exists.

All mammoths have similar cheek teeth. These consist of a series of enamel plates set parallel to one another and surrounding the hard, calcified material known as dentine. These plates are arranged, one after another, at right angles to the long axis of the tooth. Flat on top, these teeth were built for grinding, unlike the teeth of mastodons, built for crushing.

The woolly mammoth is by far the best known of these species. It is, in fact, one of the best known of all extinct Pleistocene mammals. This is because we have not just its skeletal remains, but also the beautifully preserved carcasses of the animals themselves.[287] Woolly mammoths ranged widely during the late Pleistocene, from southwestern Europe across northern Eurasia and into

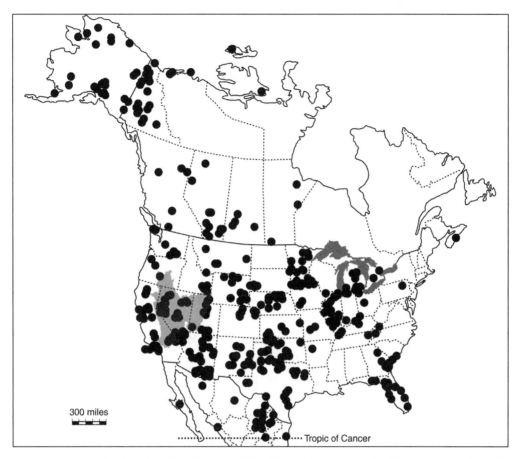

FIGURE 3.64. The late Ice Age distribution of the Columbian mammoth (*Mammuthus columbi*), the woolly mammoth (*Mammuthus primigenius*), and pygmy mammoth (*Mammuthus exilis*) in North America.

North America. In North America, they were found in the north—in eastern Beringia (Alaska and the Yukon), and across much of North America just south of glacial ice. Their preserved carcasses are known from permafrost deposits in both Alaska and Siberia. In some instances, those carcasses were so protected by long-term refrigeration that their flesh was eaten by scavengers when, 10,000 years or more after their death, they were exposed on the surface. As a result, we know about this animal's hair, its ears, its trunk, its brain, what it was eating just before it died, and even the thickness of its subcutaneous fat.[288] It does not hurt that the woolly mammoth coexisted with talented artists in Eurasia who left its depictions on cave walls, on carved bone, antler, and tooth, and even in the form of sculptures.[289]

That is how we know that the woolly mammoth really was woolly. It had a dense undercoat of fine yellowish hairs some 4 to 8 inches thick, with long

guard hairs up to 3 feet long that emerged from it. With the exception of the
soles of the feet, the woolly mammoth was woolly all over, including the
trunk, ears, and tail, with the longest hair on the shoulders. That hair hung
along the sides of the animal, providing an appearance similar to that of
today's skirted muskox. Many of the preserved guard hairs from Siberia are
dark in color, often reddish brown. People have often wondered whether this
color might not have resulted from degradation, given that the youngest of
the carcasses had been buried for about 10,000 years and the oldest for over
40,000. That, however, turns out not to have been the case. Genetic analysis of
a 43,000-year-old woolly mammoth bone showed that this animal's genome
had two different versions of a gene that controls for hair color, meaning that
some individuals had hair that probably ranged from reddish to yellowish.[290]
The preserved hair of Siberian mammoth carcasses was not lying.

The dense underwool and long guard hairs were just one of a series of
adaptations to the cold climatic regimes that marked the areas and times in
which these animals thrived. Beneath the 1–inch-thick skin that supported
all that hair, they had a thick insulating layer of subcutaneous fat about
3.5 inches thick. Their ears were small, about 14 inches from top to bottom and
11.5 inches from side to side, dramatically reducing the heat loss that larger ears
would have allowed. This is about 10 percent of the size of the ears of African
elephants and 20 percent of the size of the modern Asian version. The ears
of African elephants help shed heat; the ears of woolly mammoths helped
preserve it. Tails shed heat as well, so it is no surprise that woolly mammoth
tails were shorter than those of their modern relatives. Modern elephants have
about 30 caudal vertebrae, the bones that support the soft tissue that forms the
outside of the tail. Woolly mammoths had about 20 of them. They even had
genetically modified hemoglobin that helped ensure that sufficient amounts
of oxygen would be carried to the nether regions of their bodies.[291] All this is
enough to make you believe in evolution.

These adaptations to cold came mounted to a body that was, in general,
elephant-like, and particularly similar to that of the Asian elephant. However,
even if you shaved all the hair off and didn't notice the ears, you would still
not be likely to mistake a woolly mammoth for a modern elephant. Woolly
mammoths had a very high-domed forehead, as well as shorter limbs and a
longer body than modern elephants (fig. 3.65). The shorter limbs and longer
body, elephant expert Henryk Kubiak has suggested, may have been an adaptation to a diet that depended more on grazing than that of modern elephants.

Then there were those enlarged upper second incisors—the tusks. These
were enormous, though more so in males than in females. They emerged from

the skull quite close to one another, angling downward and outward, only to twist upward and inward. In old males, they could grow so long and turn so far inward that their tips actually crossed one another. The biggest could be over 13 feet long and weigh 250 pounds or more. Compared to the far straighter and shorter tusks of modern elephants, they would have made the woolly mammoth immediately recognizable even when shaved.

Even if they didn't, the woolly mammoth's profile sets it apart from modern elephants, especially the African versions. From a shoulder height of about 10 to 11 feet, within the range of modern African elephants, the back of the woolly mammoth sloped more sharply to its hips than that of modern elephants, African or Asian. A close look at a woolly's trunk would also show that this was not a modern elephant, since the tip of that trunk had a fingerlike projection at the top coupled with a flattened projection at the bottom. More complex than the tip of an elephant's trunk, these projections would have provided a grasping mechanism to help its owner pluck low vegetation as it fed.

We know that woolly mammoths plucked vegetation low to the ground because the materials they fed on have been preserved in the gastrointestinal tracts of frozen carcasses. Many scientists have studied that material, so we know that these animals fed on a very diverse variety of plants, with grasses, sedges, and the twig tips of willows, alders, larch, and spruce looming high on the dietary list. They also ingested mosses and, in the case of the 18,500-year-old Yukagir mammoth from northern Siberia, the dung of other mammoths.[292] That these animals ate mammoth dung has been shown only for the Yukagir individual but does not come as a surprise, since it is well known that modern herbivores ingest the dung of other herbivores, perhaps for its vitamin content.

Woolly mammoths seem to be of northeastern Siberian origin, appearing there by about 400,000 years ago and then dispersing as far as Europe to the west and North America to the east.[293] Turnabout being fair play, the analysis of ancient DNA has shown that the latest Ice Age Siberian woollies came from North America after about 50,000 years ago, replacing the Siberian natives that had been there before them. How and why this happened is not at all clear, but it is clear that it did happen.[294]

It is also clear that something didn't happen on some of the islands in the Bering Land Bridge area that did happen in mainland Siberia and North America. Mammoths were gone from mainland Siberia by about 9,700 years ago[295] and from North America by 10,000 years ago (chapter 4). Remarkably enough, however, they survived far later on islands in the general Bering Strait region. On Siberia's Wrangel Island, on the northern edge of the Bering Land

FIGURE 3.65. The Columbian mammoth (*Mammuthus columbi*). Drawing by Wally Woolfenden.

Bridge, they survived until about 3,700 years ago.[296] As far as we know, the last mammoth on Wrangel was also the last surviving member of its genus. On Saint Paul Island, one of the Pribilof Islands in the eastern Bering Sea, they survived until about 5,700 years ago.[297] Thanks to the analysis of ancient DNA, we know that both sets of these island mammoths were descended from the mammoths that had crossed the Bering Land Bridge from east to west, replacing the Siberian natives that had been there before. Even so, the Saint Paul mammoths were so genetically distinct from all other known woolly mammoths that they seem to have belonged to a population that was living on the now-submerged Bering Land Bridge. If so, all their close relatives are now under water.[298] Manny would not have been happy had he known about this.

Columbian mammoths were the more southerly of the two species, taking over for woollies once far enough south of glacial ice (see figs. 3.65 and 3.66). These are the ones that made it as far south as Costa Rica and the ones that were found in the Great Basin. Because no preserved carcasses of these animals have ever been found, and because the early Americans who encountered them did not provide us with the richness of depictions found in Europe, we do not know what the animal looked like on the outside.

FIGURE 3.66. A Columbian mammoth under excavation by the Desert Research Institute, Black Rock Desert, Nevada. The project director, Stephanie D. Livingston, provides scale.

We do know that early Americans encountered mammoths, because there are 11 North American sites that contain mammoth remains so closely associated with artifacts, in particular with Clovis fluted points (see fig. 3.62 and chapter 6), as to leave no doubt that the two species interacted with one another.[299] In fact, one of the most common explanations of the late Ice Age extinction of North American mammoths—and of virtually all the other mammals in this chapter—is that Clovis hunters did them in (see chapter 7). Even though mammoths and people knew one another, however, there is only a single depiction of a mammoth known from North America that might have been drawn from life. One must tread carefully here since frauds are known,[300] but the engraving of a proboscidean, almost certainly a mammoth, on a bone from Vero Beach, Florida, has passed every test for authenticity to which it has been subjected.[301] If we accept it as valid, then the Columbian mammoth looked a lot like a woolly mammoth with a haircut.

This is not surprising, since, as elephant expert Gary Haynes has pointed out, the skeletons of these two different kinds of mammoths are very similar to one another, so much so that he has wondered whether they might be fairly minor variants of one another.[302] There are differences, but the prime way of telling them apart is through subtle differences in their teeth.

We know a lot about the diet of Columbian mammoths. Most directly, that knowledge comes from preserved mammoth dung found in caves and rockshelters in the Colorado Plateau of the arid American Southwest. Dung from southern Utah's Bechan Cave—the name taken from the Navajo word for, to put it politely, excrement—has been best studied, but there are other sites as well.[303] They all show pretty much the same thing. Columbian mammoths in this part of the world were heavily dependent on grasses and sedges for their diet, with the twigs of such woody plants as saltbush, sagebrush, water birch (*Betula occidentalis*), and spruce eaten as well. The Great Basin has pitched in here as well, since we have the stomach contents from a mammoth found in this region. This animal, from the Huntington Mammoth site on the Wasatch Plateau of central Utah (chapter 5), had eaten grasses or sedges, or both, along with the needles of subalpine fir. All that we know of the diet of the Columbian mammoth suggests that while the particular species of plants it ate might have differed, its diet was, in general, similar to that of the woolly mammoth.

The second kind of dietary information we have for Columbian mammoths comes from chemical analyses of their bones. This approach does not tell us precisely what the animals were eating but confirms that grasses formed a major component of their diet, both in the arid southwestern United States and in Florida.[304] As most paleontologists had concluded long before this kind of information became available, mammoths were primarily grazers.

Although some have suggested that mammoths might have migrated vast distances during their lives, the information we now have available suggests that they did not. This information again comes from the chemical content of their skeletons, which scientists can match with the chemistry of the water they were drinking. The results of this research suggest that while some mammoths did migrate, the distances weren't that huge—a straight-line trek of 120 miles or so—and that many of them had home ranges that required much less long-distance walking than this.[305]

The third kind of mammoth known from North America during the late Pleistocene is the kind you might have wanted as a pet if large pets were your thing. This was the pygmy mammoth, *Mammuthus exilis*, known only from the northern Channel Islands off the coast of southern California. As the common name suggests, these were small for mammoths, with an average shoulder height of about 5.5 feet and an average weight of about 1,600 pounds. Because of their small size and specialized limb bones, they were able to negotiate steeper slopes than their far larger relatives.[306]

Pygmy mammoths were descendants of Columbian mammoths that swam from the California mainland to what are now the Channel Islands during the

Pleistocene. This happened sometime before about 80,000 years ago, since pygmy mammoths have been there for at least that long. Similar phenomena—dwarf versions of mammoths and other mammals evolving from large ancestors in an island setting—are well known from other parts of the world. During the Pleistocene, for instance, Sicily supported an elephant that stood about 3 feet tall and Cyprus a pygmy hippopotamus the size of a pig.[307]

Today, the closest the Channel Islands come to the southern California mainland is about 12 miles, but during the Ice Age, when glaciers had expanded on a global basis and sea levels had dropped accordingly (chapter 2), that distance had dropped to a little more than 4 miles.[308] Elephants are powerful swimmers, and the Columbian mammoth ancestors of pygmy mammoths made that swim. When they arrived, they would not have been confined to a small island since the same drop in sea level that had brought the mainland closer also created a much larger, continuous landmass out of what are now the separate northern Channel Islands. Columbian mammoths are found throughout the later Pleistocene paleontological record of these islands, perhaps reflecting multiple arrivals from the mainland through time, but the pygmy version was the common one. It lasted until the very end of the Ice Age: the latest date for a Channel Islands pygmy mammoth falls at 11,030 years ago, not that much different from the latest date for a mainland mammoth (chapter 4).[309]

Mammoths have been found throughout the Great Basin. Certainly the most notable of them all was the individual from the Huntington Mammoth site that I mentioned above and that I discuss in detail in chapter 5. Found at an elevation of 8,990 feet and dated to 11,200 years ago, this exquisitely preserved animal was identified, on the basis of its teeth, as a Columbian mammoth, as have all other Great Basin mammoths that have been identified to the species level. This animal was so well preserved that it provided not just stomach contents but DNA as well. The genetic work, done by Jacob Enk and his colleagues, showed that this animal, which had the teeth of a Columbian mammoth, was, according to its DNA, a woolly mammoth. Not only did its DNA match that of woolly mammoths, but it matched the DNA of those woolly mammoths that were native to the eastern side of the Bering Land Bridge, which I discussed earlier.[310]

This poses more than a little problem. In terms of its form, the Huntington mammoth is a Columbian mammoth, *Mammuthus columbi*. In terms of its genetics, it is a woolly, *Mammuthus primigenius*. The quality of the paleontology cannot be questioned, nor can the quality of the ancient DNA work. The only way to explain this is the way that Enk and his colleagues explain it: Columbian and woolly mammoths could and did interbreed. Gary Haynes had very good reason to wonder whether the two forms might simply be minor variants of one another.

That brings us back to Jefferson's mammoth, known primarily from the Midwest. At the outset of this section, I mentioned that some paleontologists accept this animal as a separate species.[311] Others, however, point out that Jefferson's mammoth seems intermediate in form between the Columbian and woolly forms and might, in fact, be a hybrid between the two.[312] Enk and his colleagues have shown that this may well be a fully reasonable interpretation of those animals considered to be Jefferson's mammoths.

It also gets us back to where we started. I mentioned at the outset that some consider all American mammoths to belong to a single species, *Mammuthus columbi*, and divide that species into several subspecies. This assumes that they could all potentially interbreed and produce fertile offspring, which is exactly what the DNA from the Huntington mammoth suggests.

Three Extinct Species

As if 37 genera of extinct late Ice Age North American mammals were not enough, several well-defined species of mammals that belong to genera that still exist in North America were also lost as the Ice Age came to an end.

The beautiful armadillo (*Dasypus bellus*) is essentially a much larger version of the nine-banded armadillo (*Dasypus novemcinctus*). In the United States, the nine-banded armadillo is now found from eastern New Mexico to Florida; it also occurs along the eastern and western edges of Mexico and well into South America. Although this animal is familiar to many people who live within its current range, that range is surprisingly recent. With one exception, there is not a single record for the nine-banded armadillo in the United States prior to about 150 years ago.[313] That one exception is substantial, though, since it suggests that these animals may have been living in Florida toward the very end of the Ice Age.[314]

Although large nine-banded armadillos can overlap small beautiful armadillos in size, the latter were, in general, about twice the size of the former. Other than size, the skeletons of these two animals are extremely similar, as are their distributions (fig. 3.67; there is also a record from southern Mexico). As a result, paleontologists have long wondered whether the two actually belong to the same species, with the modern form simply being smaller than the late Ice Age version. We have ancient DNA from only one beautiful armadillo specimen, but that DNA suggests that they really do belong to separate species.[315]

There is no such uncertainty about the three species of extinct mammals that belong to existing North American genera that were found in the Great

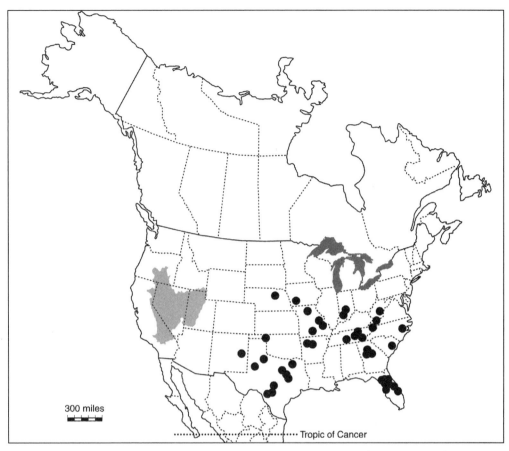

FIGURE 3.67. The late Ice Age distribution of the beautiful armadillo (*Dasypus bellus*) in North America.

Basin during the late Ice Age: the American lion, the dire wolf, and Harrington's mountain goat.

The American Lion (*Panthera leo atrox*)

I have pointed out that the American cheetah (*Miracinonyx trumani*) wasn't a real cheetah—it just looked like one. The American lion, on the other hand, was a real lion, found from Alaska and western Canada deep into Mexico and missing only from the forested east (figs. 3.68 and 3.69).

Lions, in fact, were almost literally all over the place during the late Pleistocene, from Africa throughout Eurasia, across the Bering Land Bridge into North America, and south into at least southern Mexico.[316]

Modern lions are referred to scientifically as *Panthera leo*. The Eurasian Pleistocene version is known as the cave lion, called either *Panthera leo spelaea* or *Panthera spelaea*, depending on whether one treats them as a subspecies of the modern lion or as a separate species. The North American lions that

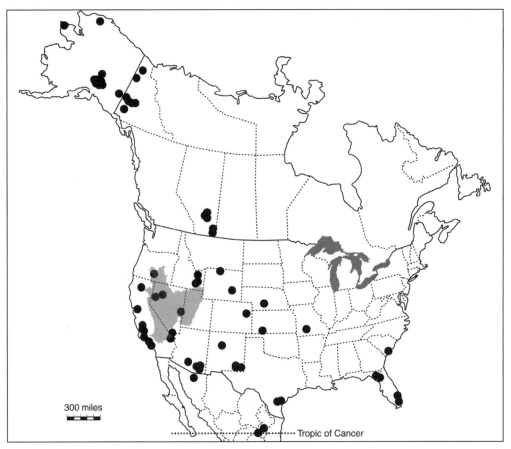

FIGURE 3.68. The late Ice Age distribution of the American lion (*Panthera leo atrox*) in North America.

were found south of glacial ice are referred to as *Panthera leo atrox* or *Panthera atrox*, again depending on how one assesses how closely related they are to African lions.

For quite some time, paleontologists argued about whether these really were lions, with many suggesting that they were a giant form of jaguar, *Panthera onca*.[317] Geneticist Ross Barnett and his colleagues, however, put an end to this debate by analyzing the DNA of modern African and Asian lions and comparing it to the DNA of the remains of the Pleistocene lions of Eurasia and North America. The results were clear. While the Eurasian and North American Pleistocene lions differ genetically from one another, and from African lions, all three are closely related. As part of this work, Barnett and his coworkers also showed that North American lions came in two varieties. Those from eastern Beringia (Alaska and the Yukon) were genetically the same as the Eurasian cave lion. Those that lived south of glacial ice in North America

FIGURE 3.69. The African lion (*Panthera leo*), Serengeti National Park, Tanzania. Photograph by James F. O'Connell.

formed their own, slightly different, group.[318] This doesn't solve the minor problem of whether to treat them as different species (*Panthera spelaea* and *Panthera atrox*, or *Panthera leo spelaea* and *Panthera leo atrox*), but that doesn't matter. Sometime in the deeper past, the Eurasian lion crossed the Bering Land Bridge and entered northern North America. At least 200,000 years ago, some of those lions made it south, where they evolved into the American lion, left their distinctive large footprints in a cave in southern Missouri,[319] and survived until the end of the Ice Age (chapter 4).

Modern lions are sexually dimorphic, with females having about 80 percent of the body weight of males. American lions were even more so, with females about 70 percent of the weight of males. Exactly what that weight was depends on how the estimates are made, but regardless, the numbers are impressive. Paleozoologists Todd Wheeler and George Jefferson estimate males to have weighed about 540 pounds, females about 390. For comparison, modern male lions are about the size of extinct American female lions. These averages don't do justice, though, to the size these animals could reach. The largest individual Wheeler and Jefferson examined came in at an estimated 1,000 pounds.[320] Longer legged than modern lions, these were formidable predators, capable of taking down prey as large as adult bison.[321]

American lions have been reported, with varying degrees of confidence, from six Great Basin sites. Oregon's Fossil Lake provided a series of fragmentary specimens that paleontologist Herbert Elftman thought might pertain to this species, but he could not be certain.[322] The record for Rye Patch Dam, along the Humboldt River in northern Nevada, is even less certain, since it consists of a single small tooth fragment.[323] We are on firmer ground with a specimen from Astor Pass in northwestern Nevada. Although this site provided only a single tooth, there is no question that it came from an American lion.[324] The Smith Creek Cave record, from the eastern edge of the Snake Range in far eastern Nevada, is based on a single but distinctive and very large toe bone.[325] The mandible that came from Tule Springs, in Las Vegas Valley, is also clearly from an American lion.[326] The final record, from Piute Valley on the California-Nevada border, has been identified as American lion but has not been described in detail.[327]

I did not discuss the American lion with the extinct genera of the North American Pleistocene because, while the species (*Panthera leo*) is extinct in North America, the jaguar (*Panthera onca*) belongs to the same genus and still occurs in the far southwestern United States.[328] In fact, the jaguar has also been reported from the Ice Age Great Basin, including Smith Creek Cave, the same site that I mentioned above as having also provided an American lion specimen, and from Fort Irwin in southeastern California's Mojave Desert.[329]

The Dire Wolf (*Canis dirus*)

To judge by the frequency with which I see coyotes (*Canis latrans*) in the Great Basin, to say nothing of my Seattle neighborhood, wild members of the genus *Canis* are thriving in North America today. "Thriving" is not the word to use for the current status of North American wolves (*Canis lupus*), but they, too, are doing well in parts of the continent. Dogs belong to the same genus, but they are so closely related to wolves, from which they descended, that they are now most commonly treated as a subspecies of wolf, *Canis lupus familiaris*. Because dogs were domesticated in Eurasia during the late Pleistocene, it is possible that they accompanied people on their journey across the Bering Land Bridge sometime prior to 12,500 years ago and were found in North America as well. If they did, though, we have no evidence for it.[330]

There was another species of this genus on the North American landscape during the late Pleistocene, the famous dire wolf (*Canis dirus*). Picture a large, heavily built wolf with substantial teeth, and you are close to picturing what a dire wolf would have looked like.

If you pet a large dog on the back of the head—a German shepherd will do nicely—you may notice the large pointy thing at the rear end of its skull. That large pointy thing is created by the intersection of two ridges of bone. One of these ridges, called the nuchal crest, is where the neck muscles attach and runs along the outer edge of the back of the skull. You can feel where the human version would be, if we had one, by flexing your neck muscles and running your fingers up the back of your neck and onto your skull to the point where you can no longer feel those muscles flexing. We don't have a nuchal crest, largely because we are bipedal, balancing our head on our upright vertebral column. Quadrupeds don't have that option and so need larger muscles to hold their heads up. When those muscles are large enough in relation to the size of the skull, a nuchal crest develops for muscle attachment.

The second ridge, called the sagittal crest, runs down the center of the skull and serves for the attachment of muscles that help close the jaw. You can feel where yours would be, if you had one, by clenching your teeth and moving your fingers upward on the sides of your head until you reach the point where the muscle seems to blend into the bony part of your skull. If your skull were to become a lot smaller, or your jaw muscles a lot bigger, those muscles would move up the sides of your head until they ran out of room to attach. The sagittal crest provides extra room for that attachment. Animals with huge sagittal crests have correspondingly large muscles for closing their jaws. The sagittal crests on dire wolves were substantial, sometimes scarily so, and often larger than those on wolves.[331] These animals had significant power in their jaws. "Nice doggy" would not have applied to the dire wolf.

Dire wolves were found throughout North America south of glacial ice (fig. 3.70), and deep into South America.[332] In North America, they came in two slightly different versions, a larger eastern one that weighed an estimated 150 pounds and a smaller western one that weighed about 130 pounds.[333] Wolves weigh, on average, about 90 pounds, though they can get much bigger than this.[334] Dire wolves were, in short, the size of very large wolves.

By analyzing the size of dire wolf skulls and teeth, paleobiologists Blaire Van Valkenburgh and Tyson Sacco have shown that these animals had about the same level of sexual dimorphism as modern wolves, with males only slightly larger than females. This, in turn, suggests that male dire wolves did not compete with one another for access to mates and so likely lived in

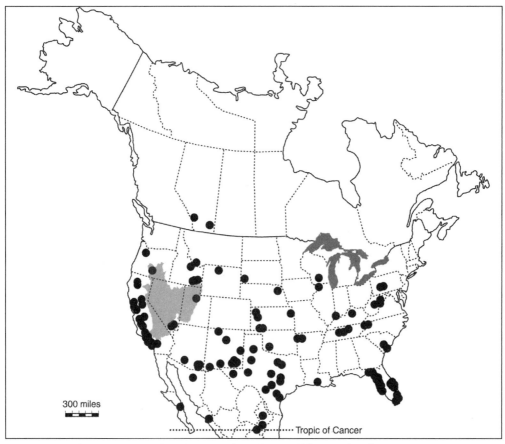

FIGURE 3.70. The late Ice Age distribution of the dire wolf (*Canis dirus*) in North America.

monogamous pairs, just as modern wolves do. Van Valkenburgh and Sacco combine this with the fact that dire wolves are by far the most common carnivore at Rancho La Brea to suggest that they were being entrapped there not as single individuals, but as groups. They conclude, as do others, that dire wolves lived in packs, much like modern wolves, and that those packs were likely composed of an alpha pair—leaders of the pack—along with their current offspring and those from previous years.[335]

Dire wolves were powerful carnivores the size of very large wolves, they had massive biting power, and they were organized into packs. All this suggests a formidable predator. Given that modern wolves can take down a moose, dire wolves could almost certainly have done better than that, including adult bison within their predatory range.[336]

Dire wolves have been reported from three, and perhaps four, sites along the edges of the Great Basin (fig. 3.70). Fossil Lake, Oregon (chapter 5), provided four specimens tentatively identified as belonging to this species. One of these was a tooth, the very careful description of which makes that identification convincing.[337] Silver Creek, above Salt Lake City in the Utah's Wasatch Range, yielded three bones tentatively identified as dire wolf, and the descriptions of these specimens again makes it very likely that this is the animal involved.[338] Most recently, geologist Josh Bonde reported a dire wolf foot bone from the Tule Springs area of southern Nevada.[339] The article in which this discovery was reported includes an excellent photograph of the specimen and claims that this is the first record for dire wolf from Nevada. In fact, it probably is. The only other possibility that has been reported comes from Gypsum Cave, also in southern Nevada (chapter 5). This material, unfortunately, has never been adequately described and might have come from either a wolf (*Canis lupus*) or a dire wolf. I have plotted it on figure 3.70, but the Tule Springs specimen does provide the first trustworthy record for this animal in Nevada.

Harrington's Mountain Goat (*Oreamnos harringtoni*)

In October 2010, a hiker was gored to death by a mountain goat (*Oreamnos americanus*) in Olympic National Park, in the far northwestern corner of Washington State.[340] Prior to 1925, hikers in the mountains of the Olympic Peninsula did not have to fear goats, since that was the year that goats began to be introduced there, the starter kit consisting of four animals transported from Alaska.[341] They have also been introduced to the mountains of central, northeastern, and southeastern Utah. Here, the starter kit was provided by animals from Olympic National Park. The reasons for the continuing introductions in Utah are purely economic: enhanced tourism and the sale of hunting permits. In 2014, if you wanted one of the latter and were not a Utah resident, you could get one for $1,518.[342]

According to the state of Utah, the native status of mountain goats in Utah is debatable and controversial.[343] In fact, if anyone is debating this, they must be talking to a mirror. Over 100,000 archaeological sites have been reported for the state.[344] Only a small fraction of these have been excavated, but not a single bone or tooth of a modern mountain goat has come from any of them. There is a record of *Oreamnos americanus* from west-central Nevada, but it hardly counts, not simply because it is so far from Utah, but because it may be several hundred thousand years old.[345]

Today's mountain goats are not native to Utah. Instead, they are native to southern Alaska and adjacent Canada, south through the central Cascades of Oregon and the northern Rocky Mountains as far south as Idaho.[346]

This is not to say that there were never any mountain goats in Utah. There were, but they belonged to a very different species and they have been extinct for about 10,000 years (chapter 4). Harrington's mountain goat (*Oreamnos*

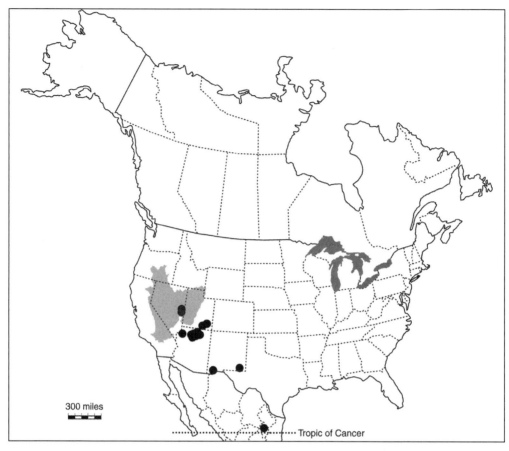

FIGURE 3.71. The late Ice Age distribution of Harrington's mountain goat (*Oreamnos harringtoni*) in North America.

harringtoni) is known from the central Great Basin south into northern Mexico, with the majority of sites known from the Grand Canyon area (fig. 3.71). The analysis of DNA taken from both modern and ancient versions shows them to belong to closely related but separate species.[347]

That is the same conclusion that had been drawn from the analysis of Harrington's mountain goat bones and teeth. In fact, that was the conclusion that was drawn when the animal was first described, in 1936, from material that had been excavated in eastern Nevada's Smith Creek Cave.[348] All that we have learned since has confirmed that initial decision.

Fortunately, this is an animal we know a lot about, much of it as a result of the work of paleontologist Jim Mead and his colleagues. It doesn't hurt that Harrington's mountain goat frequented the dry caves of the southern Colorado Plateau, where a wide variety of its remains have been beautifully preserved. Those remains include not just bones and teeth, but such things

FIGURE 3.72. The skull of Harrington's mountain goat (*Oreamnos harringtoni*), Tse-An Bida Cave, Grand Canyon. Photograph by Kenneth L. Cole.

as horn sheaths and hair. We know what the animal looked like and we know what it ate. We even know at least some of the places it pooped, since its dung pellets have been preserved in abundance in some of those dry caves.

Compared to existing mountain goats, Harrington's mountain goats were about 30 percent smaller and peered out from a face that was more slender and elongated (fig. 3.72). The horns that projected up and back from behind their eye orbits tended to be slimmer than those of today's mountain goats but at times got as long as those now carried by females. The long bones of the lower ends of their limbs—the metacarpals and metatarsals—were shorter than those of *Oreamnos americanus*, but they were also wider. These animals were well adapted to nimbling their way around places like the steep-walled Grand Canyon. That is a good thing, because they certainly weren't camouflaged; hairs preserved in Grand Canyon caves show them to have been as white as their modern counterparts.[349]

Thanks to those preserved dung pellets, and to those who have spent the time analyzing them, we know what the Harrington's mountain goats that were living near the Colorado River in southern Utah and northern Arizona were eating. Simply put, it was pretty much everything on the landscape that they could eat, from grasses and shrubs to parts of trees, including bark.[350]

Harrington's mountain goat is known from two sites in the Great Basin: Smith Creek Cave and Labor-of-Love Cave, both in eastern Nevada. Smith Creek Cave provided the material—skull parts, horn cores, and other specimens—that paleontologist Chester Stock used to define the species in the first place. As we will see in the next chapter, the location of this site seems better fitted to mountain goats than it would appear to be to people, though prehistoric peoples lived there for significant amounts of time. Labor-of-Love Cave provided a single foot bone, identified after the initial report on this site appeared.[351] Given the mountainous terrain of so much of the Great Basin, it would be surprising if the remains of more of these animals were not found here in the future.

THE EXTINCT LATE PLEISTOCENE BIRDS
OF THE GREAT BASIN

Even though this is a book about the extinct Pleistocene mammals of the Great Basin, the now-extinct birds that were found here, and that often depended on those mammals for lunch, were so remarkable that I can't ignore them. I also can't ignore them because some of those birds first became known to science through work done in this region.

We know far less about the birds of the late Ice Age Great Basin than about the mammals because very few scientists specialize in the paleontology of birds, and very few of them have worked in the Great Basin. Much of the most interesting and important work was done decades ago, a good deal of it as a result of the efforts of avian paleontologist Hildegarde Howard (I was fortunate to have been able to work with her when I was a graduate student).

There are about 20 genera of extinct birds known from the North American late Pleistocene, of which 8 are known from the Great Basin (table 3.4). While the list of extinct late Ice Age mammals known from the Great Basin is probably reasonably complete, there is no reason to think that this is the case for the birds. Not only have few avian paleontologists worked here, but the number of carefully excavated Great Basin sites known to have provided the remains of Pleistocene birds is very small. I would guess, for instance, that the extinct vulture *Neophrontops*, known from northern Mexico to California and Wyoming,[352] also occurred in the Great Basin but has simply yet to be found here, or if found, has yet to be identified by someone with the necessary skill. Its time will likely come.

Even as it stands, however, the list shows that the Great Basin late Pleistocene avifauna was as astonishing as its mammalian counterpart. In addition

TABLE 3.4. Extinct Late Pleistocene genera of North American birds (genera marked with an asterisk survive outside North America; those in bold are known from the Great Basin [from Grayson 2011])

Family	Genus	Modern relatives	References
Ciconiidae	*Ciconia**	Storks	4
Phoenicopteridae	*Phoenicopterus**	Flamingos	9
Anatidae	*Anabernicula*	Shelducks	10
Cathartidae	*Breagyps*	Condors, vultures	3, 7, 17
Teratornithidae	*Aiolornis*	(Extinct teratorn)	2
	Teratornis	(Extinct teratorn)	2, 6, 7, 17
	Cathartornis	(Extinct teratorn)	17
Accipitridae	*Spizaetus**	Hawks, eagles	7, 17
	Amplibuteo	Hawks, eagles	5, 7, 17
	Neophrontops	Old World vultures	7, 17
	Neogyps	Old World vultures	7, 17
Falconidae	*Milvago**	Caracaras	4
Phasianidae	*Neortyx*	Quails	4, 8
Burhinidae	*Burhinus**	Thick-knees	11
Charadriidae	*Belonopterus**	Lapwings	1, 4
Icteridae	*Cremaster*	Cowbirds, blackbirds	4, 13, 16
	Pandanaris	Cowbirds, blackbirds	4, 13, 15, 16
	Pyeloramphus	Cowbirds, blackbirds	12, 14
Corvidae	*Protocitta*	Jays	13, 16
	Henocitta	Jays	4, 13, 16

References: (1) K. E. Campbell 2002; (2) K. E. Campbell, Scott, and Springer 1999; (3) Emslie 1988; (4) Emslie 1998; (5) Emslie and Czaplewski 1999; (6) Fox-Dobbs et al. 2006; (7) Hertel 1995; (8) Holman 1961; (9) Howard 1946; (10) Howard 1964a, 1964b; (11) Howard 1971b; (12) Howard and Miller 1933; (13) Hulbert and Becker 2001; (14) A. Miller 1932; (15) A. Miller 1947; (16) G. Morgan 2002; (17) Van Valkenburgh and Hertel 1998.

to the eight genera of extinct birds that are known from here, there are also three species of birds that belong to genera that still exist and that are very much worth mentioning. These are the Western Black Vulture (*Coragyps occidentalis*) and the Fragile Eagle (*Buteogallus fragilis*), both of which are extinct, and the California Condor (*Gymnogyps californianus*), which is still with us, albeit barely.[353]

The Ciconiidae: Storks

Storks come in many forms, but the ones that matter here belong to the genus *Ciconia*. Most of these are Old World in distribution, found from southern Africa north through Europe and Asia. The only exception is the Maguari Stork (*Ciconia maguari*), which is South American. During the late Pleistocene,

North America had its own, widely distributed member of this genus, the extinct Asphalt Stork, *Ciconia maltha*. First described from Rancho La Brea (hence the common name), and about the same size as the Maguari Stork, this bird is known from California to Florida.[354] It is also known from the Mojave Desert's Lake Manix Basin, in deposits that appear to be about 20,000 years old.[355] Today, the only stork found in the Great Basin is the Wood Stork (*Mycteria americana*), and it is here only occasionally during the summer months after it has finished breeding elsewhere. There is a single sighting of the Marabou Stork (*Leptoptilos crumenifer*) from the Great Basin, but since this is an African species, the chances are good that it was an escaped captive.[356]

The Phoenicopteridae: Flamingos

The American Flamingo (*Phoenicopterus ruber*) today breeds as far north as the Caribbean and sometimes wanders into the eastern United States. There are flamingo sightings from the Great Basin, but these almost certainly also represent escaped birds.[357] There were, though, two species of flamingos in the late Ice Age Great Basin. The first of these, Cope's Flamingo (*Phoenicopterus copei*), was slightly larger than the American Flamingo, with more robust legs. First identified from Fossil Lake, Oregon, well over a century ago,[358] Cope's Flamingo is known to have bred here; Hildegarde Howard showed that the Fossil Lake collections contain the bones of immature individuals of this species (chapter 5).[359] Cope's Flamingo is also known from the Lake Manix Basin, but so is a second, smaller species of flamingo, *Phoenicopterus minutus*.[360] Including the Manix Lake Basin flamingos on our list may be cheating, though, since the bottom part of the deposits from which this material comes is about 300,000 years old.[361]

The Anatidae: Ducks, Geese, and Swans

Rancho La Brea may be the most famous of southern California's "tar pits," but it is not the only asphalt deposit in this area to have provided rich hauls of late Pleistocene vertebrates. McKittrick, located west of Bakersfield, has not provided the wealth of material that Rancho La Brea has, but what has come from there has been extremely important. That includes the remains of a slender, lightly built, goose-like bird that was placed in a new genus and species, *Anabernicula gracilenta*. This bird, Hildegarde Howard pointed out, was "too small to be assigned to any living species of goose, and unlike any species of duck."[362] Although the closest living relatives of this waterbird are debated, most consider those relatives to be shelducks of the genus *Tadorna*.[363] Today, shelducks occupy primarily temperate regions in Europe, Asia, Africa, and the Pacific, but this was clearly not the case during the late Pleistocene, since *Anabernicula* was fairly common from California to the southwestern United States and in adjacent Mexico.[364] It is also known from two sites in the Great Basin: Oregon's Fossil Lake and eastern Nevada's Smith Creek Cave,[365] both of which are discussed in detail in the next chapter.

The Cathartidae: New World Vultures

The rest of our extinct birds were raptors, making their living by preying on live animals or scavenging the remains of dead ones. There were six kinds, organized into three distinctly different families (table 3.4).

North America now has three species of New World vultures, or Cathartidae, all of which are scavengers: the Turkey Vulture (*Cathartes aura*) and California Condor (*Gymnogyps californianus*) in the West, and the Black Vulture (*Coragyps atratus*) in the East. Of the four species that occur farther south, the Andean Condor (*Vultur gryphus*) is certainly the best known, famous for a wingspan that can reach 10 feet or more.[366]

In 1910, Loye Miller reported that Rancho La Brea had yielded a single large foot bone (tarsometatarsus) from a vulture the size of a California Condor. Since it seemed most similar to the same bone from an Andean Condor, this was the genus to which he assigned it.[367] By 1937, enough material from this bird had been found to make it clear that this large scavenging condor from Rancho La Brea deserved not only its own species but also its own genus: *Breagyps clarki*, or Clark's Condor.[368]

Most bird experts continue to agree that, among living species, Clark's Condor is most closely related to the Andean Condor.[369] Compared to the California Condor, with a wingspan that can reach over 9 feet, Clark's Condor was slightly larger, with a longer and more slender beak, broader and deeper chest, and a larger wingspan.[370] Best known from the asphalt deposits of southern California, it has also been reported from one site in central Mexico[371] and one in the Great Basin—Smith Creek Cave in eastern Nevada's Snake Range (chapter 4). Smith Creek Cave, however, provided at least six individuals of this substantial scavenging bird. One of those was immature, suggesting that the bird was breeding in the Snake Range or nearby.[372]

For decades, it was thought that Clark's Condor was not the only condor to have left its remains in Smith Creek Cave. In 1952, Hildegarde Howard tentatively identified five specimens of the California Condor from this site, but she later realized that this material had actually come from *Breagyps clarki*.[373] California Condors are, however, known from late Ice Age deposits at Antelope Cave, located in southeastern California's Mescal Range,[374] and from Gypsum Cave, in southern Nevada (chapter 5).[375] Thanks to paleontologist Steve Emslie, we have radiocarbon dates for the Antelope and Gypsum Cave condors; they are 11,080 and 14,720 years old, respectively.

It might come as a surprise to learn that California Condors once lived in the Great Basin, since we think of these huge scavengers as birds of the Pacific Coast. During the late Pleistocene, however, they were found from California to Florida and New York.[376] As Emslie has shown, the distribution of these birds seems to have retracted dramatically as the Ice Age came to an end; the latest dates for them in the Southwest, for instance, fall at about 9,600 years ago.[377] Their restricted distribution during historic times is, from a deeper perspective, very recent.

People who spend any time in the Great Basin today have seen Turkey Vultures (*Cathartes aura*) teetering high in the sky as they search for food. The Turkey Vulture is the only member of the cathartid family to be found here today, but it is found in significant numbers.

In North America, the Black Vulture (*Coragyps atratus*) is primarily a bird of the eastern United States, coming no closer to the Great Basin than southern Arizona. There is, however, an apparently extinct version of this species. Common at Rancho La Brea and known from a variety of other late Ice Age sites from northern Mexico to California, the Western Black Vulture (*Coragyps occidentalis*) would have looked very much like today's Black Vulture, except that it was larger and had longer wings and shorter, stouter legs.[378] Like its close relative, the Western Black Vulture was a terrestrial scavenger.[379] The real question about the Western Black Vulture, similar in form and in diet to the Black Vulture, is whether it is actually extinct. Some think that the Black Vultures that are so common in parts of North America today are simply the slightly smaller descendants of the Ice Age versions.[380]

Even though the Turkey Vulture is common in the Great Basin today, it is unknown from any Ice Age site in this region. The apparently extinct Western Black Vulture, however, is known from eastern Nevada's Smith Creek Cave, although from only two specimens.[381] This is something of an oddity, since both the Western Black Vulture and the Turkey Vulture are known from Rancho La Brea, though the former is far more common than the latter.[382]

The Teratornithidae: Teratorns

The most impressive birds of the North American late Pleistocene were surely the teratorns. Closely related to both vultures and storks, teratorns are so distinctive that they are placed in their own family, the Teratornithidae.

They became known to science in 1909, when Loye Miller described a huge, condor-like bird that had been recovered from Rancho La Brea and that he called Merriam's Teratorn, *Teratornis merriami*. With a wingspan of 12 to 13 feet and weighing at least 30 pounds, this giant bird lived from southern California through northern Mexico to Florida.[383] Along the way, it also lived in the Great Basin, where its remains have been identified from the late Ice Age deposits of Tule Springs in southern Nevada, and, somewhat less securely, from Crystal Ball Cave in western Utah.[384]

As deeply impressive as it would have been to see Merriam's Teratorn sailing though the sky, it would have been even more impressive to see the Incredible Teratorn (*Aiolornis incredibilis*) doing the same thing. As I discuss in more detail in chapter 5, this bird was first identified by the equally incredible Hildegarde Howard on the basis of a single wrist bone that had been excavated from Smith Creek Cave. She called it the Incredible Teratorn because it was so big, with a wingspan estimated at 16–18 feet, about twice the wingspan of a California Condor. The remains of this bird have rarely been found, but by 1999, enough were known to realize that it was different enough from Merriam's Teratorn to merit its own genus. In 1999, paleontologist Kenneth Campbell and his colleagues gave it the name *Aiolornis incredibilis*, which translates to the very appropriate "Incredible God of the Winds Bird." The Smith Creek Cave specimen remains the only one known from the Great Basin.[385]

Both of these teratorns seem to have been fairly agile on the ground, but better adapted for walking than running. An obvious question is how they got off the ground, but it seems likely that they could run well enough to take off by getting a running start. If they couldn't, it seems unlikely they would have lasted as long as they did, since the earliest known specimen is over three million years old.[386] What they were doing on the ground seems clear enough. Although it has been suggested that they were active predators, stalking and taking small vertebrates, chemical analyses of their bones makes it seem far more likely that these birds made their living by scavenging.[387]

In case you are having trouble believing any of this, I will now make things worse by saying that the Incredible Teratorn is not the largest known member of the teratorn family. That honor belongs to *Argentavis magnificans*, the Giant Teratorn ("Incredible Teratorn" had already been taken), which lived in South America about six million years ago, weighed about 150 pounds, and had a wingspan of about 23 feet. It was, as paleobiologist Sankar Chatterjee and his colleagues have pointed out, "about the size of a Cessna 152 light aircraft."[388] It is also the largest flying bird known to have existed.

The Accipitridae: Kites, Hawks, Eagles, and Old World Vultures

When paleontologists work with ancient bones long enough and hard enough, sooner or later they encounter something they can't quite believe. It has happened to me on a few occasions, and it has happened to most paleontologists I know. It happened to Loye Miller around 1915, when he was working with the birds of Rancho La Brea. He had come across the bones of what looked like an Old World vulture—not the cathartids he was used to, like the Western Black Vulture, but a real Old World vulture, the kind you see in the Serengeti today. He was so surprised by this that he set those bones aside for several years. During those years, he reported, "the problem has been taken up, worked out, and then laid aside many times."[389] Not until he was utterly convinced that he was right did he publish the results: Rancho La Brea had yielded the remains of two distinct Old World vultures. Not only was Miller right, but the record for Old World vultures in North America now extends back some 20 million years, into the early Miocene.[390]

Noting the similarity of one of his Rancho La Brea vultures to the Egyptian Vulture (*Neophron percnopterus*), Miller called the first one the American Neophron (*Neophrontops americanus*). As I noted earlier, it is now known from northern Mexico to California and Wyoming (an earlier form is known from Florida), but it has yet to be reported from the Great Basin.[391]

The second one he called the Errant Eagle (*Neogyps errans*). This bird was about the size of the familiar Golden Eagle (*Aquila chrysaetos*), though with longer wings and relatively short legs. Its powerful beak and strong feet suggest it was far more capable of taking live prey than are the living Old World Vultures.[392] The results of detailed analyses of its skull and mandible agree: this was a bird that seemed almost equally at home taking live prey animals and scavenging dead ones.[393]

Remains of the Errant Eagle have been found from southern California and northern Mexico north to Colorado and the Great Basin. Within the Great Basin, it is known from two specimens from Smith Creek Cave, detected (of course) by Hildegarde Howard.[394]

Miller struggled with the discovery of an Old World vulture in the New World's Rancho La Brea, but he had no problem with a single leg bone from that site that he found similar, but not identical, to the corresponding bone of the South American Black-chested Buzzard-Eagle (*Geranoaetus melanoleucus*). This, at least, was a bird that was known from the Americas. He assigned the Rancho La Brea material to an extinct species of the Buzzard-Eagle genus, *Geranoaetus grinnelli*. Not until two decades later had enough material been discovered to show that this bird is actually far more closely related to the

Hawk-Eagles, found from central Mexico deep into South America.[395] Miller's extinct species from Rancho La Brea is now placed in the Hawk-Eagle genus and is known as Grinnell's Eagle (*Spizaetus grinnelli*).

Two extinct species of this genus have been reported from the Great Basin, *Spizaetus pliogryps* from Fossil Lake[396] and *Spizaetus willetti* from Smith Creek Cave.[397] The modern relatives of these birds are active predators, feeding on a wide array of smaller vertebrates. Analyses of the skeleton of Grinnell's Hawk-Eagle suggest that it did the same thing, and there is no reason to think the Great Basin versions were any different.[398]

Our last raptor, and last extinct bird, was also first detected by Loye Miller from Rancho La Brea material.[399] He called it the Fragile Eagle because the leg bones he had available were so long and slender. As additional remains of this bird became available, it became clear that he was right: this was a large, long-legged, and lightly built eagle. It also become obvious that the Fragile Eagle was very similar in form to today's Black Hawks, of the genus *Buteogallus*. As a result, the Fragile Eagle is now known scientifically as *Buteogallus fragilis*.

A close relative of the Fragile Eagle, the Common Black Hawk (*Buteogallus anthracinus*), now breeds as far north as the Utah-Arizona border and has been seen a few times in far southern Nevada.[400] The Fragile Eagle itself is known from central Mexico and from a number of places in the western United States, including New Mexico, the asphalt deposits of southern California, and from Hawver Cave in northern California. Much earlier examples have also been reported from Florida.[401] There is a single record for this bird in the Great Basin, from Mineral Hill Cave, in northeastern Nevada, which I discuss in the next chapter.[402]

Why So Many Raptors?

If you know the Great Basin well enough, picture a dry valley—the Snake Valley on the border between eastern Nevada and western Utah, or the Fort Rock Basin in south-central Oregon, or just about any other. Close your eyes and picture the Incredible Teratorn, Clark's Condor, and California Condor jostling for position around the carcass of a dead camel or horse near the shore of a lake dotted with flamingos and storks and perhaps even a shelduck. From a modern perspective, it is hard to imagine that this could have been the case, but it was.

Today, the Turkey Vulture is the only large scavenging bird in the Great Basin. During the late Pleistocene, however, this same area supported at least five species of vultures, condors, and teratorns, all of which were scavengers:

Clark's Condor, the Western Black Vulture, the California Condor, the Incredible Teratorn, and Merriam's Teratorn. We can probably add the Errant Eagle to this list, since it is very likely that it scavenged as well. That gives us six large scavenging birds in a region that today supports only one. If it turns out that the Turkey Vulture was here as well, that would give us seven. There are only seven species of vultures and condors in all of the Americas today,[403] but there may have been this many in the late Ice Age Great Basin alone.

How could this region have supported all these birds? Obviously, the late Ice Age mammal fauna here must have been rich enough not just in species but also in individual animals to support them. That, in turn, might seem to suggest that the Great Basin's late Pleistocene landscape teemed with large mammals. As we will see in chapter 6, however, this seems unlikely, and the explanation for the diverse set of large scavenging birds known from the Great Basin Pleistocene is likely to lie elsewhere.

Explaining how the Great Basin was able to support so many large scavenging birds during the late Ice Age is just one of two important challenges these birds present. We can flip that question around and ask why this region has been able to support only one such bird, the Turkey Vulture, during what seems to have been the past 10,000 years or so.

The answer might seem simple. All those large scavengers became extinct because the terrestrial meat market collapsed at the end of the Ice Age. When the large mammals became extinct, the large birds that depended on them for meals also became extinct.[404] To some extent, this is clearly true. Two research teams, one led by C. Page Chamberlain and the other by Kena Fox-Dobbs, have shown, through analyses of bone chemistry, that Merriam's Teratorn and the Western Black Vulture both had diets focused on scavenged terrestrial herbivores, while the California Condor fed both on such herbivores and, in coastal settings, on scavenged marine mammals. California Condors living near the coast of the state for which they were named survived because the marine component of their diet survived the Ice Age; those condors entirely dependent on terrestrial herbivores succumbed.[405]

Clearly, the loss of so many large terrestrial herbivores at the end of the Ice Age played a role in the loss of large scavenging birds. It is hard to see, though, how it can account for the full range of raptors that were lost. After the Ice Age ended, the Great Basin still had bighorn sheep and pronghorn in numbers. There were also elk, bison, and mule deer, albeit in far lesser numbers.[406] Where Black Vultures and Turkey Vultures occur together, they have no problem feeding and roosting together. Why, then, did the Western Black Vulture, so closely related to today's Black Vulture, become extinct while the Turkey Vulture survived? Why did all but the Turkey Vulture become extinct when

TABLE 3.5. Extinct mammal genera shared by North and South America and those extinct in the north that still exist in the south

Extinct genera shared by North and South America	Genera extinct in North America but extant in South America
Cuvieronius	*Hydrochoerus*
Equus	*Tapirus*
Eremotherium	*Tremarctos*
Glyptotherium	
Hemiauchenia	
Holmesina	
Mixotoxodon	
Neochoerus	
Palaeolama	
Pampatherium	
Smilodon	

there is every reason to think that there was enough food on the landscape to support more than one such scavenging bird? The same question has been raised before, and in other geographic contexts,[407] and we still have no answer for it. However, the climates to which these birds were adapted disappeared as well, and a full explanation of the loss of such a rich array of Pleistocene raptors will likely have to take that into account.

HOW MANY GENERA OF MAMMALS WERE LOST TOWARD THE END OF THE AMERICAN PLEISTOCENE?

The loss of so many raptors toward the end of the North American Pleistocene raises an important question we cannot fully answer, though it is clear that the loss of so many large terrestrial herbivores played a significant role.

We cannot get even this close to explaining why so many mammals were lost in the Americas, including the Great Basin, toward the end of the Ice Age. I will discuss the explanations that exist for those losses in the last chapter of this book. To understand the magnitude of the explanatory problem posed by these losses, it helps to have a feel for how many genera of mammals were lost in the Americas as the Ice Age came to an end. Without knowing that, it becomes easier to be convinced by explanations for the extinctions that actually aren't convincing at all.

Table 3.1 lists the genera of late Ice Age mammals that were lost in South America and indicates all the extinct South American forms that were also found in North America. There are 11 of these (table 3.5). In addition to these, three genera became extinct in North America but continue to exist in the

TABLE 3.6. The surviving genera of American terrestrial mammals weighing more than 100 pounds (genera found in both North and South America are in bold)

North America (13 genera)	South America (10 genera)
Alces (moose)	*Blastocerus* (marsh deer)
Cervus (elk)	*Hippocamelus* (guemal)
Odocoileus (deer)	**Odocoileus (deer)**
Rangifer (caribou)	*Hydrochoerus* (capybara)
Oreamnos (mountain goat)	*Lama* (llama, guanaco)
Ovibos (muskox)	**Panthera (jaguar)**
Ovis (mountain sheep)	**Puma (cougar)**
Bison (bison)	*Priodontes* (giant armadillo)
Antilocapra (pronghorn)	*Tapirus* (tapir)
Canis (wolf)	*Tremarctos* (spectacled bear)
Panthera (jaguar)	
Puma (cougar)	
Ursus (bears)	

south: the capybara (*Hydrochoerus*), the Andean bear (*Tremarctos*), and the tapir (*Tapirus*). Given that there were 54 genera that became extinct in South America and 37 in North America, subtracting the 14 genera that were either shared by both continents or that continue to exist in one of them gives us the total number of mammal genera that were lost in the Americas as the Pleistocene came to an end. If we are ever to have an adequate explanation of the extinction of American mammals toward the end of the Ice Age, that explanation is going to have to account for the loss of 77 genera of mammals. Of these, 20 occurred in the Great Basin.

THE LOSS OF THE LARGE

At the beginning of this chapter, I pointed out that Alfred Russel Wallace was impressed by the fact that the "hugest" animals were lost when the world became, in his words, zoologically impoverished. He was right to be.

Of the 37 genera of mammals that were lost in North America as the Pleistocene came to an end, 32, or 86 percent, weighed more than 100 pounds (table 3.2). In South America, it was 47 of 54 genera (87 percent; table 3.1). Of the 77 that became extinct in the Americas as a whole, 65, or 84 percent, weighed more than 100 pounds. Genus-level extinction targeted the largest of American mammals.

There is another, equally impressive, way of looking at this. Table 3.6 shows the existing genera of American terrestrial mammals that contain species that weigh more than 100 pounds. There are 13 of these in North America and 10 in South America. Since three occur in both places, there are now 20 genera of mammals in the Americas that make this large-mammal grade. All of these were also here during the late Ice Age, although, as I have discussed, elk and moose were very late entrants from northeastern Asia. Adding these to the late Ice Age faunas of North and South America tells us that the Americas as a whole supported 85 genera of mammals that weighed over 100 pounds. Of these, 65, or 76 percent, failed to survive.

The Americas lost three-fourths of their large land mammal genera as the Ice Age came to an end. Whatever explains the American extinctions must explain not only the loss of 77 genera of mammals but also why so many of them were so large. As we will see in chapter 7, getting that explanation has not proved to be easy.

4 Dating an Ass

As I discussed in the last chapter, the concept of extinction moved from the realm of the impossible to an established fact around the year 1800. As the nineteenth century accumulated, so did our knowledge of the extinct mammals of the relatively recent North American past—the helmeted muskox (*Bootherium*) in 1825, the giant beaver (*Castoroides*) in 1838, the flat-headed peccary (*Platygonus*) in 1848, the tapir (*Tapirus*) in 1849, the camel (*Camelops*) and the giant bear (*Arctodus*) in 1854, and on and on.

However, in the absence of any accurate means of providing ages in years for the fossil remains of these animals, it was impossible to place them in time accurately. Gypsum Cave provides a good example. When, in 1931, Chester Stock discussed "problems of antiquity presented in Gypsum Cave, Nevada," he was concerned only with how they might relate in time to such things as the Pleistocene lakes of the Great Basin. To him, there was simply no way of assessing their ages in years. The paleontologist Richard Lull, who brought the Aden Crater Shasta ground sloth to a broad scientific audience in 1930 (chapter 3), threw his hands up in chronological despair when discussing the age of that animal: "It must be thousands of years old, but how old one is utterly at a loss to say."[1]

Archaeologist Mark Harrington was bolder in this realm. As I discuss in the next chapter, Harrington excavated southern Nevada's Gypsum Cave in the early 1930s, uncovering the remains of Shasta ground sloths and a rich haul of artifacts. Some of those artifacts seemed to be associated with the sloth remains, suggesting that the two had been in the cave at the same time. Deeply interested in estimating the age of his sloths, he took a close look at the layering of one of the parts of the cave he had dug, an area he called Room 1. The highest layer in Room 1, he noted, contained the remains of the archaeological cultures referred to as Pueblo and Basketmaker. He estimated that the bottom of this layer was 3,500 years old. He then measured down to what he thought was a fireplace beneath sloth dung. If, he reasoned, the bottom of Layer 1 was 3,500 years old, and the rate of deposition through time was fairly steady, then "man met the sloth in Gypsum Cave about 8500 BC," or 10,500 years ago.[2]

HOW TO DATE AN EXTINCTION

Although Harrington's fireplace was not a fireplace (chapter 5), his estimates were remarkably accurate. That 3,500-year estimate for the bottom of the Basketmaker stratum? It was dead-on. We know that because archaeologist Amy Gilreath and her colleagues recently returned to Gypsum Cave to conduct additional excavations. Of the 16 new radiocarbon dates they obtained, 13 fall between 2,400 and 3,900 years ago, and two fall almost exactly at 3,500 years ago.[3] The youngest certainly reliable date for the ground sloth dung layer falls at 11,000 years ago. Harrington's guesswork had provided him with an estimate that we now know was very close to the truth.

But the operative word here is "guesswork," since that is what it was. Not until the development of radiocarbon dating did a means of dating late Ice Age paleontological and archaeological deposits become available. That happened in 1949, when Willard F. Libby of the University of Chicago and his colleagues published the first radiocarbon dates. Two years later, the first radiocarbon dates for the Gypsum Cave sloth dung appeared. Those dates suggested that the dung was between 8,000 and 8,900 years old, dates Harrington welcomed with justifiable pride. His estimate had, he noted, been "pretty close, as such guesses go" (see tables 4.1 and 5.6).[4]

We now know that those first radiocarbon dates were too young, but to understand how there can be good and bad radiocarbon dates requires that we know something about the radiocarbon dating method.

The logic is simple. Three isotopes of carbon occur naturally. Two of these are stable: carbon-12 (^{12}C) and carbon-13 (^{13}C). The third, carbon-14 (^{14}C), is unstable, created when cosmic rays hit the atmosphere and produce neutrons, which in turn interact with nitrogen-14 to produce radioactive carbon-14.

Once produced, carbon-14 combines with oxygen to form carbon dioxide, which is then distributed throughout the atmosphere. Some of it ends up in plants and animals. Although there are some tricky exceptions, as long as those plants and animals remain alive, the amount of carbon-14 they contain remains in equilibrium with the amount in the atmosphere. As soon as the organism dies, the carbon-14 is slowly lost, decaying into a beta particle and nitrogen-14. By slowly, I mean that half of it disappears about every 5,730 years (a slightly different half-life was used during the early days of radiocarbon dating), another half after another 5,730 years, and so on. After enough years pass, the amount of carbon-14 that is left is so negligible that it is hard to measure accurately. Given enough time, it will be entirely gone. Because of this, radiocarbon dating can get us back to about 50,000 years ago.

When carbon-14 atoms decay, they emit beta particles. The conventional method counts the number of beta particles emitted as radiocarbon decays in a sample. Because of this approach, and because carbon-14 atoms are rare to begin with, this means of counting requires large samples and long counting times. There have also been multiple methods of obtaining "conventional" radiocarbon dates. Very early in the history of radiocarbon dating, solid carbon from a target sample was dated, but it was soon realized that this approach often gave inaccurate results, and other, more accurate methods were quickly developed.[5] The original dates on the Gypsum Cave sloth dung, published in 1951, were produced using the solid carbon approach. These are the dates that so pleased Harrington, but they can no longer be considered trustworthy.

Today, tiny samples can be dated using a method called accelerator mass spectrometry (or AMS or accelerator dating). Rather than counting the beta particles emitted as carbon-14 decays, this method counts the number of carbon-14 atoms in a sample directly. Either way, radiocarbon dates come associated with an error term. This term takes into account the random error that is always present in radiocarbon measurements; the error term represents one standard deviation around the age estimate. For instance, the radiocarbon date of 11,005 ± 100 years for one of the Gypsum Cave sloth dung samples (see table 4.1) means that the best estimate of the age of this sample is 11,005 radiocarbon years ago, and that there is a 68 percent (one standard deviation) chance that the actual age lies between 10,905 and 11,105 years ago. If we instead used two standard deviations for this date (± 200 years), then there would be a 95 percent chance that the actual age lies between 10,805 and 11,205 years ago. We are actually all familiar with error terms of this sort because pollsters use them all the time. If, for instance, a Gallup poll using a two-standard-deviation error term were to report that 90.2 ± 3 percent of Americans recognize that dogs are superior to cats as pets, what the poll is telling us is that there is a 95 percent chance that between 87.2 and 93.2 percent of Americans feel this way. Radiocarbon dates are read the same way.

Conventional radiocarbon dates can give very accurate results. However, certain kinds of organic materials are easily contaminated as they sit in the ground, and such contaminated samples can provide dates significantly older or younger than the time of death of the organism itself. Bone provides

TABLE 4.1. Radiocarbon dates for extinct Pleistocene mammals in the Great Basin (bone, soft tissue, and dung dates only; trustworthy last appearance dates are in bold)

Genus	Site	Date	Method and material	References
Arctodus	Labor-of-Love Cave, NV	5,320 ± 120	Conventional collagen	1
Arctodus	**Huntington Mammoth site, UT**	**10,870 ± 75**	**AMS bone**[a]	**2**
Arctodus	Huntington Mammoth site, UT	10,976 ± 40	AMS bone	3
Arctodus	Monroc, UT	12,650 ± 70	AMS bone	3, 4
Arctodus	Duck Flat, NV	13,830 ± 50	AMS bone	5
Bootherium	Logan City, UT	7,080 ± 160	Conventional bone	6
Bootherium	Monroc, UT	11,690 ± 190	Conventional collagen	6
Camelops	Tabernacle Crater, UT	11,075 ± 255	Conventional bone	7
Camelops	**Sunshine Locality, NV**	**11,330 ± 60**	**AMS bone**	**8**
Camelops	Sunshine Locality, NV	11,340 ± 50	AMS bone	8
Camelops	Paisley Cave 5, OR	11,795 ± 30	AMS bone	9
Camelops	Sunshine Locality, NV	11,910 ± 50	AMS bone	8
Camelops	Paisley Cave 5, OR	12,460 ± 70	AMS bone	9
Camelops	Jessup Playette, NV	12,690 ± 60	AMS bone	10
Camelops	Jessup Playette, NV	13,070 ± 60	AMS bone	10
Camelops	Harrelson 2, Pyramid Lake, NV	22,270 ± 330	AMS bone	11
Camelops	Harrelson 2, Pyramid Lake, NV	22,400 ± 320	AMS bone	11
Camelops	Harrelson 2, Pyramid Lake, NV	22,810 ± 320	AMS bone	11
Camelops	Harrelson 2, Pyramid Lake, NV	22,890 ± 320	AMS bone	11
Camelops	Harrelson 2, Pyramid Lake, NV	23,250 ± 360	AMS bone	11
Camelops	Harrelson 2, Pyramid Lake, NV	24,290 ± 380	AMS bone	11
Camelops	Harrelson 2, Pyramid Lake, NV	25,230 ± 390	AMS bone	11
Camelops	Harrelson 2, Pyramid Lake, NV	25,370 ± 420	AMS bone	11
Camelops	Harrelson 2, Pyramid Lake, NV	25,860 ± 590	AMS bone	11
Camelops	Harrelson 2, Pyramid Lake, NV	25,870 ± 590	AMS bone	11
Camelops	Harrelson 2, Pyramid Lake, NV	25,470 ± 230	Average of the ten Harrelson 2 dates above	11
Camelops	Mineral Hill Cave, NV	44,600 ± 3,000	AMS bone	12
Camelops	Mineral Hill Cave, NV	46,550 ± 1,100	AMS bone	12
Camelops	Mineral Hill Cave, NV	48,900 ± 3,100	AMS bone	12
Equus	Paisley Cave 5, OR	11,130 ± 40[b]	AMS bone	9, 13
Equus	**Fishbone Cave, NV**	**11,210 ± 50**	**AMS bone**	**5**

TABLE 4.1. (cont'd.) Radiocarbon dates for extinct Pleistocene mammals in the Great Basin (bone, soft tissue, and dung dates only; trustworthy last appearance dates are in bold)

Genus	Site	Date	Method and material	References
Equus	Fishbone Cave, NV	11,350 ± 40	AMS bone	5
Equus	Paisley Cave 5, OR	11,365 ± 35 [b]	AMS bone	9
Equus	Paisley Cave 5, OR	11,420 ± 35 [b]	AMS bone	9
Equus	Paisley Cave 5, OR	11,435 ± 35 [b]	AMS bone	9
Equus	Paisley Cave 2, OR	11,740 ± 25	AMS bone	9, 14
Equus	Paisley Cave 5, OR	11,810 ± 40 [c]	AMS bone	9, 14
Equus	Paisley Cave 5, OR	11,820 ± 40	AMS bone	9, 14
Equus	Paisley Cave 2, OR	11,980 ± 35	AMS bone	9
Equus	Fishbone Cave, NV	12,280 ± 520	AMS bone	15
Equus	Paisley Cave 2, OR	12,340 ± 35	AMS bone	9
Equus	Paisley Cave 2, OR	12,340 ± 35	AMS bone	9
Equus	Paisley Cave 5, OR	12,340 ± 25	AMS bone	14
Equus	Paisley Cave 5, OR	12,385 ± 30 [d]	AMS bone	14
Equus	Paisley Cave 5, OR	12,410 ± 35	AMS bone	9
Equus	Gypsum Cave, NV	13,070 ± 55	AMS bone	16
Equus	Gypsum Cave, NV	13,310 ± 210	Conventional hoof keratin	17
Equus	Crystal Ball Cave, UT	18,820 +1,510/ −1,270	Conventional bone	18
Equus	Gypsum Cave, NV	25,000 ± 1,300	Conventional hoof keratin	17
Equus	Mineral Hill Cave, NV	31,130 ± 200	AMS bone	12
Equus	Mineral Hill Cave, NV	35,080 ± 280	AMS bone	12
Equus	Mineral Hill Cave, NV	42,420 ± 820	AMS bone	12
Equus	Mineral Hill Cave, NV	45,700 ± 1,000	AMS bone	12
Equus	Crystal Ball Cave, NV	> 23,000	Conventional apatite	18
Equus	Mineral Hill Cave, NV	> 46,400	AMS bone	12
Euceratherium	**Falcon Hill Cave, NV**	**11,950 ± 50**	**AMS bone**	**5**
Hemiauchenia	**Mineral Hill Cave, NV**	**36,320 ± 320**	**AMS bone**	**12**
Hemiauchenia	Mineral Hill Cave, NV	39,230 ± 1,330	AMS bone	12
Hemiauchenia	Mineral Hill Cave, NV	50,190 ± 1,940	AMS bone	12
Mammut	Sinkhole Mastodon site, UT	7,080 ± 200	Conventional apatite	19
Mammut	Sinkhole Mastodon site, UT	7,590 ± 100	Conventional apatite	19
Mammut	Sinkhole Mastodon site, UT	7,650 ± 100	Conventional apatite	20
Mammut	Sinkhole Mastodon site, UT	10,800 ± 250	Conventional collagen	19, 20
Mammuthus	Sandy Mammoth, UT	5,985 ± 210	Conventional collagen	21
Mammuthus	Sandy Mammoth, UT	7,200 ± 190	Conventional collagen	21
Mammuthus	Sandy Mammoth, UT	8,815 ± 100	Conventional collagen	21
Mammuthus	**Pyramid Lake, NV**	**10,340 ± 40**	**AMS bone**	**5**
Mammuthus	Pyramid Lake, NV	10,360 ± 50	AMS bone	5
Mammuthus	Wallman Mammoth, NV	10,630 ± 70	AMS bone	5

TABLE 4.1. (cont'd.) Radiocarbon dates for extinct Pleistocene mammals in the Great Basin (bone, soft tissue, and dung dates only; trustworthy last appearance dates are in bold)

Genus	Site	Date	Method and material	References
Mammuthus	Wallman Mammoth, NV	11,080 ± 300	Apatite	11
Mammuthus	Huntington Mammoth site, UT	11,220 ± 110	AMS bone	2
Mammuthus	City Creek Canyon, UT	14,150 ± 800	Conventional tusk	21
Mammuthus	Sandy Mammoth, UT	14,150 ± 800	Conventional bone	6
Mammuthus	Silver Creek, UT	18,150 ± 950	Conventional collagen	21
Mammuthus	China Lake, CA	18,600 ± 4,500	Conventional tusk	22
Mammuthus	Escalante Valley, UT	28,670 ± 260	AMS organic material	23
cf. *Miracinonyx*	Mineral Hill Cave, NV	> 52,200	AMS bone	12
Navahoceros	**Mineral Hill Cave, NV**	**37,750 ± 440**	**AMS bone**	**12**
Navahoceros	Mineral Hill Cave, NV	49,800 ± 1,700	AMS bone	12
Nothrotheriops	Gypsum Cave, NV	8,051 ± 450	Solid carbon dung	24
Nothrotheriops	Gypsum Cave, NV	8,692 ± 500	Solid carbon dung	24
Nothrotheriops	Gypsum Cave, NV	8,838 ± 430	Solid carbon dung	24
Nothrotheriops	Gypsum Cave, NV	10,075 ± 550	Solid carbon dung	24
Nothrotheriops	Gypsum Cave, NV	10,902 ± 440	Solid carbon dung	24
Nothrotheriops	**Gypsum Cave, NV**	**11,005 ± 100**	**AMS dung**	**25**
Nothrotheriops	Gypsum Cave, NV	11,080 ± 90	AMS dung	25
Nothrotheriops	Gypsum Cave, NV	11,290 ± 70	AMS dung	15
Nothrotheriops	Gypsum Cave, NV	11,360 ± 260	Conventional dung	26
Nothrotheriops	Newberry Cave, CA	11,600 ± 500	Conventional collagen	27
Nothrotheriops	Gypsum Cave, NV	11,690 ± 250	Conventional dung	26
Nothrotheriops	Gypsum Cave, NV	19,500 ± 205	AMS dung	25
Nothrotheriops	Gypsum Cave, NV	19,875 ± 215	AMS dung	28
Nothrotheriops	Gypsum Cave, NV	21,470 ± 760	Conventional dung	29
Nothrotheriops	Gypsum Cave, NV	23,700 ± 1,000	Conventional dung	17
Nothrotheriops	Gypsum Cave, NV	27,810 ± 455	AMS dung	25
Nothrotheriops	Gypsum Cave, NV	29,205 ± ?	AMS dung	25
Nothrotheriops	Gypsum Cave, NV	33,910 ± 3,720	Conventional dung	29
Platygonus	**Franklin, ID**	**11,340 ± 50**	**AMS bone**	**30**

References: (1) Emslie and Czaplewski 1985; (2) Madsen 2000a; (3) Schubert 2010; (4) Nelson and Madsen 1983; (5) Dansie and Jerrems 2005; (6) Nelson and Madsen 1980; (7) Nelson and Madsen 1979; (8) Beck and Jones 2009; (9) D. Jenkins et al. 2014; (10) K. Adams and Wesnousky 1998; (11) Dansie, Davis, and Stafford 1988; (12) Hockett and Dillingham 2004; (13) M. Gilbert et al. 2008; (14) D. Jenkins et al. 2012; (15) Stafford et al. 1987; (16) Gilreath 2009; (17) Long and Muller 1981; (18) Heaton 1985; (19) W. Miller 1987; (20) Gillette and Madsen 1992; (21) Madsen, Currey, and Madsen 1976; (22) E. L. Davis 1978a, 1978b; (23) P. Larson 1999; (24) Arnold and Libby 1951; (25) Hofreiter et al. 2000; (26) A. Long and Martin 1974; (27) C. Davis and Smith 1981; (28) Poinar et al. 1998; (29) Thompson et al. 1980; (30) H. McDonald 2002.

[a] AMS = accelerator mass spectrometry. See the discussion of AMS dating in this chapter.

[b] These four dates are from the same specimen. D. Jenkins et al. (2014:502) note that the three earlier dates, all of which suggest an age of about 11,400 years ago, are from samples that underwent more extensive pretreatment to remove potential contaminants than did the sample that provided the 11,130 ± 40 date. I follow D. Jenkins et al. (2014) in rejecting this date and so do not use it as the last appearance date for *Equus* in the Great Basin.

[c] Standard deviation given as ± 35 in D. Jenkins et al. (2012).

[d] Standard deviation given as ± 25 in D. Jenkins et al. (2012).

only one of many examples, but it is most appropriate for our purposes here. During the early days of radiocarbon dating, portions of whole bones might be dated, or just the mineral (apatite) portion of the bone, or just the collagen from that bone. All three different kinds of samples, we now know, are easily contaminated by groundwater introducing older or younger material. As a result, dates obtained in this way are no longer trusted. With AMS dating, however, the protein fraction of a bone can be extracted, the individual amino acids extracted from that protein, and those amino acids dated directly. To be even more certain that a date is accurate, individual amino acids from a given sample can be extracted, each one dated, and the results compared. If skeletal material is to be dated, the gold standard for doing this is now provided by AMS dating of the protein fraction of that material.

Conventional dates of such things as animal soft tissue can provide good results, as can conventional dates of animal dung, including sloth dung. Great care has to be used to remove potential contaminants when obtaining such dates, but in the right contexts with the right treatment, those dates can be trustworthy.

Finally, I mentioned at the very outset of this book that all the dates I am using here are expressed in "radiocarbon years." It has long been known that the amount of radiocarbon in the atmosphere has not been constant through time, nor has the nature of the exchange of carbon dioxide between the atmosphere and the oceans. Because of that, radiocarbon years differ from calendrical years, and scientists have put a lot of effort into developing ways to convert radiocarbon years into calendrical ("calibrated") years. This has been done by radiocarbon dating material that has also been dated in other ways: radiocarbon-dated tree rings provide an obvious example. Radiocarbon years can now be converted into calendrical years back to about 50,000 years ago, although the further back in time we go, the less trustworthy those calibrations are. For instance, the sloth dung from Gypsum Cave that I mentioned above as dating to 11,005 radiocarbon years ago actually dates to about 12,900 calendrical years ago.[6]

Here, I have used radiocarbon years for some very simple reasons. First, and least important, the older literature provides dates only in radiocarbon years. This is least important because I could have calibrated them myself, using freely available software. Second, and far more important, as our ability to translate radiocarbon years into calendar years improves, the older calibrations become outdated. The underlying radiocarbon dates, however, remain the same. That is why all the dates in this book are in radiocarbon years. Appendix 1 provides the relationship between radiocarbon dates and calendar years for 10,000 to 25,000 years ago, the period of most interest to us.

DATING AN ASS

In the early 1950s, archaeologist Phil Orr excavated a series of caves on the eastern edge of the dry Winnemucca Lake Basin. One of those sites, Fishbone Cave, provided him with the remains of extinct camel (*Camelops*) and horse (*Equus*), including two horse mandibles (lower jaws). He reported that one of those mandibles was closely associated with archaeological material, including a bed of shredded plants. That bed, when radiocarbon dated, turned out to be 10,900 ± 300 years old.[7]

Three decades later, dating expert Tom Stafford and his colleagues decided to include one of the Fishbone Cave horse mandibles as part of an important project to improve the accuracy of radiocarbon dating through the use of accelerator mass spectrometry. They found the specimen to have the physical and chemical properties of modern bone and got a date of 12,280 ± 520 years ago for it.[8]

Recently, archaeologist and bone specialist Amy Dansie and her colleague Jerry Jerrems returned to the Fishbone Cave specimens as part of a much larger project dealing with the archaeological, paleontological, and geological history of Pleistocene Lake Lahontan (chapter 2).[9] Dansie identified the horse mandibles from Fishbone Cave as having come from onagers, or wild Asian asses. Because one of the mandibles had already been accelerator dated, they decided to date the other one. The date they got for this second specimen was 11,350 ± 40 years ago, about 900 years younger than the date Stafford and his colleagues had gotten for the first mandible.

In theory, there is absolutely nothing wrong with an age discrepancy of this sort, especially with material from a site excavated the way many sites were being excavated in the 1950s, including Fishbone Cave. In practice, though, Dansie and Jerrems were surprised by the difference and so decided to get a second date on the specimen analyzed by Stafford and his colleagues. This time, they got a date of 11,210 ± 50 years ago, directly on amino acids.

So, two dates on a single specimen, 900 years apart. What to do about this? Dansie and Jerrems decided to reject the Stafford date and accept the new one. Their reasoning was simple: the first date was obtained many years ago, while their new one used state-of-the art methods. Although they didn't point it out, the standard deviations themselves would suggest that the date Dansie and Jerrems obtained is to be preferred. Their date comes with a standard deviation of 50 years; the original date, obtained some 20 years earlier, has a standard deviation of 520 years. They made the right decision. Taking into account the standard deviations on both specimens, their asses are almost identical in age.

DATING LAST APPEARANCES

As I discussed in chapter 3, late Ice Age extinctions swept away 37 genera of mammals in North America. If we are ever going to be able to understand why these extinctions occurred, we need to understand when they occurred.

There are actually many "when" questions that need to be answered. To understand the cause of the extinctions, we need to have a strong chronology of extinction for each genus of mammals (and birds) that was lost. We need to have enough trustworthy dates for each of those genera to allow us to plot them in space and through time across North America. If we could do that, then we would be able to tell whether any given mammal was lost at the same time from all places, or whether the loss of a genus was "time transgressive," occurring earlier in one region than in another. We could compare the chronology and geography of the extinction of each kind of mammal to that of all the others. That would dramatically increase our ability to understand why the extinction of any given mammal might have occurred, thereby increasing our ability to understand why all of them occurred.

We aren't close to being able to do that for the vast majority of the North American mammals that were lost, and we aren't able to do it for any of the sadly neglected birds. In some rare instances, we can take a reasonable stab at it, but that tells us primarily how much we still need to know. The giant bear *Arctodus* provides an example. As figure 4.1 shows, we have what appear to be reliable radiocarbon dates for *Arctodus* specimens from 14 North American sites. Those dates appear to show that this huge bear may have become extinct in the Yukon and Alaska long before it did so south of glacial ice.[10] The dates also show that the bear lasted in some places until the very end of the Pleistocene. But to know that giant bears really were lost earlier in far northwestern North America would require far more dates than we have, and to know that they lasted until roughly 11,000 years ago everywhere in North America would require far, far more dates than we have.

Until we have enough dates to be able to do this for every single one of the extinct mammals, we have no chance of understanding why the extinctions occurred. We are instead stuck trying to figure out the date after which each of these genera no longer existed. This is an ancient enterprise. Scientists have been doing it since the 1950s, when the insightful Paul Martin gave it the first shot.[11] That we still do not have enough dates to do more than this in any meaningful way is frustrating, but we don't.

We are, in short, stuck trying to figure out what are called "Last Appear-

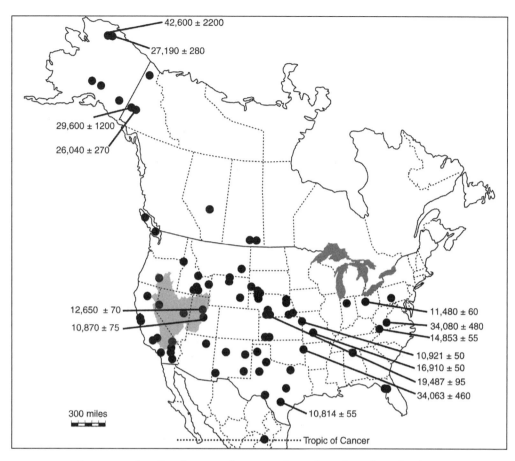

FIGURE 4.1. The geographic distribution of trustworthy radiocarbon dates for the giant bear (*Arctodus*). The radiocarbon dates are from Schubert (2010) and Mann et al. (2013); where more than a single date is available from a given site, I have used the youngest trustworthy date.

ance Dates," or LADs. An LAD is simply the most recent reliable date available for any given extinct organism. There are two tricks to working with LADs, one of which has now become trivial but the other of which may never become so.

The trivial trick is that bad dates can fool us into thinking that an animal became extinct more recently than it actually did. Paul Martin's early work provides a very good example. In his 1958 attempt to decipher the timing of the North American losses, his list of dates included many that fell after the end of the Ice Age, including some as recent as 2,000 years ago. When he tried again in 1963, this time focusing only on the arid Southwest, the extinctions seemed to have continued to as late as 7,000 years ago.[12] We now know that all the dates younger than 10,000 years ago were bad—either because they were obtained during the early years of radiocarbon dating, when we were less aware of the problems that can enter into the dating process, or because associations between the dated material and the extinct mammals were incorrectly interpreted.

I said that this problem with dealing with LADs is trivial because we quickly learned what the problems were and how to deal with them. If we are interested in getting radiocarbon dates on an extinct mammal, then the best way to proceed is to date the remains of that mammal itself. In most instances, dating organic material that happens to be lying near those remains is far more likely to mislead us than it is to tell us when the animal died. For example, no one thinks the horse mandible lying near the archaeological material at Fishbone Cave means that the two were deposited at the same time, even though Orr did at the time. Dating the remains themselves removes any questions about the relationship between what we are dating and the animal itself. Those remains can be soft tissue or dung from the animal, or properly prepared proteins from the bones or teeth. Conventional dates from carefully prepared soft tissue or dung are fine. They are not fine for dating bones and teeth. For this, as I discussed earlier, we need accelerator dates.

The second problem with obtaining LADs—the one that hasn't, and may not, go away—is easy to understand. As I write this, the chance of winning $1,000 in Washington State's "Jackpot" lottery is 1 in 27,100. This, of course, makes the state quite happy, although even it doesn't explain why people play the game (yes, I have, and no, I didn't win). If we turn the 27,100 into mammoths and array them through time so that each mammoth lives at a different time, then the chances of finding the latest one are, all other things being equal, the same as the chances of winning $1,000 in this particular lottery game.

Obviously, not all mammoths lived at different times (they couldn't have and still have produced more mammoths), but getting an LAD is akin to playing the lottery. The more dates you get, the more you can approach a true LAD. That, in turn, means that as time goes on, you can expect LADs to get younger. Because this matters a lot, I return to it below.

THE LAST APPEARANCE DATES: BIRDS

It would be wonderful to be able to talk about the LADs that we have for the eight genera of extinct Pleistocene birds known from the Great Basin (chapter 3). It would be wonderful, but it can't be done. There is not a single trustworthy radiocarbon date available for any of these birds. The mammals have gotten all the attention. This is a project waiting to happen. Because it hasn't happened, all my attention here has to be on the mammals.

THE LAST APPEARANCE DATES: MAMMALS

Table 4.2 provides the reliable LADs that are available for the 37 genera of North American mammals that were lost during the Ice Age. Count them up and you get 24. That means that 13 of the 37 genera have never been reliably dated by the radiocarbon method. Since they haven't been reliably dated by any other method, either, that means we simply don't know when they were lost. There are no dates for pampatheres, none for glyptodonts, none for the four-horned pronghorn. That is part, but only part, of what I meant when I said we have a long way to go in the dating realm.

For unfortunate reasons that I discuss in chapter 7, most scientists believe that all the North American mammal extinctions occurred around 11,000 years ago. If we broaden this by 1,000 years on either side, to between 12,000 and 10,000 years ago, it turns out that only 17 of the 37 genera of mammals can be shown to have lasted this late. These 17 are in bold in table 4.2. If we broaden our time frame just a bit to reach back to the earliest evidence of people in the Americas—Monte Verde, Chile, at 12,500 years ago, and the Paisley Caves, Oregon, at 12,400 years ago[13]—we add only *Saiga* to the list.

Since LADs can only get younger as more and more dates become available, and since 17 genera have dates that fall very close to the end of the Pleistocene, or 10,000 years ago, is it not possible that as more and more dates become available, we will discover that many of these animals lasted well into the Holocene? After all, exactly that has happened in Europe with such animals as the woolly mammoth (*Mammuthus primigenius*) and the giant deer (or "Irish elk," *Megaloceros giganteus*). Both were once thought to have been lost as the Pleistocene came to an end, but we know now that mammoths continued to exist on Wrangel Island, northeastern Siberia, until at least 3,700 years ago and that the giant deer lasted until at least 6,900 years ago in western Siberia.[14] Why could such things not have happened in North America?

The answer is that they could have. However, as Paul Martin pointed out years ago, thousands of archaeological sites known from North America are under 10,000 years old, and none of them have yielded the remains of any of the extinct mammals in a secure context.[15] That would seem to suggest that while some of them might have outlasted the Pleistocene itself, they must have been vanishingly rare by then. It also suggests that whatever the cause of the extinctions might have been, that cause must lie before 10,000 years ago.

TABLE 4.2. Trustworthy last appearance radiocarbon dates (LADs) for North American late Pleistocene extinct mammal genera (dates that fall between 12,000 and 10,000 years ago are in bold)

Genus	Common name	LAD	Reference
Megalonyx	**Jefferson's ground sloth**	**11,235 ± 40**	1
Eremotherium	Laurillard's ground sloth	38,860 ± 1,300	2
Nothrotheriops	**Shasta ground sloth**	**10,500 ± 180**	3
Paramylodon	Harlan's ground sloth	20,450 ± 460	2
Arctodus	**Giant bear**	**10,870 ± 75**	2
Smilodon	**Sabertooth**	**11,130 ± 275**	2
Homotherium	Scimitar cat	22,250 ± 130	4
Miracinonyx	American cheetah	19,765 ± 80	5
Castoroides	**Giant beaver**	**10,150 ± 50**	6
Equus	**Horses**	**10,370 ± 350**	2
Tapirus	**Tapirs**	**10,940 ± 90**	2
Mylohyus	**Long-nosed peccary**	**11,860 ± 40**	2
Platygonus	**Flat-headed peccary**	**10,750 ± 50**	6
Camelops	**Yesterday's camel**	**10,370 ± 350**	2
Hemiauchenia	Large-headed llama	36,320 ± 320	7
Palaeolama	**Stout-legged llama**	**10,890 ± 130**	2
Navahoceros	Mountain deer	37,750 ± 440	7
Cervalces	**Stag-moose**	**10,800 ± 45**	8
Saiga	Saiga	12,220 ± 130	1
Euceratherium	**Shrub ox**	**11,630 ± 150**	2
Bootherium	**Helmeted muskox**	**10,980 ± 80**	2
Cuvieronius	**Cuvier's gomphothere**	**11,550 ± 60**	9
Mammut	**American mastodon**	**10,032 ± 40**	10
Mammuthus	**Mammoths**	**10,340 ± 40**	2

References: (1) H. McDonald, Stafford, and Gnidovec 2015; (2) Faith and Surovell 2009; (3) Thompson et al. 1980; (4) Widga et al. 2012; (5) Williams 2009; (6) Feranec and Kozlowski 2010; (7) Hockett and Dillingham 2004; (8) R. Guthrie, Sher, and Harington 2001; (9) Sanchez et al. 2014; (10) Woodman and Athfield 2009.

Notes: Faith and Surovell (2009) is the most widely cited source of LADs for extinct North American Pleistocene mammals. Readers familiar with that paper may notice that my dates at times differ from theirs. There are multiple reasons for this. The date they used for *Nothrotheriops* was rejected by Thompson et al. (1980) because it could not be replicated. Faith and Surovell did not include a date for *Saiga*, but a reliable date exists and I have included it. In four instances (*Castoroides, Cervalces, Mammut, Platygonus*), the dates provided here became available after Faith and Surovell (2009) appeared. In the four other instances (*Hemiauchenia, Homotherium, Miracinonyx, and Navahoceros*), they used dates they did not consider reliable. These have been replaced with dates that are trustworthy, two of which are from after Faith and Surovell appeared (*Homotherium, Miracinonyx*).

LAST APPEARANCE DATES IN THE GREAT BASIN

Table 4.1 provides every radiocarbon date of which I am aware that comes directly from the bones, soft tissue, or dung of an extinct Ice Age mammal in the Great Basin. I say "of which I am aware" because this is the first collation of these dates to have been made in many years. The literature is scattered and publications easy enough to miss, and I have probably missed some of them.

Even so, there are 98 dates on my list. These come from 12 of the 20 genera of extinct Ice Age mammals known from the Great Basin. Obviously, 8 have never been dated at all. A close look at the list shows that many of these dates are conventional bone dates of the sort that are not to be trusted (these are the ones listed as "bone," "collagen," and "apatite"). If we drop these, we lose the dates for the helmeted muskox (*Bootherium*) and the mastodon (*Mammut*). That means that we have reliable dates for only 10 of the 20 Great Basin genera.

These are shown in table 4.3. Of the 20 genera, we can show that seven have LADs between 12,000 and 10,000 years ago. In fact, those for mammoth (*Mammuthus*) and giant bear (*Arctodus*) are the youngest trustworthy dates available for these animals from anywhere in North America.

We have already seen that in North America as a whole, 17 of 37 genera, or 46 percent, are known to have LADs between 12,000 and 10,000 years ago. In the Great Basin, the comparable number is 7 of 20, or 35 percent. This percentage difference might seem important, but it isn't. A simple statistical test (chi-square) shows that this difference is so small that it can be expected to occur by chance about 40 percent of the time. Given current knowledge, the same proportion of mammal genera were lost during the last 2,000 years of the Ice Age in the Great Basin as were lost during that period in North America as a whole.

There are two obvious explanations for this similarity in proportions.

The first possibility is that it reflects the abundances of the extinct mammals on the landscape at the end of the Ice Age. Figure 4.2 shows the number of occurrences of late Pleistocene mammals known from North America as a whole. Figure 4.3 shows the same thing for the Great Basin alone. Although these data were drawn from a database that is now showing its age,[16] which explains why the Aztlán rabbit (*Aztlanolagus*) is not on figure 4.3 (it was first reported in 2007), these figures make the point well. The Great Basin and North American Ice Age mammal faunas have similar abundance structures, with a relatively small number of genera very abundant and a larger number fairly rare. It is the abundant ones that we have most easily been able to place at the end of the Pleistocene—the results of playing that lottery game

TABLE 4.3. Trustworthy last appearance radiocarbon dates (LADs)
for the extinct Pleistocene mammals of the Great Basin

Genus	Site	Date
Mammuthus	Pyramid Lake, NV	10,340 ± 40
Arctodus	Huntington Mammoth site, UT	10,870 ± 75
Nothrotheriops	Gypsum Cave, NV	11,005 ± 100
Equus	Fishbone Cave, NV	11,210 ± 50
Camelops	Sunshine Locality, NV	11,330 ± 60
Platygonus	Franklin, ID	11,340 ± 50
Euceratherium	Falcon Hill Cave, NV	11,950 ± 50
Hemiauchenia	Mineral Hill Cave, NV	36,320 ± 320
Navahoceros	Mineral Hill Cave, NV	37,750 ± 440
Miracinonyx	Mineral Hill Cave, NV	> 52,200

References: See table 4.1.

I discussed above.[17] That the Great Basin and North America as a whole have similar proportions of genera placed in the last 2,000 years of the Ice Age might simply be a result of this fact.

The second possibility is quite different. It could be that some or many of the mammals that we have been unable to date to the very end of the Ice Age were simply gone by then, and that the similarities between the Great Basin and North America as a whole reflect this history. For instance, the youngest trustworthy date for the huge ground sloth *Eremotherium* in North America falls at about 39,000 years ago; for the mountain deer (*Navahoceros*) at 38,800 years ago; for Harlan's ground sloth (*Paramylodon*) at 20,500 years ago. And, of course, many of the mammals have no trustworthy dates at all. Perhaps we have been able to date only about 45 percent of North America's extinct Pleistocene mammals to the very end of the Ice Age because this is close to all that was left by then.

My bet is that both of these explanations are true. In this view, the animals that we can place at the very end of the Ice Age are those that were the most common on the landscape at that time. Because of this, we have been able to win the dating lottery with them. As we get more and more dates on more and more animals, we will probably learn that others also lasted this late. Some, however, were probably long gone by then. Unfortunately, it may take decades of work before we really know the time by which each of these animals was gone.

How about the three extinct species of mammals that belong to genera that still exist in North America and that occurred in the Great Basin: the

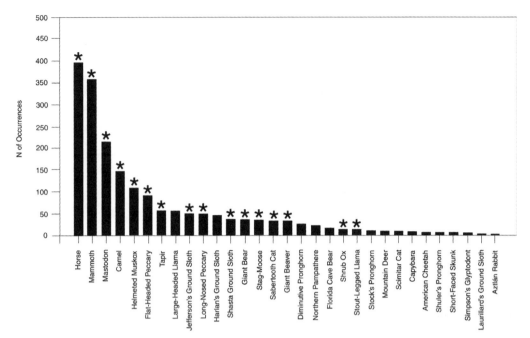

FIGURE 4.2. The number of occurrences of extinct late Pleistocene mammal genera known from North America (from FAUNMAP Working Group 1994 and Faith and Surovell 2009). Asterisks mark genera that have been securely dated to between 12,000 and 10,000 years ago.

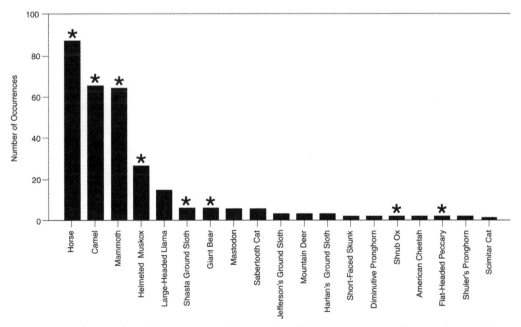

FIGURE 4.3. The number of occurrences of extinct late Pleistocene mammal genera known from the Great Basin (from FAUNMAP Working Group 1994). Asterisks mark genera that have been securely dated to between 12,000 and 10,000 years ago.

TABLE 4.4. Trustworthy last appearance radiocarbon dates (LADs) for North American dire wolf, American lion, and Harrington's mountain goat

Genus and species	Site	Date	Reference
Canis dirus	Wasden Cave, Idaho	11,280 ± 50 (AMS bone)	1
Panthera leo atrox	Ikpikpuk River, Alaska	11,290 ± 50 (AMS bone)	2
	Pit 48, Alberta	11,355 ± 55 (AMS bone)	3
Oreamnos harringtoni	Stanton's Cave, Arizona	10,870 ± 200 (conventional dung)	4

References: (1) Suzann Henrikson, personal communication, November 5, 2013, with permission of the Idaho Museum of Natural History; (2) Mann et al. 2013; (3) Barnett et al. 2009; (4) Mead, O'Rourke, and Foppe 1986.

American lion, the dire wolf, and Harrington's mountain goat? There is not a single trustworthy date for any of these animals from the Great Basin. There are good dates for them from elsewhere, however, and those dates show that all three made it to at least 11,000 years ago or so (table 4.4). The LADs for the dire wolf and Harrington's mountain goat are from sites so close to the Great Basin—southern Idaho and the Grand Canyon, respectively—that it would be a little surprising if they did not last that long here as well.

In short, all we really know is that in North America as a whole, 17 of the 37 extinct mammal genera lasted until the final 2,000 years of the Ice Age, to between 12,000 and 10,000 years ago. In the Great Basin, 7 of 20 genera did so. Since people were in the Great Basin by 12,400 years ago, it is very probable that they encountered some or all of these seven. Horses, camels, Shasta ground sloths, flat-headed peccaries, mountain deer, shrub oxen, giant bears; any or all might have seen and been seen by the early peoples of the Great Basin. I return to this issue in chapter 6.

5

A Stable of Ground Sloths

The dots on the maps in chapter 3 that show the distribution of the extinct Ice Age mammals of the Great Basin often reflect the discovery of the remains of a single distinctive tooth or of a bone or two. At times, such things have been found exposed on the surface of the ground. At other times, they have been discovered as the result of the careful excavation of sites done for archaeological, not paleontological, goals. Dennis Jenkins's exacting excavations in south-central Oregon's Paisley Cave 5, for instance, provided human coprolites (dried fecal material, to be nice about it), the oldest of which date to 12,400 years ago. By so doing, he has also provided us with the earliest secure evidence for people in North America. This site is old enough to contain the remains of extinct Pleistocene mammals, but the yield was scanty: a few specimens of horse (*Equus*) and camel (*Camelops*), and a few coprolites from the American lion (*Panthera leo atrox*).[1] And that's it (see fig. 5.1 for the location of sites discussed in this chapter, and chapter 6 for more on the Paisley Caves).

Hidden Cave, in western Nevada's Carson Sink, was excavated with equal care by Dave Thomas of the American Museum of Natural History and provided, in addition to rich archaeological assemblages, nearly 5,000 bones of birds and mammals that could be identified. Since the deposits here extended back into the Pleistocene, the site could have provided rich assemblages of extinct mammals, but it didn't. Instead, it yielded four specimens of horse and one of the extinct camel (*Camelops*).[2] And that's it.

Then there is the Sunshine Locality in eastern Nevada's Long Valley. This area was the focus of fastidious excavations by Charlotte Beck and Tom Jones of Hamilton College, reported in perhaps the best monograph ever published on the late Pleistocene and early Holocene archaeology of the Great Basin. Their work also reached Pleistocene deposits, but the haul of late Pleistocene

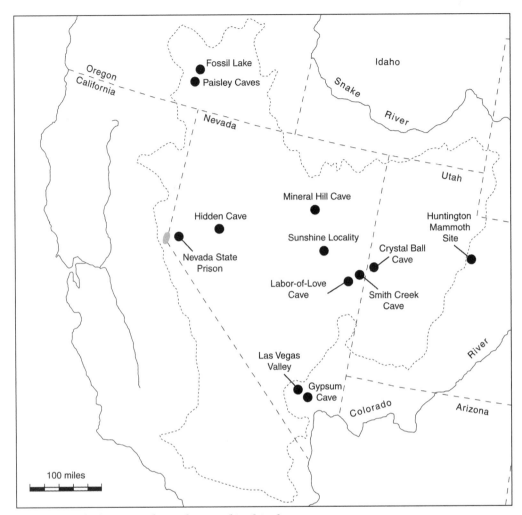

FIGURE 5.1. The location of sites discussed in this chapter.

mammals from their efforts was meager: 13 specimens of yesterday's camel (*Camelops hesternus*) and a single horse tooth.[3] And that's it.

Fortunately, the Great Basin has provided some extremely rich Ice Age paleontological sites, a few of which have international fame. In the first chapter, we encountered one of these sites imprisoned, a site famous not simply because of its history but also because fossil footprints are so rare. In this chapter, I look at five more of these sites.

The selection of these five was not easy, but my criteria were simple. I chose sites from which we have learned a lot about the Ice Age mammals of the Great Basin, that have intriguing histories, and that have importance outside the Great Basin itself.

I also chose sites that are well-enough analyzed and published that those

FIGURE 5.2. The Fossil Lake area. The association of archaeological materials with extinct Ice Age mammals in this area was caused by wind deflating the kinds of sand dunes seen here. Artifacts that were deposited on these dunes were then dropped onto the remains of the animals found by Sternberg.

who read this book in the coming years can be assured that what they are reading is still current. It was for this reason that I did not include one extremely important location, southern Nevada's Upper Las Vegas Wash and, within it, the famous Tule Springs Site. Tule Springs has a rich late Pleistocene fauna and an equally rich history,[4] and it has been the scene of important paleoclimatic work.[5] It is for very good reasons that nearly 23,000 acres of the Upper Las Vegas Wash have been set aside as the Tule Springs Fossil Beds National Monument.[6] However, I have not included this area in what follows because so much recent and unpublished work has been done there that whatever I might have to say would be quickly outdated.[7]

There is no clever order to what follows. I start with the sites that are most complex and end with the ones that are least so.

FOSSIL LAKE, OREGON

A Brief History

South-central Oregon's Fossil Lake may well be the Great Basin's most famous paleontological site (figs. 5.1 and 5.2). In part, this is because it has provided the remains of a very broad variety of Ice Age vertebrates, from snakes and fish to birds and mammals. But it is also famous because it was discovered so early, and because that early discovery put it firmly on the Ice Age map.

Fossil Lake is located in the eastern part of the Fort Rock Basin, the site of one of the massive Pleistocene lakes that I discussed in chapter 2. Today, the "lake" that gave Fossil Lake its name is itself a fossil, only occasionally holding sufficient water to merit the name.

We have no way of knowing who first discovered the bones that once covered the surface of the ground here, but we can be sure that the discovery was made by Native Americans during prehistoric times, since the basin in which Fossil Lake sits is also rich in archaeological sites. Those sites, in fact, played a significant role in interpreting the paleontology that this area has provided.

We do know something about the first non–Native Americans who discovered the site and roughly when they discovered it. This discovery has at times been attributed to the first governor of the state of Oregon, John Whiteaker.[8] Whiteaker served as governor from 1859 to 1862, but at the time he first visited Fossil Lake, in September 1876, he was apparently on a campaign tour in his successful attempt for a seat in the Senate.[9] It was Whiteaker, geologist Thomas Condon tells us, who first "noticed some fossil bones on the surface of the open prairie and shortly after this brought some fragments to the writer of these pages for examination."[10]

This, though, is not quite true, as others realized. Instead, the site ended up being brought to the attention of scientists after it was discovered by cattlemen. We don't actually know when that first happened, but an article that appeared in the *San Francisco Chronicle* on May 8, 1877, reported that three local cattlemen had discovered large numbers of bones and teeth about 30 miles east of Silver Lake, Oregon, at the edge of a small alkali lake. The *New York Times* found that report interesting enough to republish it on June 1 of that year.[11] It even named names. The cattlemen were Rufus Dullard, Jack Parton, and Andrew Foster.

Normally, the discovery of a bunch of ancient bones in a remote part of the country would not merit the attention of the *New York Times*, but this discovery did not seem normal. "Now," the anonymous author reported, "comes one of the strangest of this strange bone deposit. Vast quantities of stone arrowheads, such as were used by Indians, are found throughout these bones." Even though the reporter had no idea the bones of extinct animals were to be found here, the very fact that artifacts had been found associated with "eighty acres of fossil remains" made this story irresistible.

Whiteaker learned of this site from local ranchers, the same way archaeologists and paleontologists often learn of important sites.[12] After his initial visit in 1876, he visited the site again in mid-June 1877, along with a group of local residents. While there, he gathered some 200 pounds of fossil material that he then gave to Condon, chair of the Natural Science Program at the newly opened University of Oregon.[13] Whiteaker also provided Condon with a team of horses and convinced his son, Charles, a student of Condon's at the university, to guide Condon there.[14]

Condon arrived at Fossil Lake on August 4, 1877, accompanied by Charles Whiteaker and George Duncan, Silver Lake's first postmaster.[15] He spent the day gathering fossils, finding the remains of elephants, camels, horses, birds, and fish. At Condon's request, Duncan sent a collection of Fossil Lake specimens to the distinguished paleontologist Edward Drinker Cope to let him know of the riches to be found there.[16] Later that year, Condon wrote Cope to tell him that he would be sending him some of what he had collected as a loan. This he did, but to Condon's great annoyance, Cope never returned the specimens in spite of being asked to do so. Loaned in 1877, they were not returned to the University of Oregon until 1926, 19 years after Condon's death.[17] Ironically, Condon had chosen to notify Cope, and not Cope's archenemy, famed paleontologist O. C. Marsh, because Marsh had not returned specimens from Oregon's John Day fossil beds that Condon had loaned him in 1870.[18] No matter who actually first found the site, Condon was the first scientist to visit it, collect from it, and begin to make it known to the wider scientific community.

Once Cope learned of this location, he sent his field collector, Charles H. Sternberg, to make his own collections there. Sternberg arrived in late August[19] with clear instructions. Not only was he to collect fossils there, but he was to "look for human implements mingled with the extinct animal remains I was sent in search of; to write at once if I found any, their manner of occurrence, etc."[20] Guided to the site by George Duncan, Sternberg found exactly what Cope was hoping for, making the largest collection of fossil bones he had ever been able to make in a single day and shipping off over a thousand pounds of specimens to his employer.[21] In splendid romantic rhetoric, Sternberg reported that he had found the remains of extinct horses, camels, mammoths, and the ground sloth *Paramylodon*, and that they had been covered by deposits that had been laid down as a result of volcanic activity. He quickly understood, incorrectly, what must have happened:

> A few days exploration convinced me that great numbers of animals had been destroyed at once, and an examination of the volcanic sand and ashes that had covered them proved that they had sought shelter from a fierce storm of sand and ashes, from an active volcano in the vicinity, and beasts of all descriptions forgot their natural instincts, and

rushed together to the cooling waters of the lake. The gigantic elephant, the horse, llama, deer, wolf and other smaller animals awaited with fear the storm that must overtake them and bury them beneath the accumulating *debris.* ... What a commotion there must have been on that fearful day, when all the beasts of Southern Oregon found death and burial.[22]

As if fire and brimstone were not enough, Sternberg also found artifacts intermingled with the bones of the extinct animals he had discovered. Given the instructions Cope had provided him, he sat down that evening and wrote to tell Cope what he had encountered.

We know exactly what Sternberg told Cope. In an act of remarkable audacity, in 1878 Cope published Sternberg's letter in both the *American Naturalist* and the *Proceedings of the American Philosophical Society* without mentioning Sternberg's name.[23] In the latter version, the text was enclosed in quotation marks, but this did not matter since there was no reason to think that Cope was not the author of the text. Because Cope felt that the remains he was describing dated to the Pliocene (the geological epoch immediately prior to the Pleistocene and now dated to between 5.3 and 2.6 million years ago), he entitled the paper "Pliocene Man." The *American Naturalist* version begins with the announcement that "Prof. Cope has recently made an important addition to our knowledge of this subject in North America."[24] The important addition was that Cope had received from Oregon a collection of the remains of such extinct mammals as mammoth, horse, camel, and ground sloth and that

> mingled in the same deposit in undistinguishable relation, were found numerous flakes with arrow and spear heads of obsidian, many of them much tarnished by long erosion. All were lying mingled together on the surface of a bed of clay, which was covered by a deposit of "volcanic sand and ashes" of from fifteen to twenty feet in depth.[25]

Four years before the supposed human footprints were discovered in deposits of great antiquity at the Nevada State Prison (chapter 1), Cope was reporting the commingled remains of extinct mammals and human artifacts in the Oregon desert. And he was doing it by plagiarizing Sternberg's work.

It had been easy for Cope to get away with this. In December 1877, the *American Naturalist* announced that its home base was moving from Boston to Philadelphia, and it was now to be edited by A. S. Packard Jr. and E. D. Cope. It was moving because Cope had bought it. In many ways, this important scientific outlet had become Cope's house journal, and he could do with it pretty much what he wished. Between 1878 and 1897, Cope published an astonishing 776 papers in his journal, compared to none by Sternberg.[26] There was little Sternberg could do about this except resent Cope for mistreating him in this fashion and wait for Cope's demise to right the wrong that had been done to him.

And wait he did. In 1881, Sternberg reported that he had found not only artifacts intermingled with the bones of extinct animals at Fossil Lake, but also "an old Indian village" in the dunes not far from the lake. Since the dunes were younger than the fossil deposits, he recognized that the artifacts he found with the fossils on the surface of the ground might represent materials of very different age that had become commingled with one another. Nonetheless, he concluded that "who knows but what man was a witness of the scenes we have described, and perished amidst the storm of burning ashes."[27]

In 1884, he came closer to suggesting the obvious, noting that "we had thought at first" that the artifacts might be contemporaneous with the bones, but he still did not suggest that the association was accidental.[28] That had to wait until Cope died in 1897 and Sternberg no longer had to fear the wrath of a person Henry Fairfield Osborn properly described as a "militant paleontologist," a person whose motto was "war whatever it cost."[29]

Shortly after Cope died, Sternberg strongly and publicly rejected any possibility that Fossil Lake provided secure evidence that people had seen the extinct beasts the site had provided in such abundance. The artifacts, he wrote, were deposited in dunes that had covered the bones. "A powerful wind… drifted the sand away from over the bones. Arrows, being heavy, would drop down until they became mingled with the bones, and may not have been over a hundred years old."[30] Figure 5.2 shows the process in action today.

Cope would have been furious, but his obituary had already appeared in the pages of the *American Naturalist*, and not even he could harm Sternberg from the grave.[31] We can be certain that Cope would not have been pleased by Sternberg's apostasy because Cope believed until his dying day that Fossil Lake showed that the ancestors of modern Native Americans had walked the Fossil Lake basin at the same time as mammoths, sloths, camels, and horses. Not only had he said that in the paper he had stolen from Sternberg's letter, but he had repeated it in the report he had written—in the pages of the *American Naturalist*, of course—on his own trip to Fossil Lake, taken in 1879.

Cope was guided there by Charles Whiteaker, just as Condon had been two years earlier. He found the most abundant fossils to belong to horses and camels, discovered alongside the bones of mammoths. "Scattered everywhere in the deposit were the obsidian implements of human manufacture,"[32] he carefully observed. He also observed that it was impossible to prove that the artifacts and fossils were contemporaneous but made it clear that he had no doubt on this score: "Had they been other than human flints, no question as to their contemporaneity would have arisen." He repeated the point time and again throughout his career, often, but not always, in the pages of the *American Naturalist.*[33]

Although Cope often noted that the Fossil Lake evidence for very early people in North America was not "absolutely conclusive,"[34] there can be little doubt that he fully accepted this association at face value. In 1893, he observed that the European stone implements that were assigned to the Mousterian cultural complex, and that were correctly thought to have been manufactured by Neandertals, "are of the same form as the obsidian implements which I collected at Fossil Lake, Oregon, with the bones of extinct llamas, horses, elephants, etc."[35] Because this was the case, he argued, it was possible that Neandertals, or someone like them, would be found in North America as well. Cope returned to this issue two years later—in the pages of the *American Naturalist*—noting yet again that he had found obsidian artifacts "in profusion mixed with the bones of the extinct fauna at Fossil Lake, Oregon," and concluding that it was very likely that people and extinct mammals had walked together on the North American landscape.[36]

This was Cope's last comment on Fossil Lake as an archaeological site. He died on April 12, 1897.[37] Soon thereafter, Sternberg started working to set matters straight.

It is important to recognize how significant a matter this was. As I discussed in chapter 3, until about 1860, and for reasons that were more theological than scientific, western scientists did not believe that people had coexisted with such extinct Ice Age mammals as woolly mammoths and woolly rhinos in Europe, or mammoths and ground sloths in North America. That changed dramatically around 1860, when it became obvious, from work done in France and England, that people really had existed that far back in time. Soon, earlier and earlier archaeological sites were being reported from western Europe, ultimately reaching back into the Oligocene, now pegged at about 30 million years ago.[38] The European sites associated with the extinct Ice Age mammals were real, and everyone accepted them. The earlier sites were not real, and most scientists recognized that.[39]

The situation in North America was quite different. The European archaeological record suggested that people might have been here during the Ice Age as well, and Cope was even willing to consider the possibility of Miocene-aged human ancestors in what is now Nebraska (the Miocene is now dated to between 23 and 5 million years ago).[40] But although many sites had been put forward as documenting an ancient human presence in North America, none had proved compelling to a broad variety of scientists, triggering a debate that archaeologist David Meltzer has called "The Great Paleolithic War." In fact, no such site was to be accepted until the discovery and verification of the Folsom site in New Mexico in 1927.[41]

This context explains why Cope recounted the Fossil Lake situation on so many occasions. This was a big deal, and he knew it. It is this context, along with Sternberg's deep resentment of Cope, that explains why Sternberg took great pains to set the matter straight on multiple occasions after Cope's demise. He first did this in 1898, immediately after the death of his former employer, and then again in 1903 and 1909:

> I will now pass on to "Pliocene man." An article under this head appeared in the *American Naturalist* in 1877. The whole is enclosed in quotation marks and was a copy of a letter I wrote to Prof. E. D. Cope.... As one part of the story of the Pliocene man was told in the article, and as I was really the author of it, though my name does not appear, I have often felt it my duty to tell the end also, which, in this instance at least, proved to my mind that I was entirely mistaken in regard to man being contemporary with the Pliocene animals, birds, and reptiles that I found so abundant at Fossil Lake. I was young, and anxious to be able to say that I was the first collector to find traces of man so far back in the world's history, associated with the extinct horse, llama, elephant, etc.
>
> I concluded...that the bones of the extinct animals had been covered with sand, protected by sage-brush. It would naturally be a good place for game, on account of the water. The Indians would lose many arrows and spears. A powerful wind...drifted the sand away from over the bones. Arrows, being heavy, would drop down until they became mingled with the bones, and may not have been over a hundred years old.
>
> This experience taught me caution in accepting the first evidence of early man that comes along. Paleontologists wish to add to their reputation by discoveries in the remote past, and it is natural for them to take the first evidence that comes to hand, so they may have the credit if it proves of value. If they are, after all, mistaken they either stick to their first opinion with a determination that seems foolish to an outsider, or back out as gracefully as possible.[42]

Sternberg's memory had failed him a bit—Cope's papers had been pub-
lished in 1878, and the version in quotations was published by the *American
Philosophical Society*—but this does not matter. He was backing out gracefully.
Cope, he hoped, was now the one who looked foolish.

And how about the name "Fossil Lake" itself? Sternberg routinely claimed
that the name was his invention,[43] and it is true that it came into general use
not long after his visit there. Cope was using the term by 1883;[44] others followed suit.[45] However, it was Whiteaker whose use of the name first made it
into print. In conjunction with his trip to the site in June 1877, he referred to
the location as "Fossil Lakes."[46] It seems likely that the name was in local use
before Sternberg visited the site late in 1877 and that he simply borrowed it.
The name has been in use ever since.

An Archaeological Aside

Since Cope's demise, no one has accepted the juxtaposition of artifacts and
the remains of extinct animals that he reported as suggesting that the former
were deposited while the latter still existed on the hoof. The Fossil Lake area
is rich in archaeological sites,[47] and, just as Sternberg speculated, it is almost
certain that erosional processes caused the two to end up lying on the ground
together. If not, we will never know.

More recently, however, archaeologists Rick Minor and Lee Spencer excavated the remains of an extinct camel, unidentified as to genus, from the
Fossil Lake area. On the basis of associated stone artifacts, they maintain that
this animal was killed by people.[48] This would be exciting if true, since possible associations between people and extinct camels are vanishingly rare in
North America. Only the Wally's Beach site, in southwestern Alberta, provides one.[49]

Cope would be pleased if this were true, but there is no reason to think
that it is. Minor and Spencer report that the camel remains occurred in three
groups, two of which had been disturbed by erosion. They claim to have found
five artifacts in "direct association" with the camelid remains, of which three
fit back together to form a single projectile (spear or dart) point. Of the three
point fragments, one was found amid the erosion-disturbed bones, one with
no bones at all, and the third with what was apparently an undisturbed part
of the camel's remains.

Because there are no drawings of the relationship between the projectile
point fragments and the bones, it is hard to know what the relationship between all these things actually was, but there is no reason to think that a more
detailed report would help here. That is because the point fragments were
collected before Minor and Spencer arrived to do their excavations.[50] Of the

TABLE 5.1. Collecting expeditions to Fossil Lake that resulted in analyzed collections of fossils

Collector	Date	Reference
J. Whiteaker	1876	Cope 1878a
J. Whiteaker	1877	"Wonderful fossil beds" 1877; Clark 1989
T. Condon	1877	McCornack 1928
C. H. Sternberg	1877	Sternberg 1909
E. D. Cope	1879	Cope 1889a
I. C. Russell	1882	Howard 1946; Packard 1952
W. Day	1883	Howard 1946
A. Alexander	1901	Elftman 1931
C. Stock and E. L. Furlong	1923–1924	Howard 1946
C. Stock and E. L. Furlong	1927–1931	Howard 1946
I. Allison	1939–1941	Howard 1946
R. R. Miller and E. R. Hampton	1963	Jehl 1967
R. E. Wilson	1965	Jehl 1967
E. C. Gibson and R. Minor	1977	Gibson and Spencer 1977; Minor and Spencer 1977
J. Martin	1977; 1989–2015	J. Martin et al. 2005

Note: Many other collections of Fossil Lake material are known to have occurred. For instance, Jackman and Long (1964:131–32) reported that in 1902, a group of students took a trip to the site and collected "a four-horse wagonload" of material that they thought may have ended up at the Smithsonian. If so, and if that material was analyzed and published, the collectors have not been identified and so are not included in this table.

two remaining artifacts, one was found with erosion-disturbed bones. That leaves only a small stone flake, excavated from fine sands along with part of the camel. Given the nature of these sediments, and given that the bones themselves bear no indication that they had ever been touched by early human hands, there is no reason to think that this camel met its demise at the hands of Ice Age people. I would follow Sternberg's advice and back out gracefully from this one.

Unfortunately, the kind of camel involved has never been identified. Minor and Spencer sent specimens of it to camel expert John Breyer of Texas Christian University, but Breyer tells me that he has no idea what the genus might be.[51] It would be nice to know, since it has been dated to 10,275 ± 95 years ago,[52] making it the youngest well-dated member of the camel family known from North America (chapter 4).

The Fossil Birds of Fossil Lake

Over the years, a substantial number of fossil-collecting expeditions to Fossil Lake have resulted in the remains of a wide variety of vertebrates ending up in the hands of professional paleontologists, of whom Cope was just the first (see table 5.1 for a list of the major expeditions). Cope began the process of publishing all this;[53] many others have followed since that time.

Of the major categories of vertebrates that have been reported from Fossil Lake—fish, birds, and mammals—the birds have been the most fully described, the fish the least. In fact, while the Fossil Lake fish attracted attention early on and there has been more recent work done on them, no modern analysis of them has been done and I do not discuss them here.[54]

At first, it might seem somewhat odd that the Fossil Lake birds have been so well studied, since paleontologists with expertise in mammals are far more common on the North American scientific landscape than are those specializing in birds, and they always have been. However, from the very beginning, it was clear that the Fossil Lake fauna was rich in bird remains; in 1912, the collection was reputed to be the largest collection of fossil bird remains from anywhere in the world.[55] It is, as a result, not surprising that paleontologists with the ability to analyze fossil birds have flocked to the collections provided by this site.

Cope himself was the first to tackle any of the Fossil Lake birds. In 1878, he described ten species of them drawn from the collections made by Whiteaker, Condon, and Sternberg.[56] Ultimately, however, he turned the birds over to Robert W. Shufeldt, an army surgeon and racist of almost unimaginable virulence.[57] Shufeldt then produced an important and detailed set of publications of the Fossil Lake birds directed toward both popular and scientific audiences.[58] This work was based not only on the collections that had been made by Condon and Cope, but also on those made by two US Geological Survey personnel, Israel C. Russell and William C. Day, in 1882 and 1883. By the time he was done, Shufeldt had described 74 species of birds from Fossil Lake, of which 21 were said to be extinct. Among the latter was a species of flamingo that he named in honor of Cope, *Phoenicopterus copei* (see chapter 3). Whatever his personal flaws, and they were vast, Shufeldt put the Fossil Lake birds on the map.

Additional collections of Fossil Lake bird fossils were made in 1901 by Annie M. Alexander, the founder of the Museum of Vertebrate Zoology and the Museum of Vertebrate Paleontology at the University of California, Berkeley.[59] The birds she collected from there were then analyzed by paleontologist Loye H. Miller, whose important work we encountered in chapter 3.[60]

By the mid-1940s, more bird remains from Fossil Lake had become available thanks to visits made by paleontologists Chester Stock and E. L. Furlong and geologist Ira S. Allison. This new material led Hildegarde Howard (chapter 3) to revisit the full run of fossil material that had been collected from there and to produce a monograph that remains one of the most important scientific works to have been published on the site.[61] She provided the first accurate counts of the size of the bird collection—2,500 bones identified to at least the family level, and about 1,800 to species. Using a much larger comparative collection of modern birds, she was also able to clarify the identifications that Shufeldt had made.

By the time Howard was done, she had described 66 securely identified bird species from Fossil Lake, 16 of which were extinct. Because of the care she brought to her analysis, Howard was also able to point out that 91 percent of these were waterbirds—such as American Coots (*Fulica americana*), Western Grebes (*Aechmophorus occidentalis*), and a wide variety of ducks, geese, shorebirds, and gulls.

In addition, Howard was the first to document that Fossil Lake contained not only an extinct species of flamingo, *Phoenicopterus copei*, just as Shufeldt had said, but also the remains of juvenile members of this species, meaning that the bird actually bred there. She also noted the presence of an extinct jaeger, which she named after Shufeldt (*Stercorarius shufeldti*). Jaegers are seabirds, but Howard noted that it was possible that the single individual of this species at Fossil Lake might simply have been a vagrant rather than a resident or migrant. At the time, there were no records for jaegers from eastern Oregon; today, there are, making it very possible that she was correct.[62]

Others have continued to work on the Fossil Lake birds,[63] the most recent of which is Jennifer Hargrave. In her PhD dissertation, Hargrave identified some 290 bird specimens collected by paleontologist James Martin, who began working in the Fossil Lake area during the 1970s. Hargrave added a number of species to the Fossil Lake list, the most notable of which are the Anhinga (*Anhinga anhinga*) and the Scarlet Ibis (*Eudocimus ruber*).[64] Today, the Anhinga is found no closer to Fossil Lake than the southeastern United States; the Scarlet Ibis, no closer than Trinidad and northern South America, although there have been occasional spottings to the north.[65]

The extinct birds that have been recognized at Fossil Lake are presented in table 5.2, following the scientific nomenclature in use today. I have not listed the Fossil Lake birds that are still with us, but they would not be out of place at the Malheur National Wildlife Refuge, some 90 miles to the east.

The Fossil Mammals of Fossil Lake

One of the most famous Ice Age paleontological sites in the world, southern California's Rancho La Brea, has yielded thousands of specimens of a broad variety of vertebrates. Within months of the initial discovery and excavation of this critically important site, scientists began describing first the mammals

TABLE 5.2. The extinct Pleistocene birds of Fossil Lake (extinct genera are in bold; extinct species are marked by *)

Scientific name	Common name	Reference
*Podiceps (= Colymbus) parvus**	Extinct Grebe	Howard 1946
*Phalacrocorax macropus**	Extinct Cormorant	Howard 1946
Anhinga anhinga	Anhinga	Hargrave 2009
*Phoenicopterus copei**	Extinct Flamingo	Howard 1946
Eudocimus ruber	Scarlet Ibis	Hargrave 2009
*Cygnus (= Sthenelides) paloregonus**	Extinct Swan	Howard 1946
*Branta propinqua**	Extinct Goose	Howard 1946
Anabernicula oregonensis	Extinct Shelduck	Howard 1964a
*Buteogallus sodalis**	Extinct Eagle	Howard 1946
Spizaetus pliogryps	Extinct Eagle	Howard 1946
*Falco oregonus**	Extinct Falcon	Howard 1946
*Dendragapus gilli**	Extinct Grouse	Howard 1946; Jehl 1967
*Dendragapus lucasi**	Extinct Grouse	Howard 1946
*Dendragapus nanus**	Extinct Grouse	Howard 1946
*Stercorarius shufeldti**	Extinct Jaeger	Howard 1946
*Larus robustus**	Extinct Gull	Howard 1946
*Larus oregonus**	Extinct Gull	Howard 1946

Note: The Anhinga and Scarlet Ibis, though not extinct, are found at distances so great from Fossil Lake that I have included them on the list.

and then the birds that it was providing, a process that has continued ever since.[66] If a bird or mammal is known to occur at Rancho La Brea, it has probably been described and studied.

This is also the case for the Fossil Lake birds, but it is not so for the mammals (table 5.3). Cope's deepest interests were in those mammals and he invested his greatest Fossil Lake efforts in them.[67] Although the importance of the Fossil Lake mammals was recognized early on,[68] no one returned to them in detail until Herbert Elftman in 1931.[69] The discussion he provided remains the state of the art. Indeed, it was Elftman's work that geologist Ira Allison used over 30 years later in the only thorough review of the Fossil Lake site that has ever appeared.[70] This situation will change when the results of the more recent work by James Martin become available.

As it stands now, Fossil Lake has provided the remains of seven extinct genera of Ice Age mammals: giant bear (*Arctodus*), Harlan's ground sloth (*Paramylodon*), mammoth (*Mammuthus*), horse (*Equus*), flat-headed peccary (*Platygonus*), large-headed llama (*Hemiauchenia*), and yesterday's camel (*Camelops*). The record for dire wolf provides one of the few for the Great

TABLE 5.3. The extinct Pleistocene mammals of Fossil Lake
(extinct genera are in bold; extinct species are marked by *)

Scientific name	Common name
Canis cf. *dirus**	Dire wolf
Felis sp.	Large cat
Arctodus sp.	Giant bear
Cynomys sp.	Prairie dog
*Ondatra oregonus**	Extinct muskrat
Paramylodon (**harlani?**)	Harlan's ground sloth
Mammuthus (**columbi?**)	Columbian mammoth
Equus sp. (small)	Extinct horse
Equus pacificus	Extinct horse
Platygonus cf. vetus	Extinct peccary
Platygonus sp. (small)	Extinct peccary
Hemiauchenia sp.	Extinct llama
Camelops hesternus	Yesterday's camel

Reference: Allison 1966.
Note: The prairie dog *Cynomys* still exists in North America but at
distances so great that I have left it on the list. The large cat may include remains that belong to the American lion, *Panthera leo atrox.*

Basin as a whole, and the large cat remains may include those of the American
lion (*Panthera leo atrox*), though these would need to be studied in detail to
know for sure. There are no prairie dogs (*Cynomys*) in Oregon today, or even
close to it, but Fossil Lake contained them.[71] Although there are other Ice Age
records for prairie dogs outside their modern distribution,[72] none come from
as far away as Fossil Lake.

One other mammal that has been reported from Fossil Lake is worth mentioning: the giant beaver (*Castoroides*). Cope appears to have never identified
this animal from the site or, if he did, never published that identification. He
certainly had not identified it by 1883, since he did not mention it in the discussion of *Castoroides* he published in that year.[73] To my knowledge, the first
person to say that the remains of giant beaver had been found here was the
paleontologist William D. Matthew in 1902. He noted that his compilation
of the Fossil Lake mammals was "revised from Prof. Cope's list,"[74] but he did
not say what that list was. In 1904, William J. Sinclair followed Matthew, as did
renowned paleontologist Oliver P. Hay in 1927. Elftman found no evidence for
it, nor has anyone else since. If there really was *Castoroides* at Fossil Lake, this
would be quite remarkable, since there is not a single record for it west of the
Rocky Mountains in the United States (fig. 3.28).[75]

How Old Are the Fossil Lake Fossils?

Estimates of the age of the Fossil Lake fossils have varied dramatically since the site was first reported in the scientific literature. As I have mentioned, Cope felt that the remains he was describing dated to the Pliocene, but by the end of the century it was coming to be recognized that these materials were Pleistocene in age, perhaps even from the later Pleistocene.[76]

Today, we know that Fossil Lake is Pleistocene in age. The more recent problem has been figuring out where in the Pleistocene all the site's various animal remains fit. The discovery of rich assemblages of fossils held by the asphalt deposits ("tar pits") of southern California helped here. Rancho La Brea,[77] McKittrick,[78] and Carpinteria[79] all provided large collections of bird bones, which, once identified, could be compared to those known from Fossil Lake. Comparisons of the proportions of the extinct birds from all four of these sites suggested that the assemblages had been deposited reasonably close to one another in time.[80] Since that was the case, and since it was accepted by the mid-1950s that Rancho La Brea was late Pleistocene in age, it followed that Fossil Lake was as well.

This all began to change with a series of important technical advances. The first of these was the development of radiocarbon dating (chapter 4), with its ability to date organic material deposited within the last 50,000 years or so. The second was the development of methods to identify volcanic ashes, or tephras, by their chemical composition, which in turn allows them to be correlated across space. If an ash has been dated anywhere, you can apply that date to all the deposits in which the ash is found. The famous Mazama ash provides a great example. Crater Lake, Oregon, was formed in the huge bowl, or caldera, created when Mount Mazama exploded, spewing about 12 cubic miles of volcanic ash over a huge region. Radiocarbon dating tells us that this cataclysmic eruption occurred about 6,730 years ago.[81] Whenever you find Mazama ash in an undisturbed deposit, you know how old that part of the deposit is. Since volcanic ashes well beyond the range of radiocarbon dating can now be placed in time using other methods, this general approach can be extended deep into the past.

Ira Allison began to put all this into play at Fossil Lake.[82] He worked at a time when radiocarbon dating was not nearly as trustworthy as it is today, and the means of identifying volcanic ashes not nearly as precise as they are now. Nonetheless, by identifying the various layers of deposits at the site and showing that they contained a series of volcanic ashes that could be correlated with ashes elsewhere, he began to make chronological sense of the site's deposits. By obtaining a single radiocarbon date, of 29,000 years ago, he could

TABLE 5.4. The volcanic ashes from Fossil Lake

Tephra layer	Age (years ago)
Rye Patch Dam	ca. 646,000
Dibekulewe	ca. 610,000
Tulelake T64/Llao Rock Pumice Castle-like 1	ca 95,000/71,000
Mt. St. Helens C (Marble Bluff)	47,000
Trego Hot Springs	23,000

References: J. Martin et al. 2005; Retrum 2010.

argue that at least some of the Fossil Lake deposits really were late Pleistocene in age, just as others had surmised in the absence of such direct evidence. Even though we now know that Allison's date was incorrect,[83] it doesn't really matter. His most fundamental contribution was to show that all this could be done.

Those who have followed him have now begun to do it in earnest. Paleontologist James Martin and his colleagues have shown that the Fossil Lake deposits contain five tephras. Because these have been dated elsewhere, he was able to show that the oldest deposits at Fossil Lake that he studied date to about 646,000 years ago, the youngest, to 23,200 years ago. In addition to these dates, we also have the date from the Fossil Lake camel that I mentioned above: 10,300 years ago.[84]

Geologist Julie Retrum has provided an even deeper look at the Fossil Lake depositional sequence. Following in Martin's tracks, she was able to reidentify one of the ashes Martin had uncovered and was also able to show, in exquisite detail, that the Fossil Lake Basin has been the scene of a series of lakes that came and went during the last 650,000 years, with one of the deepest and coldest occurring at the very end of the Pleistocene.[85]

Table 5.4 shows the dated tephras that Martin and his colleagues, on the one hand, and Retrum, on the other, have identified in the Fossil Lake deposits. It is obvious that there is a glaring gap between the tephra that falls at 610,000 years ago and the next one up in the sequence, at 71,000 years ago. This is not because volcanoes stopped doing their explosive thing during the middle of the chronological sequence at the site, but because the deposits that might have contained those tephras are not present. They might never have been here in the first place, or they might have been eroded away. Either way, they are absent.

What does this tell us about the age of the fossils from Fossil Lake? Jennifer Hargrave has begun to answer this question for the birds. She has shown that many of the birds excavated by Martin at the site came from the deposits

FIGURE 5.3. Gypsum Cave.

associated with the tephra dated to about 610,000 years ago. Those oldest
birds include the remains of both the Anhinga and Scarlet Ibis that I mentioned above. Many others came from the geological unit dated to somewhere
between 71,000 and 95,000 years ago, while a small sample came from the
47,000-year-old unit. Unfortunately, since the sample Hargrave worked with
did not include the remains of Cope's flamingo, we have no hint as to when it
might have lived, and bred, here.[86]

We can say even less about the chronology of the Fossil Lake mammals,
other than the 10,300-year-old camel I discussed above. The results of James
Martin's work are not available yet, and there is, at least for now, no way to tell
how old the collections made by all those who came before him might be.
All we can really say is that Fossil Lake most certainly contains deposits and
fossils that are late Pleistocene in age, but that it also contains deposits and
fossils that are much earlier.

GYPSUM CAVE, NEVADA

Including southern Nevada's Gypsum Cave in the Great Basin requires a small
geographic stretch (figs. 5.1 and 5.3). The site is located in the limestone rocks
at the eastern edge of the Frenchman/Sunrise Mountains, about six miles
east of Las Vegas.[87] The water that falls here, when it falls, can fill the nearby
Gypsum Wash, which in turn can drain into the Colorado River. So Gypsum

Cave is not actually within the Great Basin, but it is so close that this does not matter. If you did not know the local hydrology in detail, you would have no reason to think that you were on the wrong edge of the Colorado River drainage.

The site was first excavated by Mark R. Harrington, of the Southwest Museum. Harrington became aware of it in 1924 while touring the Las Vegas area with Nevada governor James Scrugham. He visited the site the following year and was likely impressed by its size alone. The cave is some 70 feet across and 15 feet high at its mouth, opening to a breadth of about 280 feet, though that breadth is divided into a series of five major "rooms," as Harrington later called them.[88] While there, he stuck a shovel in the ground and came up with what he knew to be dung, though he did not know what animal had left it there.

When he returned in 1929, he surmised that the animal that had left it was likely extinct, and that it might have been a ground sloth. He sent a sample to his friend, paleontologist Barnum Brown at the American Museum of Natural History in New York.[89] This was a good choice, because Brown knew his ground sloths, and he knew of the ground sloth dung that had been described from Patagonia's Mylodon Cave in the late 1800s.[90] In addition, a dung ball had been found with a Shasta ground sloth skeleton in Aden Crater, in central New Mexico. This had been described in some detail by R. S. Lull in that same year, and Brown must have known about this as well.[91] As a result, he had no trouble identifying what Harrington had sent him: Gypsum Cave contained the dung of extinct ground sloths.

Once Harrington knew he had found the remains of extinct animals, he decided this was a place worth exploring in more detail and began planning a field project, receiving the necessary Antiquities Act permit on October 21, 1929.[92] He was at the site and ready to go on January 20, 1930.[93] In his own words, he was about to excavate "a stable of ground-sloths."[94]

From a modern perspective, it is hard to believe that an archaeologist or paleontologist could plan an excavation, assemble the needed team, find the needed funds, and be in the field in a few months. This, though, was 1929, when field techniques were, from our perspective, crude, equipment inexpensive, and permitting procedures simple and fast. Harrington could do it, and he did.

The Gypsum Cave project was not cheap. Harrington began work at the site on January 20, 1930, and continued until June 30 of that year. Then, because of the heat, he broke camp for four months, returning on November 1 and continuing until January 17, 1931. During this interval, and excluding the 17 days of 1931, for which figures are not available, the Gypsum Cave fieldwork

FIGURE 5.4. Bertha "Birdie" Parker Pallan (Smithsonian
Institution Archives, Image 2009-0779).

cost $12,170.91, of which $8,026.78 went to salaries. That doesn't sound like
much, but in terms of contemporary dollars, it amounts to $172,777.36.[95]
Today, that would be a huge grant for a single year's archaeological project.

Luckily, the Southwest Museum did not have to shoulder the financial
burden on its own. The California Institute of Technology chipped in nearly
$5,300, and the Carnegie Institution of Washington, about $1,600. In the end,
the Southwest Museum paid only about 40 percent of the costs.[96]

Other institutions were eager to participate in the Gypsum Cave project
because of what Harrington and his crew began to find almost immediately.
Among that crew was Bertha "Birdie" Parker Pallan, the daughter of Harrington's friend Arthur C. Parker (fig. 5.4) and, like Harrington, an employee
of the Southwest Museum.[97] A Seneca Indian, Arthur Parker was the first
New York state archaeologist, the first president of the Society for American
Archaeology, and, at the time, the director of the Rochester (New York) Museum of Arts and Sciences. Bertha later married the actor Oscar "Iron Eyes"

Cody and, with him, was active in matters relating to Indian affairs, just as her father was. Since Harrington had married Bertha's aunt in 1927, she was also his niece.[98]

I say all this because it was Birdie Parker Pallan who, on January 30, found the skull of "a strange animal" beneath a rock slab.[99] The skull was strange enough that they sent it off to the Southwest Museum to be identified. The museum's director, James Scherer, said that he called in Chester Stock, California Institute of Technology paleontologist (see chapter 1), who immediately identified it as the skull of a Shasta ground sloth (*Nothrotheriops shastensis*).[100] Harrington, on the other hand, reported that he had the skull taken directly to Cal Tech, where it was identified.[101]

Either way, this was a remarkable find, so much so that Stock, his paleontologist colleague E. L. Furlong, and Scherer visited the site on February 12. Not only did the Cal Tech agreement to participate in the project follow quickly, but the institute sent paleontologist James E. Thurston to become part of the excavation team.[102] Thurston had come to the institute only a year earlier, having worked for Charles Sternberg, of Fossil Lake fame, for some four years. He was an experienced field collector, and his addition to the field team provided a much-needed professional paleontological presence. Unfortunately, he died in 1932, a year after the close of the Gypsum Cave excavations. He was only 27 years old.[103]

Harrington found sloths to be "a queer massive race of beasts, awkward and stupid,"[104] but their remains would have been more than enough to make Gypsum Cave an enticing target of excavation even if it had contained nothing else. However, the site also yielded artifacts and other indications of a human presence that appeared to establish that people had been using the cave at about the same time as the sloths. On May 13, 1930, for instance, Harrington's team found "a wooden foreshaft for a dart imbedded in gypsum, under a four-inch layer of sloth manure, which in turn was capped by a gypsum layer."[105] On May 16, they found charcoal and ashes "under an unbroken layer of sloth manure nearly eight feet below the surface of the ground."[106] These, they were sure, represented the remnants of a campfire.

It was around this time that John C. Merriam, president of the Carnegie Institution of Washington and a member of the Southwest Museum's Advisory Council, visited and came away deeply impressed. On May 31, 1930, he wrote to the museum's director to tell him that Gypsum Cave "ranks among the most interesting discoveries in archaeology in America."[107] No surprise, then, that the Carnegie Institution of Washington, with its deep interest in the antiquity of people in the Americas, also decided to help fund the project.[108]

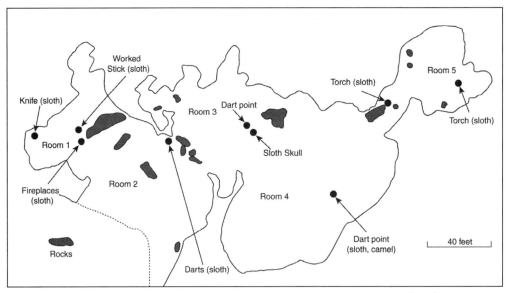

FIGURE 5.5. Harrington's Gypsum Cave floor plan, redrawn from Harrington (1933) and Gilreath (2009). The arrows mark the locations of what he felt to be associations between ancient human activity and extinct mammals.

As I mentioned, Harrington divided the cave into five major separate rooms (fig. 5.5). In Room 1, he found the remains of campfires beneath sloth dung (fig. 5.6). In Room 2, he discovered painted atlatl dart shafts beneath the dung (fig. 5.7). Room 3 provided a distinctive atlatl dart point, now known as a Gypsum point (fig. 5.8), in a layer of sloth dung (fig. 5.9). Room 4 yielded a second dart point and atlatl dart fragments between two sloth dung layers and near sloth and camel bones (fig. 5.10). In the passage between Rooms 4 and 5, he uncovered a small pit-like excavation containing 12 small sticks burned on one end and capped with "an unbroken stratum of solid well-preserved sloth dung 17.5 inches deep" (fig. 5.11).[109] He even found sloth bones that he thought showed cut marks indicating they had been butchered by people. In fact, he thought that some of the animals he found had been killed by people. "That the sloths were the victims," he said, "is hinted by scratches on bones which might have been made with flint implements in cutting off the flesh."[110]

There was more, but this is enough to convey the main point. Gypsum Cave seemed to provide evidence that people and extinct sloths had walked the earth at the same time. Exactly when Harrington thought that time was varied across the years. In 1932, he compared the Gypsum Cave projectile points to stone tools found in the French Paleolithic archaeological complex known as the Solutrean (to which they bear no resemblance), and which he

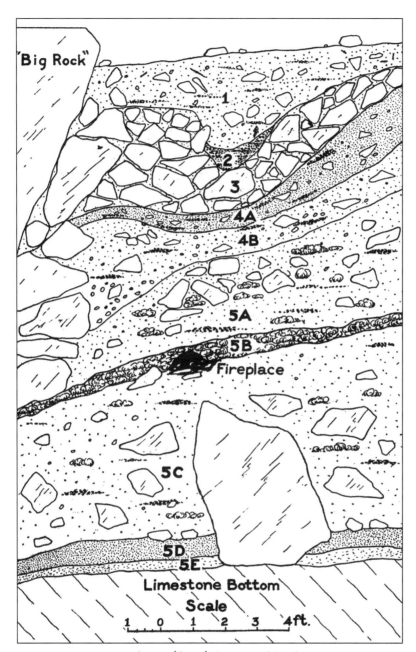

FIGURE 5.6. Harrington's 1933 (fig. 11) drawing of the deposits in Gypsum Cave Room 1. According to Harrington, Stratum 5A consisted of ground sloth dung, stones, dust, and bighorn dung, while stratum 5B was composed of "solid sloth dung" (1933:30). Note the location of the fireplace beneath the sloth dung.

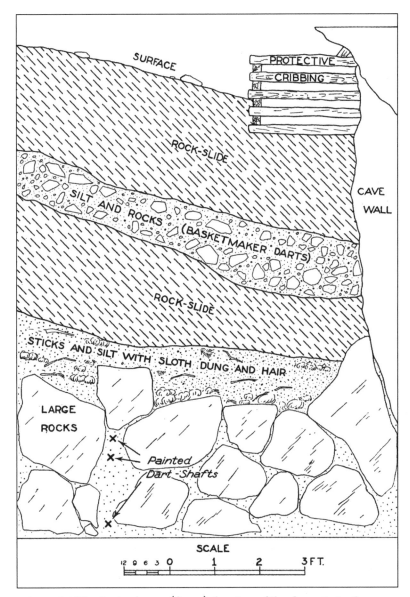

FIGURE 5.7. Harrington's 1933 (fig. 15) drawing of the deposits in Gypsum
Cave Room 2. Note the painted dart shafts beneath the sloth dung.

estimated to be between 18,000 and 20,000 years old.[111] The next year, he
estimated this material to be about 10,500 years old; this is the estimate I discussed in chapter 4.[112] A few years later, while still comparing the Gypsum
Cave projectile points to Solutrean artifacts, he estimated an age of about
9,500 years.[113]

The age estimates really didn't matter, since at the time and in the absence
of any secure dating technique, no one knew the age of the Solutrean, or

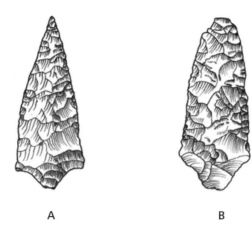

A   B

FIGURE 5.8. Gypsum points from Gypsum Cave. *A,* Found in Room 3 beneath "partly burned sloth dung" (Harrington 1933:42); Southwest Museum specimen 6F485, illustrated in Harrington (1933, fig. 19a), and Gilreath (2009, fig. 12j). *B,* Found in Room 3 in a rock crevice unassociated with the remains of extinct mammals (Harrington 1933:107); Southwest Museum specimen 6F185A, illustrated in Harrington (1993, fig. 19b), and Gilreath (2009, fig. 12i). Both points are 2.54 inches long. Drawings by Peggy L. Corson.

Shasta ground sloths, or anything else that far back in time. Gypsum Cave became famous because it seemed to show that people and ground sloths had been around at the same time, regardless of exactly when this had been.

The site is also famous for the extinct mammal bones and teeth that it provided to Harrington and his crew. Analyzing that material fell to the California Institute of Technology, with the results reported by both Chester Stock and Harrington.[114] In addition to the Shasta ground sloth, the extinct mammals they reported included yesterday's camel (*Camelops*), horse (*Equus*), large-headed llama (*Hemiauchenia*), and, perhaps, dire wolf (*Canis dirus*; see chapter 3).[115]

Other animals of Pleistocene age have been identified from the Gypsum Cave deposits, including the California Condor (*Gymnogyps californianus*) and a wide range of reptiles and amphibians.[116] Most famous of all, however, are the Gypsum Cave Shasta ground sloths. The site provided the skeletal remains of these animals, including a nearly complete foot with soft tissues still in place, along with sloth hair and skin. It also provided a substantial collection of Shasta ground sloth dung, which scientists have studied in impressive detail.

In 1929, Lull had provided a very general analysis of the single Aden Crater Shasta ground sloth dung ball, with the conclusion that that animal

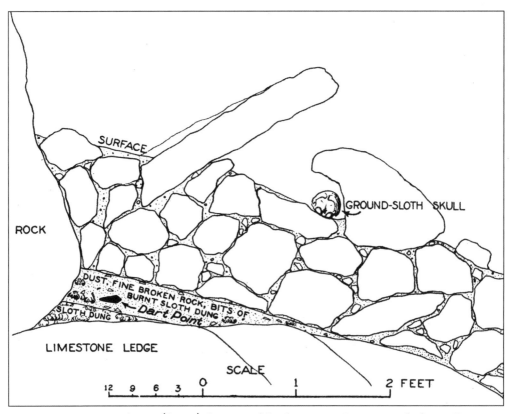

FIGURE 5.9. Harrington's 1933 (fig. 20) drawing of the deposits in Room 3, including a Gypsum point in a layer of sloth dung. The point found here is illustrated on the left in figure 5.8.

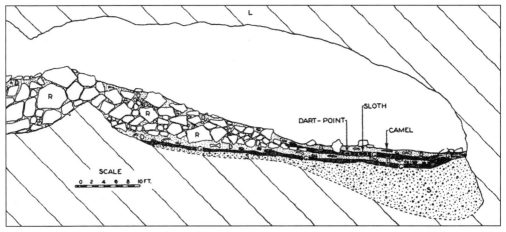

FIGURE 5.10. Harrington's 1933 (fig. 32) drawing of the deposits in Room 4. The dart (Gypsum) point in this illustration was found between layers of sloth dung near a sloth bone and beneath the remains of a camel.

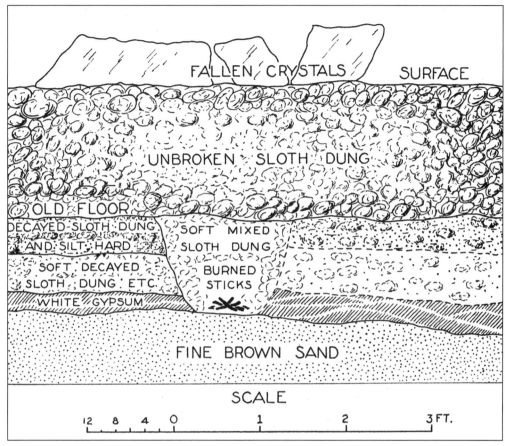

FIGURE 5.11. Harrington's 1933 (fig. 43) drawing of the deposits at the entrance to Room 5. Note the burned sticks beneath what is identified as an unbroken layer of sloth dung.

"apparently was feeding upon low-growing, hairy, desert-scrub vegetation, such as at present grows in the southwestern United States."[117] Botanist Arthur Eames soon followed, in 1930, with a far more detailed analysis of the Aden Crater coprolite. Stock was very much aware of these analyses and of the potential of the plant material preserved in the Gypsum Cave dung to tell us about sloth diet, about the plants that grew in the region when the sloths were munching on them, and about the climate that then prevailed.[118]

Harrington got the sloth dung ball rolling in this realm, reporting that Arthur D. Howard, a volunteer on the project, had found the remains of Joshua tree, saltbush (*Atriplex*), perhaps Mormon tea (*Ephedra viridis*), and either rushes (*Juncus*) or grasses in the dung he examined.[119] This work, however, was already being eclipsed by that done by two scientists who, at the request of Chester Stock, were examining the Gypsum Cave sloth dung housed at the California Institute of Technology.

Jerome D. Laudermilk and Philip A. Munz were colleagues at the Clare-mont Colleges in Claremont, southern California. The combination was per-fect. Laudermilk was a chemist who analyzed such things as the constituents of varnish on desert rocks and the tiny remnants of burned fabrics. Munz was a botanist who had helped Eames identify the plant material from the Aden Crater coprolite. Later, he was to produce one of the most famous works on the botany of California ever published.[120] Completed over 70 years ago, their analysis of the Gypsum Cave sloth dung remains important, with good reason.

Although it is the joint work of Laudermilk and Munz that is now cited, the first publication of this research came in 1933, when Stock reported the preliminary results of the microscopic analysis of the Gypsum Cave sloth dung. Those results, he noted, included the presence of Joshua tree, banana yucca (*Yucca baccata*), Mormon tea, desert holly (*Atriplex hymenelytra*), Utah agave (*Agave utahensis*), juniper (*Juniperus*), and others plants, including grasses, identified at a more general level. "An account of these studies," Stock concluded, "will be published by Mr. Laudermilk."[121]

That remarkable publication appeared the following year, now with Munz's name attached. Laudermilk and Munz had succeeded in identifying 14 species of plants from the Gypsum Cave dung (table 5.5), most, though not all, of which still grow in the area today. Of those 14 kinds of plants, the most abundant were the yuccas and agave. Importantly, the yuccas included Joshua tree, which is not found here now. The Shasta ground sloth, Laudermilk and Munz concluded, "was preeminently a yucca-feeder, while living in the vicin-ity of the cave."[122] They discovered that this beast was not a particularly careful chewer, since they found large chunks of the stems of plants like saltbush and jointfir, chunks that were up to nearly 1.5 inches long. They found that the sloths didn't seem to worry much about the pointed and potentially mouth-piercing tips of yucca and agave leaves, since their dung was full of those. The sloths didn't confine themselves to the leaves and stems of plants; they also chowed down on the seedpods and seeds of Joshua trees.

People who are prone to developing kidney stones learn to stay away from plants that are heavy in calcium oxalate—spinach and rhubarb are good examples. Many plants produce this chemical for a variety of reasons, includ-ing its ability to provide a defense against against herbivores, from insects to peccaries.[123] Yuccas and agaves can be particularly rich in calcium oxalate, so much so that those who gather and process tequila agave (*Agave tequilana*) to make tequila refer to the skin irritation that comes from touching this plant,

TABLE 5.5. Plants in the dung of the Gypsum Cave ground sloths

Scientific name	Common name
Juniperus osteosperma	Utah juniper
Ephedra nevadensis	Nevada jointfir
Typha sp.	Cattail
Aristida purpurea	Blue threeawn
Yucca brevifolia	Joshua tree
Yucca baccata	Banana yucca
Yucca schidigera	Mojave yucca
Agave utahensis	Utah agave
Atriplex sp.	Saltbush
Larrea tridentata	Creosote bush
Sphaeralcea ambigua	Desert globemallow
Petalonyx sp.	Sandpaper plant
Camissonia multijuga	Froststem suncup
Ericameria nauseosa	Rubber rabbitbrush

Reference: Laudermilk and Munz 1934.

and the calcium oxalate crystals it contains in such abundance, as *mal de aga-veros*, or "agave worker's disease."[124] Because the modern Native peoples of arid western North America make heavy use of agaves, it is not just those involved in tequila production who have this problem.[125] Ethnobotanist and agave expert Wendy Hodgson spoke with a Native southwestern basket maker who reported that making agave fibers was a miserable enterprise that gave her a significant rash by the end of the day. Hodgson herself recalls that the first time she handled Grand Canyon century plant (*Agave phillipsiana*), she "writhed with pain for 15 long minutes."[126]

Since the Gypsum Cave Shasta ground sloths were eating significant amounts of yucca and agave leaves, it is not surprising that Laudermilk and Munz found their dung balls to be loaded with this chemical. In fact, they noted that when they analyzed the "fine yellowish powder" the dung balls contained, half of it consisted of calcium oxalate.[127]

That raises a series of questions. Eating calcium oxalate–laden leaves can cause painful swelling of the mouth in people, and some herbivores avoid plants loaded with this chemical.[128] Why did the sloths not avoid them as well? Were they equipped with a means of detoxifying agaves and yuccas? The ingestion of large amounts of calcium oxalate can cause bladder and kidney stones, which modern sloths are known to get.[129] Did Shasta ground sloths get these stones? We don't know the answer to either of these questions.

However, since the form of bladder and kidney stones is distinctive, it would be worth taking a close look for them the next time a complete Shasta ground sloth skeleton is found.

The analysis of Gypsum Cave sloth dung did not stop with Laudermilk and Munz. Organic chemist Fiona Gill and her colleagues analyzed the chemical content of a single Gypsum Cave sloth coprolite and found that the chemistry confirmed what had been concluded from a previous analysis of the same coprolite—that this dung ball had been produced by an animal that had been chowing down on yucca.

Specialists in the analysis of ancient DNA—Michael Hofreiter, Heinrik Poinar, and their colleagues—have also begun to look at the DNA contained within the Gypsum Cave sloth dung. By doing so, they showed that the dung attributed to the Shasta ground sloth really did come from that sloth. We pretty much knew that already, of course (remember the Aden Crater skeleton and dung ball), but it was reassuring to have it confirmed.[130]

They also looked at the plant DNA preserved within six dung balls from Shasta ground sloths from Gypsum Cave. This allows the identification of plants that are so badly fragmented that they cannot be identified from the bits and pieces that have been munched and passed through the sloth's innards. Unfortunately, the state of the art is such that the identifications don't go below the order or family level, so something that could be identified as, say, Joshua tree from a seed or other big part could be identified from the DNA of a fragment only as Liliales, the plant order to which Joshua trees belong.

In addition to looking at the DNA from these six dung balls, the DNA team also radiocarbon dated each one of them (see table 5.6). As a result, they were able to figure out not only what these sloths were eating, but also when they were eating it. Their ground sloths from 28,500 years ago had focused their diets on plants belonging to the pine family, and probably on plants from the pine genus (*Pinus*) itself. They had also been feeding on plants belonging to the mulberry family, perhaps Texas mulberry (*Morus microphylla*), on plants belonging to the order that contains mustards and capers, and on grasses. Other plants were present in smaller abundances, but the yuccas and agaves were simply not there.

That all changed in the 20,000-year-old sample, in which yuccas and/or agaves were one of two very abundant sets of plants represented. The other was, just as before, the order that includes mustards and capers. Then, by 11,000 years ago, the family Chenopodiaceae, which includes the arid-adapted saltbushes, was the most abundant and yuccas and agaves were again absent.

There is not much mystery as to why ground sloth diet changed this way

TABLE 5.6. Radiocarbon dates on Shasta ground sloth dung balls from Gypsum Cave (dates in bold are from specimens analyzed for ancient DNA)

Date	Dating method	Reference
8,051 ± 450	Solid carbon	Arnold and Libby 1951
8,692 ± 500	Solid carbon	Arnold and Libby 1951
8,838 ± 430	Solid carbon	Arnold and Libby 1951
10,075 ± 550	Solid carbon	Arnold and Libby 1951
10,902 ± 440	Solid carbon	Arnold and Libby 1951
11,005 ± 100	AMS	Hofreiter et al. 2000
11,080 ± 90	AMS	Hofreiter et al. 2000
11,290 ± 70	AMS	Gilreath 2009
11,360 ± 260	Conventional	A. Long and Martin 1974
11,690 ± 250	Conventional	A. Long and Martin 1974
19,500 ± 205	AMS	Hofreiter et al. 2000
19,875 ± 215	AMS	Poinar et al. 1998
21,470 ± 760	Conventional	Thompson et al. 1980
23,700 ± 1,000	Conventional	Long and Muller 1981
27,810 ± 455	AMS	Hofreiter et al. 2000
29,205 ± ?	AMS	Hofreiter et al. 2000
33,910 ± 3,720	Conventional	Thompson et al. 1980

as the years went by. The period from 30,000 to 11,000 years ago was a time of dramatic climate change in the Great Basin, and the shifting ground sloth diets match what we know of these changes.[131] Obviously, Shasta ground sloths could alter their eating habits to survive under a wide variety of climatic settings.

Just as obviously, Chester Stock was right in thinking that the dung of the Gypsum Cave sloths could tell us what these animals were eating, which plants were near enough to the cave that they could eat them, and what the climatic conditions were under which they lived. Of course, one must wonder why the sloths became extinct if they could live through all this, but that is an issue for chapter 7.

When I discussed Fossil Lake, I pointed out that what we don't know about the chronology of this site vastly exceeds what we do know. This is decidedly not the case for Gypsum Cave. In fact, this site is one of the best-dated Ice Age paleontological sites in the Great Basin. Thanks to work done by a wide variety of scientists, we have 17 radiocarbon dates available for the sloth dung from here. These are listed in table 5.6.

Of these 17 dates, those that were produced very early in the history of radiocarbon dating, using what is known as the solid carbon method, are best ignored, since we now have much more reliable dating methods (chapter 4).

That leaves us with 12 trustworthy dates. They show us that ground sloths were using the cave between about 11,000 and 34,000 years ago, but there is a huge hole in the sloth-date sequence, between 19,500 and 11,700 years ago. This hole might coincide with the break between the two very different layers of sloth dung that Harrington described and illustrated from some parts of the site.[132] It might, though, simply reflect luck of the sloth-dung draw, especially since there is a date for one of the three specimens of California Condor known from the site that falls right in the middle of the sloth-gap: 14,720 ± 170 years ago.[133]

The dates that fall between 11,700 and 11,000 years ago coincide with a period when we know that people were in North America. This, in turn, suggests that Harrington might have been correct in arguing that Gypsum Cave provides evidence for interactions between humans and ground sloths. We know, however, that it does not, and that Harrington was wrong about this.

We have known this since 1967, when archaeologist Robert F. Heizer and radiocarbon-dating expert Rainer Berger obtained two radiocarbon dates from artifacts that Harrington claimed to be of sloth-like antiquity.[134] One of these dates was obtained from the small burned sticks that Harrington found in the pit-like feature between Rooms 4 and 5, beneath "an unbroken stratum of solid well-preserved sloth dung," which I mentioned above (fig. 5.11).[135] If Harrington had been right, these objects would pretty much have had to date to before 10,000 years ago, since this was very close to the time that Shasta ground sloths became extinct (chapter 4). The date that Heizer obtained, however, was not even close to this; it fell at 2,400 ± 60 years ago. This young date led Heizer to date one of the painted dart shafts that Harrington had found beneath sloth dung (fig. 5.7). It turned out to be 2,900 ± 80 years old. "It appears," Heizer and Berger concluded, "that the ground sloth lived in Gypsum Cave long before it was occupied by man, and that since 2400 to 2900 years ago a considerable amount of disturbance of the loose, surficial cave deposits has occurred."[136] No one has disagreed with this conclusion.

Any questions about this issue would have been answered by Amy Gilreath and her colleagues.[137] In 2004, her team returned to Gypsum Cave and carefully excavated some of the intact deposits in Room 1 that Harrington had left behind. Their important work showed that the "fireplace" that Harrington had thought to have been constructed by people during sloth times (fig. 5.6) was, in fact, natural in origin. They also showed that the site was stratigraphically far more complex than Harrington had realized, that it had been used by people between about 4,500 and 2,500 years ago and then again between about 700 and 200 years ago, and that there was no evidence whatsoever that both people and ground sloths used this site during the Ice Age. Even the sloth bones Harrington thought had been cut by human hands have been reanalyzed and shown to reflect the chewing of carnivores, perhaps dire wolf, and the effects of being trampled by other animals.[138]

Gypsum Cave might not be an Ice Age archaeological site and it might have been dug well before the development of the fine-grained excavation techniques in use today, but it is a gem of a paleontological site of great historical significance. No wonder that on July 8, 2010, it was placed on the National Register of Historic Places.[139] Fossil Lake should be there, too.

SMITH CREEK CAVE, NEVADA

As early as 1925, Harrington had known of another large, potentially productive cave, in Smith Creek Canyon on the eastern edge of the massive Snake Range (figs. 5.1 and 5.12), not far north of today's Great Basin National Park. Nevada governor Scrugham had asked him to conduct archaeological work in this area, but he declined. Instead, he sent two members of his team, George Evans and his son Willis, to excavate a small portion of the western part of what soon became known as Smith Creek Cave. In 1930, Willis Evans and James Thurston, the Cal Tech paleontologist who was part of the Gypsum Cave crew, excavated here again, but Harrington himself did not work at the site until 1932.[140] With support from the Carnegie Institution of Washington, he returned in 1934 and then again in 1936.

"The cave," he reported, "is a beauty as appearance goes, a high-arched, wide-mouthed chamber in the limestone, with plenty of daylight and air to the very back." It was also real work to reach: "It took me nearly an hour every morning to drag my perspiring and overweight person up to the cave mouth from our camp."[141] We know from his son that he camped, as we all do, along Smith Creek itself,[142] so his overweight person had to be dragged up some 1,100 feet of elevation, from about 5,600 feet at the creek to about 6,700 feet at the mouth of the cave, all across a very short distance. The hike is no easier today, as a glance at figures 5.12 and 5.13 will suggest.

Harrington dug deep holes in the site during the 1934 and 1936 field seasons, holes that went more than 12 feet down. He pulled out bones, some of which were broken in a way that suggested to him that people had split them for marrow, along with bits of what he thought was charcoal that might have been produced by ancient human campfires. He did not, however, find any artifacts, and this probably led him to abandon the site as a focus of additional work.[143]

FIGURE 5.12. Smith Creek Cave from near the mouth of Smith Creek Canyon.

FIGURE 5.13. The view from Smith Creek Cave. Archaeologist Charlotte Beck is on the left, archaeologist Michael Cannon on the right. Photograph by Lisbeth A. Louderback.

The bones, though, proved to be deeply interesting. In 1936, Chester Stock reported that the site had provided the remains of camels, horses, and at least six individuals of a species of mountain goat that had never been found before. This new goat Stock named *Oreamnos harringtoni*, Harrington's mountain goat (chapter 3), in honor of Harrington's "noteworthy contributions to the study of cave occurrences in the Southwest."[144]

The new goat was interesting in and of itself, but the birds Harrington retrieved included one new species that was, and still is, astonishing. Those birds went to Hildegarde Howard, whose work we have encountered many times before. In 1935, she reported that Smith Creek Cave had provided the remains of three extinct species of birds: the Western Black Vulture (*Coragyps occidentalis*), Clark's Condor (*Breagyps clarki*), and a bird new to science, which Howard designated as Willett's Hawk-Eagle (*Spizaetus willetti*; see chapter 3).[145]

That new species was important, but the astonishing one wasn't described until 1952, when Howard returned to the Smith Creek Cave bird material, this time with both the 1934 and 1936 collections, and about 650 bird bones, in hand. To the list of extinct bird species at the site, she was now able to add the Errant Eagle (*Neogyps errans*) and the shelduck *Anabernicula miniscula*, a close relative of the shelduck that she had identified from Fossil Lake (table 5.2).

Among the 650 Smith Creek Cave bird bones, there was a single wrist-bone, a cuneiform, that surprised Howard by its size. Based on her exhaustive knowledge of avian skeletal anatomy, both past and present, she immediately recognized that this single bone came from a teratorn. She also recognized that it had been far larger than Merriam's Teratorn (*Teratornis merriami*; see chapter 3).[146]

Historical connections abound here. Merriam's Teratorn was first described by Loye Miller in 1909, from specimens found at Rancho La Brea. We first encountered his work in chapter 3; Miller had studied under Joseph Le Conte at Berkeley, who had played a significant role in the interpretation of the supposed human footprints at the Nevada State Prison (chapter 1). Howard then studied under Miller and collaborated with him until his death at the age of 94 in 1970.[147] Howard was now dealing with a relative of Miller's huge, soaring predatory bird.

Because the Smith Creek Cave wristbone had clearly come from a teratorn, and because it was so much larger than the corresponding bone from Merriam's Teratorn, she assigned it to a new species, *Teratornis incredibilis*, the Incredible Teratorn. She estimated that it must have had a wingspan of 16–17 feet, and she toyed with the idea of assigning it to its very own genus.

This she did not do because she had only a single wristbone to work with. Twenty years later, when she returned to the Incredible Teratorn armed with small amounts of new material, she again mentioned the possibility that the bird should be placed in a different genus and made the same decision: not enough was known about it. Not until 1999 had "enough" material (six specimens) been accumulated to make it clear that Howard's interpretation, based on a single wristbone, had been correct. The Incredible Teratorn now gets its own genus and is referred to scientifically as *Aiolornis incredibilis*.[148] As I mentioned in chapter 3, the bird's wingspan is currently estimated at 16–18 feet. Howard, with only a wristbone to go on, got that right, too. Have I mentioned how good she was?

Just as Harrington left deposits behind at Gypsum Cave for others to excavate in the future, so he did at Smith Creek Cave, though here he left more of them. In 1955, Theodore Downs of the Los Angeles County Museum excavated here, the results of which have never been published. The work that put Smith Creek Cave on the archaeological map was conducted by Alan L. Bryan, who excavated here in 1968, 1971, and 1974.

Bryan was drawn to this site by the description of it published by Harrington in 1934.[149] While Harrington seems to have lost interest in Smith Creek Cave because he found no artifacts there, Bryan was struck by his mention of camel and horse bones that might have been split by human hands. Bryan also thought that the location would have appealed to any late Pleistocene people who lived in the area. Archaeologists often think that way and are often right. Bryan thought that way and he was right.

Bryan's work here uncovered archaeological materials whose dating has been somewhat controversial, but they likely date to between about 11,000 and 10,000 years ago and possibly a bit earlier.[150] The archaeology is not directly relevant to us in this context,[151] but Bryan argued that the stone tools he found were so distinct from those found in some other parts of North America from the end of the Ice Age that two separate historical origins were needed to account for them. That argument was not particularly popular among archaeologists, but as time goes on, it seems more likely that he was correct.[152]

The bones that Bryan uncovered at Smith Creek Cave were identified by paleontologist Suzanne Miller.[153] Unlike the earlier paleontological work that had taken place at the site, Miller's work paid attention to the stratigraphic placement of the materials that she identified. Because of that, we know that, with one exception, all the extinct mammal specimens from the site that can be placed stratigraphically come from the unit known as the "red silt zone." The one exception is the single specimen that might represent the American lion (*Panthera leo atrox*), which cannot be placed in the stratigraphic sequence that Bryan developed. The red silts are found beneath the archaeological layers at Smith Creek Cave and have only a single radiocarbon date available, 28,650 ± 760 years ago, a conventional radiocarbon date on collagen from a bone of an unidentified mammal.

Although conventional bone collagen dates are no longer state of the art (see chapter 4), that is all we have for the red silts. For obvious reasons, Bryan focused his radiocarbon dating program on the archaeological layers of the site, not on the deeper deposits. Until very recently, there was no indication that the red silts dated to anything but the late Ice Age, just as the sole radiocarbon date suggested.

Then, paleontologist Christopher Jass came along and messed everything up. Among other things, Jass is an expert in identifying the teeth of voles, animals that belong to the rodent subfamily known as the Arvicolinae. Many North Americans who live in rural areas have encountered these animals, since the family includes the genus *Microtus*. These abundant small mammals are often called meadow voles, perhaps because many of them are voles that like to live in meadows. In addition to being dangerously cute, the members of this subfamily have the virtue of evolving in minor ways through time. Not only do the species of voles change through time, but the form (for instance, tooth structure) of particular species can change as well. That means that if you know the time range for a given species of vole, or for a particular tooth structure within a given species of vole, you can infer from that species or that structure the age of the surrounding deposits.[154]

I say all this because Jass returned to the voles that had been collected during the archaeological excavations at Smith Creek Cave.[155] In the end, he gathered up 296 first lower molars, often the most diagnostic of all vole teeth. Of those 296 teeth, he discovered that five seemed to have come from two extinct species known only from older paleontological contexts: *Microtus meadensis* and *Microtus paroperarius*. From the information associated with the teeth, and from the red staining on them, he could be certain that they came from Bryan's red silt zone. Not only are both of these species extinct, but the youngest known date for them falls at 146,000 years ago. That date comes from Cathedral Cave, on the other side of Smith Creek Canyon from Smith Creek Cave itself, and was obtained as a result of careful work Jass did for his PhD dissertation.[156]

That, obviously, raises a problem. The red silt zone at Smith Creek Cave has a single radiocarbon date that falls at about 29,000 years ago. The youngest dates known for the two extinct species of voles Jass found at the site are 117,000 years older than this. So how old is the red silt zone and the extinct mammals it contains?

There are multiple choices here. Perhaps the two extinct species of *Microtus* from the red silt zone lasted until very late in the Pleistocene. Perhaps the 28,650-year radiocarbon date is simply wrong, and the red silt zone is vastly older than anyone has thought. Perhaps the red silt zone incorporates a huge amount of time. And, since vole teeth are wickedly variable (God may have put them on earth to punish those who study them), perhaps the five specimens identified as coming from extinct species simply represent rare versions of the teeth of modern voles. Jass considered each of these options in detail and, other than doubting that these are the teeth of voles that still exist, concluded that we really don't know how old the red silt zone is. That, of course, means that we also don't know the ages of the extinct mammals contained in that silt.

On the other hand, we at least know they came from the red silts. Someone really should revisit the Smith Creek Cave birds to see whether they, too, are stained red.

Table 5.7 lists all the extinct birds and mammals known from Smith Creek Cave. The reptiles and amphibians have also been analyzed but include no extinct forms and so are not discussed here.[157]

On that list, you will see *Martes nobilis*, the noble marten. Paleontologist Elaine Anderson defined this as a distinct species in 1970, and it was long thought to have become extinct at the end of the Pleistocene. Now, however, it seems to have lasted well into the Holocene, and it is appearing more and more likely that it is best treated as a subspecies of the very-much-alive American marten (*Martes americana*). At the moment, there is no reason to think that the animal succumbed at the end of the Pleistocene, even though the Smith Creek Cave specimen itself is from the Pleistocene-aged red silts.[158]

As we have seen, Gypsum Cave provided the hair and skin of Shasta ground sloths. Smith Creek Cave provided similar material, including hair that has been tentatively identified as coming from the large-headed llama (*Hemiauchenia*) and dated to 12,060 ± 450 years ago.[159] It would be helpful to analyze the DNA from this material to see whether it actually is from a llama. This would likely work, since we know that 9,900-year-old bighorn sheep (*Ovis canadensis*) hairs from the site have well-preserved DNA.[160]

TABLE 5.7. The extinct birds and mammals from Smith Creek Cave

Scientific name	Common name
Birds	
Anabernicula miniscula	Extinct Shelduck
Aiolornis incredibilis	Incredible Teratorn
Breagyps clarki	Clark's Condor
Coragyps occidentalis	Western Black Vulture
Neogyps errans	Errant Eagle
Spizaetus willetti	Willett's Hawk-Eagle
Mammals	
Camelops	Yesterday's camel
Hemiauchenia?	Large-headed llama
Equus sp. (large)	Horse
Equus sp. (small)	Horse
Oreamnos harringtoni	Harrington's mountain goat
Martes nobilis	Noble marten
cf. *Capromeryx*	Diminutive pronghorn
cf. *Panthera leo atrox*	American lion

References: Stock 1936a; Howard 1952; S. Miller 1979; Mead, Thompson, and Van Devender 1982.

Bryan suggested that the possible *Hemiauchenia* hairs came from an animal that had been killed by human hunters. He argued that since no one in their right mind would haul a dead llama to Smith Creek Cave (he said it better than this), the animal must have been butchered elsewhere and the hide brought back to the site. It is a long jump, though, from finding herbivore hairs in a site to concluding that those hairs had been attached to a hide that had been attached to an animal that had been killed by people. As with the other sites discussed in this chapter, there are no secure associations between artifacts and the remains of extinct mammals or birds at Smith Creek Cave.

MINERAL HILL CAVE, NEVADA

If you are getting the impression that most late Ice Age paleontological sites in the Great Basin that are rich in the remains of extinct mammals have also been claimed to provide evidence for the presence of late Pleistocene people, you are right. We encountered this phenomenon in chapter 1 with the "human" footprints found at the Nevada State Prison, and again at Fossil Lake, Gypsum Cave, Smith Creek Cave, and Fishbone Cave (chapter 4). Had I discussed

FIGURE 5.14. The entrance to Mineral Hill Cave, with paleobiologist Bryan Hockett prepared for the descent into the site.

Tule Springs in detail, we would have encountered it there as well.[161] We could have seen it at China Lake in California's Mojave Desert, at the Paisley Caves in Oregon,[162] and elsewhere.

Since we will also encounter it throughout the rest of this chapter, it is worth noting here that there is not a single site in the Great Basin that provides evidence that people interacted with any extinct Ice Age mammal in this region. I explore this issue in chapters 6 and 7, but it is worth remembering the caution that Sternberg learned from his Fossil Lake experience. Caution is appropriate when one is tempted to accept "the first evidence of early man that comes along."

With that introduction, it may be obvious that someone must have thought that our next site, Mineral Hill Cave, also provided evidence for human interactions with now-extinct mammals. In fact, this site was introduced to paleontologists with exactly the opposite argument.

Deep, dark, and multichambered, Mineral Hill Cave is located in northeastern Nevada's Sulphur Spring Range (figs. 5.1 and 5.14). It has been known since the nineteenth century, as vouched for by the name "Daniel Gayles" having been written on the walls of the cave in 1856.[163] It did not become known to paleontologists and archaeologists, however, until 1980, when Kelly

McGuire published the results of small excavations that had been conducted there in 1975.[164] McGuire reported that the site seemed to lack stratified deposits but had provided him with the remains of shrub ox (*Euceratherium*), large-headed llama (*Hemiauchenia*), and horse, along with those of a number of other large mammals that still exist in the region. Some of the bones of these large mammals had been split in such a way that they resembled bones that we know to have been split by people. In addition, he found bits of charcoal dispersed throughout the excavated sediments.

Charcoal, of course, is another possible indicator of a past human presence (recall Harrington's fireplaces at Gypsum Cave). What McGuire did not find, however, was any direct evidence that people had occupied the site in ancient times—not a single artifact, no blackened cave roof suggesting that a fire had been built in the cave, no burned bones to suggest that meals had been prepared there. McGuire concluded that the simple presence of charcoal in a site, or of bones that looked as if they could have been broken by people, did not mean that a site was archaeological.

Archaeologists Ruth Gruhn and Alan Bryan quickly protested. They believed that McGuire had not shown that the split bones had not been broken by people (they were correct), and that he had not shown that the charcoal had not resulted from human-built fires (they were correct again). McGuire responded; a few years later, archaeologist R. Lee Lyman echoed the points that Gruhn and Bryan had made. Clearly, if Gruhn and Bryan were correct, Mineral Hill Cave might well be an archaeological site, showing that people had processed the bones of extinct late Ice Age mammals.[165]

There the matter stood until paleobiologist Bryan Hockett decided to examine the cave in more detail.[166] Beginning in 1997 and continuing into the year 2000, he and his coworkers carefully excavated a series of units within the cave and revisited the materials that McGuire had retrieved. By doing so, they were able to show that the specimen McGuire had identified as shrub ox actually came from a bison, thus eliminating the shrub ox from the Mineral Hill Cave faunal list.[167] By the time Hockett and his colleagues were done, they had identified 64 genera, and 74 species, of fish, reptiles, birds, and mammals from the site. These included six genera of extinct late Ice Age mammals and, probably, one species of extinct bird, the Fragile Eagle (*Buteogallus fragilis*; see chapter 3 and table 5.8). What they did not find was any indication of a past human presence—not a single basketry fragment, not a single stone tool, not a single struck flake, not a single fireplace.

In the course of doing all this, Hockett's team confirmed McGuire's observation that the Mineral Hill Cave deposits, at least in the areas that had been

TABLE 5.8. The extinct birds and mammals from Mineral Hill Cave

Scientific name	Common name
Birds	
cf. *Buteogallus fragilis*	Fragile Eagle
Mammals	
Brachyprotoma obtusata	Short-faced skunk
Camelops hesternus	Yesterday's camel
Equus cf. *conversidens*	Mexican horse
Equus cf. *occidentalis*	Western horse
Equus sp.	Extinct horse
Hemiauchenia macrocephala	Large-headed llama
cf. *Miracinonyx trumani*	American cheetah
Navahoceros fricki	Mountain deer

Reference: Hockett and Dillingham 2004.

excavated, lacked any vestige of the stratigraphic layering that well-behaved sites so helpfully provide. Not only was such layering invisible in the walls of the units they excavated, but their extensive radiocarbon dating program showed that objects only a few hundred years old lay next to ones that were deposited tens of thousands of years ago. Eons of generations of burrowing rodents, along with other stratigraphy-destroying processes, had done their job well.

Hockett's team also showed how charcoal could get into Mineral Hill Cave without human involvement. In 2001, Hockett noted, the Sulphur Spring Range was the site of a massive forest fire. That, in turn, set woodrat (*Neotoma*) nests near the entrance of the cave on fire (see chapter 2). This probably made some woodrats very unhappy, but he didn't mention this, perhaps because that wasn't the point he was trying to make. When the rains came, charcoal and other material from these burned middens were flushed into the cave. Sometime in the future, these, too, will be incorporated into the cave's deposits. Hockett recognized that people carrying torches into the cave could also account for the charcoal flecked through the deposits of the site. That mechanism, however, is not needed to account for those flecks.

And how about the split bones that Gruhn and Bryan suggested might represent human handiwork? Hockett and his colleagues found lots of those as well. A skilled analyst of such things, Hockett saw nothing in them to suggest that they had been manipulated by people.[168]

Lacking any vestige of stratigraphic layering in Mineral Hill Cave, Hockett's team was faced with a dilemma. Given that the site was not stratified, and that charcoal dated to very recent times was found next to bone that dated to well into the Ice Age, how were they to figure out how old any of this stuff was?

TABLE 5.9. The extinct birds and mammals from Crystal Ball Cave

Scientific name	Common name
Birds	
cf. *Teratornis merriami*	Merriam's Teratorn
Bubo cf. *sinclairi**	Sinclair's Owl
Mammals	
Brachyprotoma obtusata	Short-faced skunk
Smilodon cf. *fatalis*	Sabertooth cat
Equus cf. *scotti*	Scott's horse
Equus (*conversidens?*)	Mexican horse
Camelops cf. *hesternus*	Yesterday's camel
Hemiauchenia cf. *macrocephala*	Large-headed llama

References: Heaton 1985, 1990; Emslie and Heaton 1987.
*Sinclair's Owl is probably just a larger version of the Barn Owl, *Bubo virginianus*, and so not extinct at all (Howard 1947; Emslie and Heaton 1987).

This is an unfortunately common situation. Paleontologist Tim Heaton had the same problem at Crystal Ball Cave, located east of the Snake Range, just barely on the Utah side of the Nevada-Utah border. Paleontologist Wade Miller had worked there in the 1970s, finding the remains of a broad variety of animals, including extinct horse, yesterday's camel, and large-headed llama. Heaton and Miller returned to the site in the early 1980s, and Heaton soon put the site on the Great Basin paleontological map. His excavations provided an extraordinary array of animals, from small to large, including five genera of extinct Ice Age mammals and two specimens of Merriam's Teratorn (table 5.9). They also yielded the remains of a Crested Caracara (*Caracara cheriway*). A member of the falcon family, this bird today comes no farther north than the far southern edge of the United States. During the late Ice Age, it had a far broader distribution, including, Crystal Ball Cave tells us, the Great Basin.[169]

How old was this material? Heaton described the problem well: "One problem with the Crystal Ball Cave assemblage is that it is impossible to separate fossil bones from Recent [Holocene] bones using superposition, because the sediments in which they are found are shallow and unstratified."[170] As a result, Heaton turned to radiocarbon dating, getting four dates on bones. Two on unidentified large mammal bones fell at about 13,000 and 18,600 years ago. One on a horse bone clocked in at 18,800 years ago; a second horse bone was older than 23,000 years ago.

From our current perspective, these dates don't help much, if at all. As I discussed in chapter 4, conventional radiocarbon dates on whole bone are no longer trusted because the material being dated can be so readily contaminated. Although the extinct vertebrates from this site clearly date to the late

TABLE 5.10. The radiocarbon-dated mammals from Mineral Hill Cave

Scientific name	Common name	Date
Equus sp.	Extinct horse	31,130 ± 200
Equus cf. *conversidens*	Mexican horse	35,080 ± 280
Hemiauchenia macrocephala	Large-headed llama	36,320 ± 320
Navahoceros fricki	Mountain deer	37,750 ± 440
Hemiauchenia macrocephala	Large-headed llama	39,230 ± 1,330
Equus cf. *occidentalis*	Western horse	42,420 ± 820
Camelops hesternus	Yesterday's camel	44,600 ± 3,000
Equus cf. *occidentalis*	Western horse	45,700 ± 1,000
Camelops hesternus	Yesterday's camel	46,550 ± 1,100
Camelops hesternus	Yesterday's camel	48,900 ± 3,100
Navahoceros fricki	Mountain deer	49,800 ± 1,700
Hemiauchenia macrocephala	Large-headed llama	50,190 ± 1,940
Equus cf. *conversidens*	Mexican horse	> 46,400
cf. *Miracinonyx trumani*	American cheetah	> 52,200

Reference: Hockett and Dillingham 2004.

Ice Age, we have no trustworthy indication as to exactly how old those animals are.

By the time Hockett did his work at Mineral Hill Cave, far more precise ways of getting radiocarbon dates had become available (chapter 4). Hockett managed to get 55 of them, making this the best-dated paleontological site in the Great Basin. The results for the extinct mammals are shown in table 5.10. As Hockett pointed out, the full set of dates range from very recent times to over 50,000 years ago, but of those that fall within the Pleistocene, most lie between 30,000 and more than 50,000 years ago. To know more about the ages of all the extinct animals, however, each specimen would have to be dated separately. There is no chance of that happening in the near future. It is, in fact, amazing that Hockett was able to get as many dates as he did.

At 30,000 to 50,000 years ago, we are pushing close to, or into, Neandertal times. Hardly anyone knowledgeable about the American past thinks there were people here then. McGuire, it turns out, got it right.

THE HUNTINGTON MAMMOTH SITE, UTAH

Not all paleontological sites in the Great Basin have complex collection histories, or lack stratigraphy. Not all were dug at a time when excavation and analytical techniques were not up to the task of deciphering the stratigraphy they did have and collecting the wide variety of scientifically important specimens they contained. Some, in fact, are quite simple and were excavated splendidly. The Huntington Mammoth site is one of those.

This site was discovered the way so many important sites are, through construction activity. In this case, that activity was to repair an earthen dam at Huntington Reservoir on the Wasatch Plateau of central Utah. This work exposed the bones of a huge animal in the deposits of an old lake. The Utah Division of State History was called in and emergency excavations begun immediately, under the direction of archaeologist David Madsen and paleontologist David Gillette.[171]

Mammoths are cool wherever you find them, but the Huntington mammoth is particularly cool because the elevation of the site—8,990 feet—makes this one of the highest-elevation mammoths known from the Americas (the record may be set by the mammoth remains at 9,639 feet in the Colorado Rockies).[172]

By the time Madsen and Gillette were done, they had recovered the remains of two extinct mammals: the giant bear (*Arctodus*) and the Columbian mammoth (*Mammuthus columbi*). The bear was represented by a partial skull and a rib. The mammoth, though, was represented by a nearly complete skeleton (fig. 5.15).

The mammoth was carefully excavated, but the *Arctodus* bones were not. Amazingly enough, those specimens were ripped from the site by the night guard, whose job included protecting the place from people who might want to rip bones from the site at night.[173] They were, at least, returned to the US Forest Service, and Madsen and Gillette were able to show that the sediments adhering to the bones were the same as those in which the mammoth had been embedded. All chances of determining the relationship of the two sets of bones in the ground, however, were completely lost.

Careful excavation of the skeleton revealed that the mammoth bones lay in the ground in almost perfect anatomical position, except that the ribs were not quite in the right place and part of the vertebral column had been moved. Analysis of the skeleton showed that the animal had lived for about 55 to 60 years, had heavily worn teeth, and had likely died of some combination of old age and poor nutrition. As befits the animal's advanced age, he had arthritis in his vertebrae and pathologies in other parts of his skeleton. I say "he" had arthritis because measurements of the animal's tusks, and the shape of its pelvis, showed that it was, in fact, a he.

The animal spent its last moments next to the open water of a small, shallow, marsh-edged lake. Identification of the bits and pieces of plant fragments from the lake's deposits showed that at the time the animal died, the lake

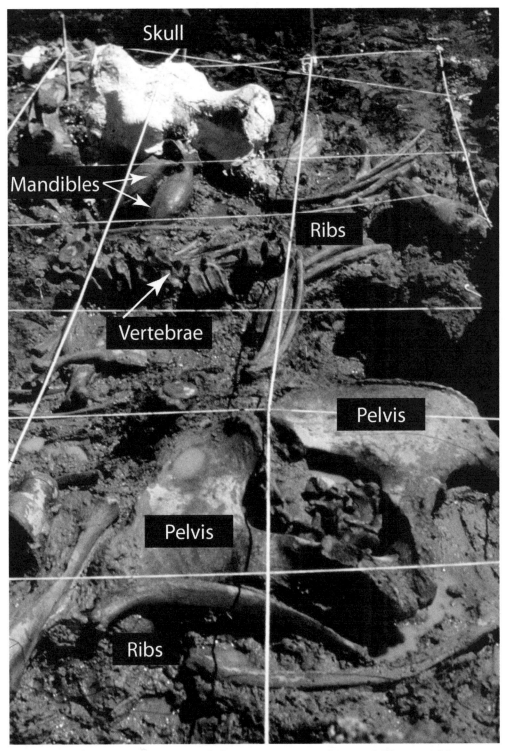

FIGURE 5.15. The Huntington mammoth during excavation and removal. Photograph by David B. Madsen.

was surrounded by sedges (*Carex*) and Engelmann spruce. Pollen from the same sediment added sagebrush to the list. The same set of plants can be seen here today.

The site also contained 11 small pieces of what appeared to be chewed-up plant material that were found amid the rib cage of the animal and that represent the remains of the mammoth's last meal, or something close to that. That material contained sedge leaves or grass leaves, or both, along with subalpine fir needles and a single sedge seed. The leaves were sufficiently mature to suggest that this meal had been eaten sometime between midsummer and late fall. Pollen from these intestinal remains came from plants similar to those identified from the plant macrofossils found in the sediments that enveloped the mammoth skeleton, but it also came from plants that today grow only at lower elevations—oak (*Quercus*), juniper (*Juniperus*), and boxelder (*Acer negundo*). Oaks and boxelder tend to produce their pollen in the spring, which doesn't match the midsummer to late fall estimate for the mammoth's last meal, but there is no reason to think that pollen in dung must coincide with the season in which that pollen was produced, or even that the animals must have been browsing on the pollen-producing plants themselves. Pollen often disperses widely across the landscape (chapter 2) and can be ingested simply because it comes to rest on something an animal eats.

Since the mammoth's intestinal contents contained subalpine fir but the deposits in which the mammoth was found contained only spruce, it is pretty clear that this animal had been eating someplace else before it came here and met its lakeside end. We know when that end came, since there is a radiocarbon date from the mammoth itself: 11,200 ± 110 years ago.

After it died, it appears that at least one other animal dined on it. There are gouges and cuts on two mammoth bones that match the canines of *Arctodus* almost perfectly, suggesting that a giant bear scavenged the carcass shortly after the animal succumbed. If this suggestion is correct, it is similar to what happened to a mammoth from the Saltville Valley of Virginia, which may also have been scavenged by *Arctodus*.[174] Even if the Huntington mammoth had been scavenged by a giant bear, that bear was probably not the same one whose bones were found at the site. That *Arctodus* has been directly dated to 10,870 ± 75 years old, much younger than the mammoth itself. This is the youngest trustworthy date available for the giant bear (chapter 4).

It is also possible, though perhaps not likely, that people scavenged this mammoth. Some of the animal's bones contain marks that might have been produced by people, but that also might have been produced in other ways. More importantly, the same fox-in-the-henhouse guard who looted the site for its *Arctodus* remains also pulled an artifact from the same deposits. That artifact appears to be a projectile point and, as with the bear bones, has the same sediments adhering to it as surrounded the mammoth bones. Because the artifact was illegally removed from its context, we have no way of knowing its relationship to the mammoth skeleton. However, the general form of the point is most consistent with artifacts that have been dated elsewhere to younger than 10,000 years ago, suggesting that the two probably had nothing to do with each other. David Madsen, who excavated the site, properly considers the evidence for human involvement with the Huntington mammoth to be "equivocal."[175]

Even the insects from the Huntington Mammoth Site have been studied. Paleoentomologist Scott Elias analyzed about 175 pounds of sediment that were taken from inside the mammoth's skull and from the deposits surrounding it. He was able to identify 36 species of insects, including 27 species of beetles, 4 of caddisflies, and 5 of ants. He also discovered that one of those species—the beetle *Bembidion gordoni*—is locally extinct, occurring today only in the mountains of the Pacific Northwest. As with the plants, the insects suggest that at about the time the mammoth died, the local environment was probably similar to what it is now, except, perhaps, for having colder winters. Unfortunately, since the full set of radiocarbon dates for the depositional unit that provided these insects spans nearly 3,000 years, it is hard to know, without dating each insect individually, exactly what to make of these conclusions.[176]

That's not all that has been studied from the Huntington mammoth skeleton. The animal was so exceptionally well preserved[177] that it provided the DNA that I discussed in chapter 3, the DNA that suggests that woolly and Columbian mammoths may have interbred. This was the mammoth whose skeleton says "Columbian" but whose DNA says "woolly."

Thanks to the exacting work of Madsen and Gillette, we know that about 11,200 years ago, a male mammoth wandered to the shore of a spruce-draped lake at 8,990 feet on Utah's Wasatch Plateau. Not long before, it had dined on subalpine fir, sedges, and perhaps grasses. Old, with worn teeth, here it died. Sometime soon after, it may have provided a meal for a giant bear that ate its meat and gouged its bones. Sometime after that, it became covered with lake sediments, not to be discovered until 1988. Today, anything else that might have been with it is buried again, a result of the repair work that led to its discovery in the first place.

DIGGING PALEONTOLOGY

Today, Harrington's descriptions of his excavations at Gypsum Cave or Smith Creek Cave, and other places not discussed here, make paleontologists and archaeologists cringe. Obviously, these sites are not renewable resources. Once dug, all that remains of them are the objects and the associated information that was gathered during the excavation process. We cringe because the excavation techniques that were used were, from a modern standpoint, akin to a dentist using a sledgehammer.

Archaeologist Don Tuohy once referred to the work conducted by George and Willis Evans at Smith Creek Cave in 1925 as "State subsidized pothunting."[178] He said "State" because this work was done at the request of Nevada governor James Scrugham. He said "pothunting" because the work was focused entirely on the discovery of interesting objects, with little attention paid to the stratigraphy of the site, and to the relationships of the objects to each other and to the deposits in which they were embedded.

From a modern perspective, Harrington's work at Smith Creek Cave wasn't all that much better. He dug, and he retrieved some cool things, like Hildegarde Howard's Incredible Teratorn and Chester Stock's Harrington's mountain goat. Today, if we want to know whether the bones of the extinct birds from this site came from the red silt zone, the best way to tell is to see whether they are stained red.

At Gypsum Cave, Harrington at least defined general stratigraphic relationships (see figs. 5.6, 5.7, and 5.9 through 5.11) and noted the relationship between the artifacts he found and his broadly defined strata, focusing primarily on whether those artifacts were above, within, or beneath sloth dung. In contrast, Amy Gilreath and her colleagues defined 14 separate depositional units in Gypsum Cave's Room 1 alone and described those strata in great detail. Harrington did not process the deposits he excavated through screens to retrieve small items; Gilreath's team passed the deposits they excavated through ⅛-inch screens. Harrington lost—though "destroyed" would be an acceptable, if harsh, term—a significant amount of important information by excavating in this fashion.

But it would be grotesquely unfair to Harrington to criticize him for this. What he was doing was standard for the time. What would be inexcusably crude today was, at the time, state of the art.

Forty years later, Alan Bryan's work at Smith Creek Cave was also state of the art. He excavated carefully, paid close attention to stratigraphy, and screened the deposits he excavated. The lengthy report he published on his work there is required reading for anyone interested in Great Basin Pleisto-cene paleontology and archaeology. Nonetheless, from a modern perspective, his excavation techniques were also crude.

For instance, Bryan used screens with a ¼-inch mesh. As a result, he, too, missed small objects of large importance. We know this is true because paleontologists interested in the small mammals that were once contained within the Smith Creek Cave deposits have gone back to the site, collected samples of the screened deposits Bryan left behind (the "backdirt"), care-fully sorted through them, and discovered that Bryan's backdirt is rich in the remains of such mammals.[179] In fact, they found the remains of a heather vole (*Phenacomys intermedius*) and a northern bog lemming (*Synaptomys borealis*) in that backdirt. These were important discoveries, since, while neither animal is extinct, neither lives in the Great Basin today.

For these reasons, contemporary excavations in sites like Smith Creek Cave and Gypsum Cave pay exacting attention to stratigraphy, and excava-tion teams often include a specialist to help unravel that stratigraphy. Some scientists use the chemical constituents of the deposits of their sites to help guide their excavations, using portable high-tech devices to monitor that chemistry as they move along.[180] Sediment samples are routinely taken from throughout the deposits and then analyzed for the pollen, plant macrofossils, and other items they might contain. Fireplaces or hearths are stratigraphi-cally excavated as if they were tiny sites, and their contents are often water-processed ("floated") for minute bits of plant material that might be hidden within. Rather than use ¼-inch screens to catch objects that might have been missed during the digging process, the standard approach in sites like these is to use ⅛-inch screens, as Gilreath's team did in their recent work at Gypsum Cave, or even ¹⁄₁₆-inch screens.[181] Those fine screens provide things like tiny bat and shrew teeth, which are missed as excavators are troweling away.

By the time Bryan excavated Smith Creek Cave, the excavation techniques Harrington used there and elsewhere were recognized to be ruinously crude, even though they were state of the art at the time. Forty years had passed in the interim. Today, we recognize that Bryan's techniques at Smith Creek re-sulted in the loss of highly important information. Forty years from now, even the most exacting excavation procedures in use today may look equally crude. Given that professional excavation destroys sites, and unauthorized digging by the untrained destroys even more of them, we can only hope that there will still be sites like Gypsum and Smith Creek Caves left to which scientists can apply whatever techniques are state of the art in the middle of the twenty-first century and beyond.

Extinct Mammals, Dangerous Plants, and the Early Peoples of the Great Basin

THE LATE PLEISTOCENE PEOPLES OF THE GREAT BASIN: GREAT BASIN FLUTED POINTS

While mammoths and sabertooth cats are the icons of the late Ice Age fauna of North America, the Clovis projectile point is the icon of the late Ice Age peoples of North America.

Clovis points, and the artifacts found with them, date from about 11,500 to 10,800 years ago.[1] These are not the earliest securely dated human remains known from North America, but they are a far better choice for an iconic emblem of early Americans than are the oldest dated remains.

That is because those oldest remains aren't artifacts at all. Instead, they are human coprolites, desiccated human fecal material that contains human DNA that could have come only from Native Americans. These were excavated from Paisley Cave 5 in south-central Oregon, on the northwestern edge of the Great Basin (see figs. 5.1 and 6.1), and have been dated to as old as 12,400 years ago.[2] This makes Paisley Cave 5 one of the two oldest securely dated archaeological sites in the Americas. The other, the Monte Verde site in far southern Chile, dates to 12,500 years ago and has provided a wide array of archaeological material.[3] The Paisley Caves coprolites, on the other hand, come securely associated only with each other, but one could not hope for a better demonstration of human antiquity in the Americas than the high-precision radiocarbon dating of material that came from the people themselves.

Many other archaeological sites from the Americas have been claimed to be older than Clovis, but none meet the rigid criteria needed to sustain such a claim.[4] Between the Paisley coprolites and the earliest Clovis sites, there is nothing convincing as of yet.

FIGURE 6.1. The setting of Paisley Cave 5 (the arrow marks the spot).

If there is to be an iconic late Ice Age North American symbol, Clovis points are better than a collection of dried-up human fecal material (I am being cautious with my wording here). It helps that Clovis points are beautifully made, high-precision tools that fill the role perfectly (fig. 6.2).

Clovis points are generally 1 to 4 inches long, but far larger ones are known. It is not their size, however, that makes them distinctive, or the care with which they were manufactured. Instead, it is their form. All Clovis points are lanceolate, with slightly concave bases, and nearly all have flutes that begin at the base and continue less than halfway toward the tip, produced by the removal of a series of thin flakes from one or both sides. There is little doubt that these points were fluted to help fit them directly into a shaft, or into a foreshaft that then fit into a spear. The base of the point that was so hafted was fixed in place by sinew. We know that, or at least we think we know that, because the bases of Clovis points are dulled by grinding them. The obvious reason for doing that is to prevent the sinew that held the point in place from being cut by the point's sharp edges. Clovis points come associated with a wide variety of other artifacts made from stone, bone, and ivory, but it is the distinctive Clovis point that generally allows a Clovis artifact assemblage to be identified.

These points are so very distinctive that archaeologists have spent a lot of time wondering where they originated. There have been four very different answers to this question.

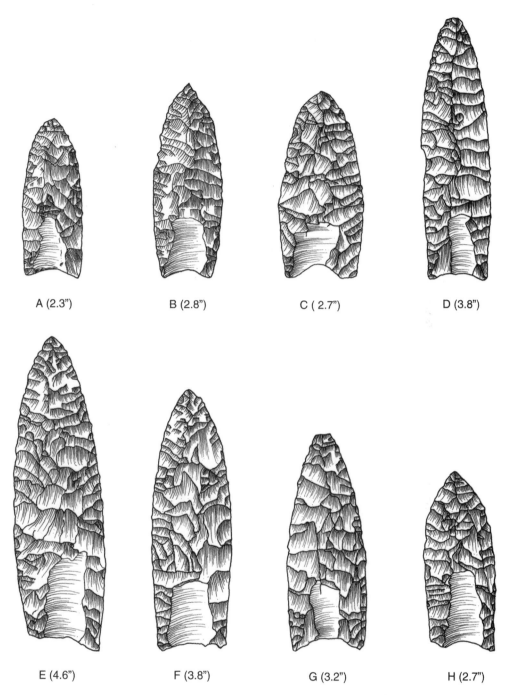

A (2.3")　　　　B (2.8")　　　　C (2.7")　　　　D (3.8")

E (4.6")　　　　F (3.8")　　　　G (3.2")　　　　H (2.7")

FIGURE 6.2. The Clovis points from the Naco site, southeastern Arizona (after Haury 1953). Specimen lengths are provided. Drawings by Peggy L. Corson.

Since it has long been assumed, for good reasons, that the earliest Americans entered the continent from Siberia by crossing over the Bering Land Bridge (chapter 2), Siberia is an obvious place to look for the technology that gave rise to Clovis. However, even though Siberia has provided bone artifacts that are very similar to forms known from Clovis sites,[5] that region has provided no fluted points, and no stone tools that suggest they might have given rise to Clovis points.

If not in Siberia, then perhaps the distinctive Clovis stone tool technology came into being on the eastern side of the Bering Land Bridge. In fact, fluted points are known from both Alaska and British Columbia, but all cases that have been dated are younger than Clovis sites to the south.[6] The idea of fluting a point seems to have moved not from north to south, but from south to north.

Then there is the odd notion, revived from time to time, that the ancestors of Clovis peoples came not from Siberia, but from Europe. In the current version of this view, Clovis is alleged to have descended from the Solutrean peoples of western Europe. The Solutrean culture existed from about 20,500 to 17,000 years ago and is marked by artifacts that are, in some ways, superficially similar to those made by Clovis peoples, though fluting was unknown to the Solutreans. Solutrean peoples, this argument maintains, boated their way west by skirting ice in the North Atlantic and living on the bounty of the sea along the way. Once in North America, they survived to give rise, thousands of years later, to Clovis.[7]

There are many archaeological reasons to reject this notion,[8] but they need not be recounted here, because ancient DNA shows unequivocally that Clovis people came from Siberia. For proof, we need look no further than the genetic material recovered from western Montana's Anzick burial site. This site provided a large series of Clovis artifacts that had been covered by red ocher and were apparently meant to accompany the male child who had been buried with them.[9] Not only are the artifacts Clovis in nature, but so is the date obtained from the remains of the child: 10,705 ± 35 years ago. With the approval of local Native people, ancient DNA was successfully extracted from those remains. The results were clear: the Anzick child was thoroughly Native American and belonged to a group that was ancestral to modern Native Americans. The closest genetic ties of this young person? They are to the late Ice Age peoples of Siberia.[10] Fringe ideas aside, the archaeology and the genetics fit perfectly.

There are no fluted points known from Siberia, those from Alaska and adjacent Canada are younger than Clovis fluted points, and genetic data show that the ancestors of Clovis came across the Bering Land Bridge from Siberia. Where, then, did Clovis points themselves come from?

The answer comes from subtraction. If they didn't come from Siberia and they didn't come from northern North America, they must have developed from a stone-tool technology south of glacial ice in North America. What that technology was we still do not know. The only securely dated pre-Clovis site in North America is Paisley. While the dated coprolites from Paisley tell us that people were there by 12,400 years ago, there are no stone tools convincingly associated with those coprolites. At least, though, we know that the ancestors of Clovis people came from Siberia, as was long surmised, and that it is highly likely that the Clovis culture developed in place, south of glacial ice.

Since, when securely dated, Clovis assemblages always fall within a few centuries of 11,000 years ago, archaeologists reasonably assume that an un-dated Clovis point or Clovis assemblage most likely falls within that same time interval. When diagnostic artifacts have been found associated with the remains of extinct mammals in North America (fig. 3.62), those artifacts have been Clovis points. The Naco site in southeastern Arizona is a great example, since this site provided Clovis points tightly associated with the remains of a single mammoth (fig. 6.2), and a series of nearby sites mirrors the Naco discovery.[11] We know these people were capable of killing elephant-sized beasts, and that they were doing it around 11,000 years ago.

Fluted points are known from the Great Basin in some number (fig. 6.3). There has been a strong tendency to call these "Clovis points," but that may be a serious mistake because it immediately leads to the assumption that they are the same age as Clovis points that have been dated elsewhere. In fact, detailed analyses of Great Basin fluted points by archaeologists Charlotte Beck and Tom Jones have shown that most, though not all, differ significantly from the classic Clovis points of the American Plains and Southwest. As a result, here I will simply refer to these early Great Basin artifacts as Great Basin fluted points, as I have long done. Doing that makes no assumptions about how old these things are.[12]

This is important because we simply don't know when Great Basin fluted points were made. While a small handful of fluted points have been found in buried settings in the Great Basin, none of these have been found in a secure stratigraphic context that has allowed them to be convincingly dated. Instead, nearly all Great Basin fluted points have been surface finds. Most of these have been in valley settings, which, during the latest Ice Age and the few thousand years that followed, would have been adjacent to now-extinct lakes and marshes (see chapter 2). Higher-elevation fluted points have been found

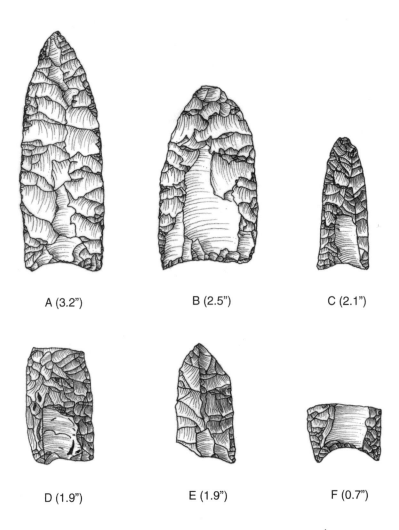

A (3.2")   B (2.5")   C (2.1")

D (1.9")   E (1.9")   F (0.7")

FIGURE 6.3. Great Basin fluted points. *A*, *B*, Alkali Lake, Oregon (after Willig 1988); *C–E*, Fort Irwin, Mojave Desert, California (after Warren and Phagan 1988); *F*, Long Valley, Nevada (after Beck and Jones 2009). Specimen lengths are provided. Drawings by Peggy L. Corson.

in the Great Basin, with the current record set by two fluted points found by Dave Rhode and his colleagues in the Pine Grove Hills of western Nevada at an elevation of 7,810 feet.[13] These, however, are extremely rare.

In other parts of North America, fluting goes out of business by about 10,200 years ago, with the demise of the cultural phenomenon known as Folsom, which is well known from sites in the American Plains.[14] Because the earliest-known fluted points date to about 11,500 years ago, and the latest to about 10,200 years ago, Great Basin fluted points almost certainly fall within this interval, but where they fall within it is simply unknown.

If all this is correct, then there is a very good chance that fluted point makers encountered at least some of the now-extinct late Pleistocene mammals of the Great Basin. As we saw in chapter 4, 7 of the 20 extinct genera of the Great Basin's late Ice Age mammals can be shown to have lasted to between 12,000 and 10,000 years ago (table 5.3), including the Columbian mammoth, American horse, and yesterday's camel. The odds are very good that people saw some or all of these animals on the hoof. Given that we know that the makers of fluted points in other parts of North America had the ability to kill very large mammals—mammoths and mastodons, for instance, for Clovis people, and bison for Folsom folk—perhaps the fluted point makers of the Great Basin were able to, and did, do the same.

No site in the Great Basin suggests that these people were hunting such animals. However, even if people began every day by hunting now-extinct large mammals, we would have no way of knowing it. Virtually all of the Great Basin fluted point sites have been found in surface settings that do not provide the kind of preservation needed to show us what these people were doing for a living. They could have been using their fluted points to hunt mammoths, to trip rabbits as they ran by, or to flip horny toads on their backs, and we would not know. (I know horny toads are actually horned lizards, genus *Phrynosoma*, but who calls them that?)

If fluted points were all there were in the Great Basin during the late Pleistocene, then Great Basin archaeologists interested in this period would have nothing to brag about other than the Paisley Caves. There is, however, something else that may be telling us important things about the early peopling of the Americas in general.

THE LATE PLEISTOCENE PEOPLES OF THE GREAT BASIN: GREAT BASIN STEMMED POINTS

The Late Pleistocene and early Holocene archaeological record of the Great Basin is marked by a variety of "projectile point" styles generally referred to as "Great Basin stemmed" points (figs. 6.4 and 6.5). I put "projectile point" in quotation marks because it is clear that these objects performed a variety of functions—cutting, sawing, chiseling, and serving as the tip to a weapon all seem to have been within their functional realm.[15] One thing they all have in common is fairly thick stems—the bottom part of the point—which usually slope gently inward to meet a rounded to square base; a shoulder separating the stem from the rest of the point (fig. 6.5D) may or may not be present. Just as with fluted points, the edges of the bottom part of these points are dulled

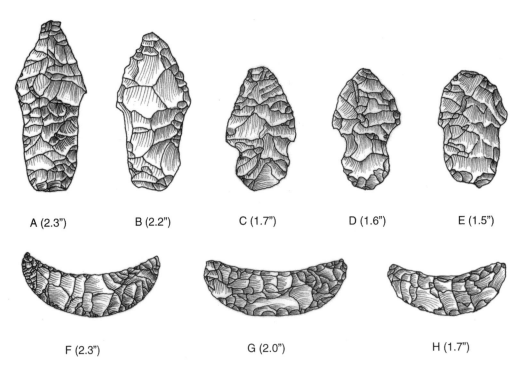

A (2.3")   B (2.2")   C (1.7")   D (1.6")   E (1.5")

F (2.3")   G (2.0")   H (1.7")

FIGURE 6.4. Great Basin stemmed points and crescents. *A, B,* Lake Mohave points; *C–E,* Silver Lake points; *F–H,* crescents (*F* after Willig 1988; all others after Beck and Jones 2009). Specimen lengths are provided (from tip to tip for the crescents). Drawings by Peggy L. Corson.

by grinding, again probably to prevent damage to the sinew that held them firmly to a shaft.

Stemmed point sites routinely have a broad variety of artifacts in addition to the points. The most distinctive are the crescents (fig. 6.4). These are generally from 1 to 2.5 inches long and about 0.5 inch wide, with the front and back edges of the midsections routinely dulled by grinding. Like the grinding on the dulled bases of fluted and stemmed points, the grinding on crescents suggests that they were hafted. That hafting provides one hint about the possible function of these artifacts; another is provided by the fact that when crescents are damaged, the damage is to their tips. Archaeologists Madonna Moss and Jon Erlandson have recently argued that crescents were used to hunt molting ducks and geese, but the truth is that we have no idea what these things were used for.[16] Figure that out and you will be applauded by every archaeologist interested in the early human history of North America.

Stemmed point sites at times contain unfluted lanceolate points of the sort illustrated in figure 6.5. These look a lot like fluted points that happen not to be fluted, but Charlotte Beck and Tom Jones have shown that, at least in

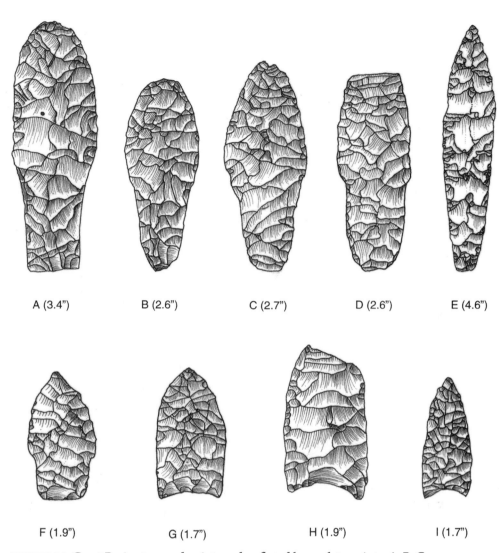

A (3.4") B (2.6") C (2.7") D (2.6") E (4.6")

F (1.9") G (1.7") H (1.9") I (1.7")

FIGURE 6.5. Great Basin stemmed points and unfluted lanceolate points. *A, B,* Cougar Mountain points; *C, D,* Parman points; *E,* Haskett point; *F,* Windust point; *G–I,* unfluted lanceolate points (*E* after Butler 1978; all others after Beck and Jones 2009). Specimen lengths are provided. Drawings by Peggy L. Corson.

eastern Nevada, these objects have widths, lengths, and thicknesses similar to those of fluted points but are flaked in a very different way. These represent a very different projectile point style, rather than being unfluted fluted points.

Most stemmed point sites have been discovered because they were lying on the surface of the ground, rather than having been buried. This, of course, is very similar to the situation with fluted point sites. The similarities are even greater, however, since fluted and stemmed points can be found in the very same sites. This is probably because the makers of both happened to live in the same kinds of places at different times—valley bottoms and valley margins adjacent to what were then shallow lakes and marshes.

However, stemmed point sites are known from a far greater range of environments than are fluted point sites, and a significant number have been found in buried contexts, including Smith Creek Cave (chapter 5). As a result, and unlike the situation for Great Basin fluted point sites, we have good dates for them. We know that stemmed points were in use as late as 8,500 years ago[17] and probably later, and as early as about 11,100 years ago and probably earlier. In fact, Paisley Cave 5, the same site that provided the noniconic 12,400-year-old human coprolite, also provided the base of a stemmed point that was found beneath material that dated to 11,070 years ago, and above material dated to 11,340 years ago. That means that the earliest stemmed points we know about were as early as true Clovis points from elsewhere in North America.[18]

Earlier in this chapter, I guessed that the makers of Great Basin fluted points were around at the same time as at least seven genera of now-extinct late Ice Age mammals. We don't have to make that guess for the makers of stemmed points in the Great Basin. Those people were clearly around when those seven genera of now-extinct mammals still existed (chapter 4).

Both fluted and stemmed point sites are routinely found in Great Basin valley bottoms and valley margins, adjacent to what would then have been shallow lakes and marshes. Archaeologists have long speculated that these are precisely the kinds of places to which large mammals would have come for both water and food.

We no longer need to speculate about the relationship between large mammals and lakeshores, and waterside settings in general, during the late Ice Age. The tracks of mammoth and camel are known from the shores of Pleistocene Lake Otero in southern New Mexico and those of camel, mammoth, and horse from the late Ice Age riverside and near-riverside settings provided by Wally's Beach, southern Alberta.[19] Even the Nevada State Prison footprints discussed in chapter 1 were likely made in a similar setting.

In short, it is now clear that people overlapped at least seven genera of now-extinct large mammals in the Great Basin both in time and in the habitats they used.

Even so, and even though we have buried sites that contain stemmed points dating to the late Ice Age, there is not a single instance of those points, or any recognizable part of the stemmed point tool kit, having been found associated with the remains of an extinct late Pleistocene mammal in such a way as to suggest that people were hunting them or scavenging their remains.

There aren't even any sites that archaeologists argue about that might contain such an association.

The lack of such sites may simply reflect the fact that so much of the Great Basin late Pleistocene archaeological record comes from the surface of the ground and that surface sites of this age are unlikely to provide compelling associations between artifacts and the remains of extinct mammals. Given that buried sites containing artifacts and the remains of extinct mammals are rare in North America as a whole, it is hardly reasonable to think that the small sample of Great Basin buried sites of the right age should have provided an association of this sort by now, if such a thing exists here.

However, the lack of these kinds of sites may also reflect the fact that large Pleistocene mammals were rare on the Great Basin landscape toward the very end of the Ice Age, and that people encountered them far less often than they did elsewhere. The challenge is to figure out whether this was the case.

TRACKING ABUNDANCES

It is fairly easy to figure out the relative abundances of different kinds of ancient animals on the landscape—which animals were most common and which least so. You tally up the number of known occurrences of each kind of organism during particular periods and assume that the results will tell you how common or rare those animals were relative to one another. Figures 4.2 and 4.3 provide examples.

There are obvious problems in doing this. The bones and teeth of a mammoth or mastodon are far more likely to survive than are those of a short-faced skunk or Aztlán rabbit. They are also much more likely to attract attention than are the remains of far smaller creatures. There aren't many people who would get excited by finding bunny bones sticking out of the ground, but there are few who wouldn't if they came across a giant tooth or leg bone. Only the latter kinds of discoveries get reported. In addition, the kinds of habitats animals occupied helped, or hindered, the preservation of their remains and our ability to find them. To make it even harder, the bones and teeth of some animals are much easier to identify than are those of others. Graphs like those shown in figures 4.2 and 4.3 should, as a result, be taken in the most general of senses. In the Great Basin, we can say that mammoths appear to have been more common than flat-headed peccaries, yesterday's camels more common than mountain deer, but that's about all we can say.

In some situations, we can track the relative abundances of single species through time in the deeper past. People provide the best example, because

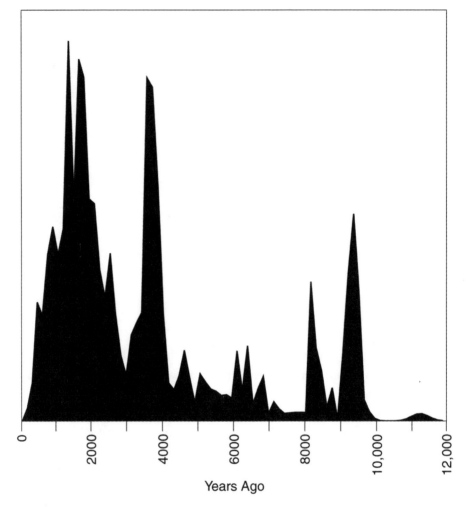

FIGURE 6.6. Changing frequencies of radiocarbon dates from archaeological deposits in the western Lahontan Basin, Nevada. The heights of the peaks indicate the chances that a site dates to that time (after Louderback, Grayson, and Llobera 2011).

archaeologists have expended so much effort in trying to do this. The approach is straightforward. For any given area, count the number of radiocarbon-dated archaeological sites, and of separate deposits in those sites, and analyze them statistically to determine how the number of those dated deposits changes through time. Figure 6.6 shows one such attempt, based on 48 sites and 210 dates from the western basin of Pleistocene Lake Lahontan in Nevada. This graph suggests that people were relatively scarce on the landscape between about 8,000 and 4,000 years ago, reflecting the fact that this was the hottest and driest period in the Lahontan Basin during the last 12,000 years or so.[20]

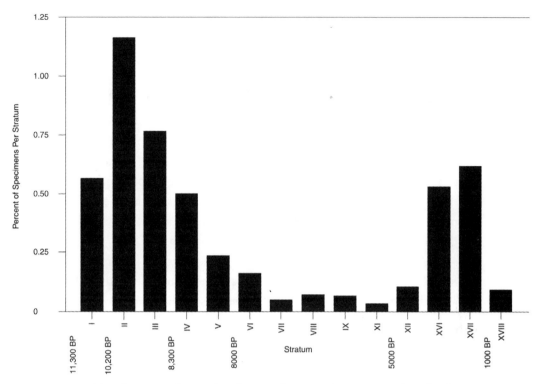

FIGURE 6.7. The changing relative abundances of western harvest mice (*Reithrodontomys mega-lotis*) at Homestead Cave, Bonneville Basin, Utah (after Grayson 2000a, 2000b). The bottom axis shows the separate layers of the site's deposits and the dates of those deposits; the left axis shows the proportion of specimens from each of those layers that came from harvest mice.

We can do the same kind of thing for nonhuman mammals. Figure 6.7 shows over 11,000 years of the changing relative abundances of the western harvest mouse (*Reithrodontomys megalotis*) at Homestead Cave in the basin of Pleistocene Lake Bonneville, just west of Great Salt Lake.[21] The remains of these animals were in the site because they had been picked off by owls who ultimately deposited the remains of their digested meals here. Obviously, harvest mice were most abundant in the Homestead Cave area between about 10,000 and 8,300 years ago. They were least abundant between about 8,000 and 4,000 years ago, after which their numbers climbed again (the sample from Stratum XVII doesn't count because it is so small). Western harvest mice prefer habitats that are relatively cool and moist and, as with people in the Lahontan Basin, they did not thrive in the hot and dry climates that marked this area during the middle of the Holocene.

Even though we can show things like this if our radiocarbon-dated record is strong enough, we still have no idea how many harvest mice were in the Homestead area, or how many people in the western Lahontan Basin, at any particular time. We can learn about the past relative abundances of various organisms this way, but translating that information into numbers of individuals is a different matter.

There have been attempts to make that translation. The most interesting example I know of was provided by Dan Mann and his colleagues. They collected about 4,000 large mammal bones and teeth from the North Slope of Alaska, the area between the northern edge of the Brooks Range and the Arctic Ocean. After identifying what they could—about 1,900 of the specimens they collected—they got radiocarbon dates for 245 of them. Those dates showed that most of their muskox and mammoth bones had been deposited more than about 40,000 years ago, and that nearly all the horse and most of the American lion specimens were younger than this.

Next, they counted the number of late Ice Age caribou specimens they had found and compared that number to the numbers of specimens of the other animals they had identified. Doing this suggested that caribou were 1.8 times more abundant on the late Ice Age landscape than muskoxen and mammoths, 1.4 times less abundant than bison, 2.6 times less abundant than horses, and so on.

To this point, what they had accomplished was a lot of work but was fairly standard. Even people like me have done such things. Their next step, though, was distinctly different. They observed that today's North Slope supports about 6.7 caribou per square mile (it also supports about eight million mosquitoes per square mile, but this is irrelevant unless you happen to be there). They reasoned that if caribou had been at least this abundant during the late Ice Age, the caribou ratios they had calculated might be turned into population estimates for the other animals they had identified. If, for instance, there were at least 6.7 caribou per square mile on the North Slope during the late Ice Age, and bison were 1.4 times more abundant than caribou, there should have been at least 9 bison per square mile on that landscape. There also should have been at least 17 horses per square mile, and so on. Doing this for all their animals led them to conclude that the Ice Age North Slope supported at least 40 or so large mammals per square mile. If this is true, then large mammals were about 6 times more abundant on this landscape during the late Ice Age than they are now.[22]

I know what you are probably thinking: this requires that modern caribou abundances on the landscape are similar to late Ice Age caribou abundances on the same landscape. You are right. You are probably also thinking that we can't assume that. You are right again. Mann and his colleagues have convinced me that, insofar as their sample is representative of the late Ice Age

North Slope, horses were more abundant than caribou here, muskoxen less abundant. Because of the assumptions they had to make, I am not convinced that something like 17 horses per square miles galloped here, or that large mammals were at least six times more abundant in this region than they are today. Mann and his coworkers recognized this problem and suggested that their huge numbers probably applied to only the best of large-mammal times on the late Ice Age North Slope. If caribou weren't as abundant as they assumed, however, all bets on these horses are off.

My point here isn't to disagree with Dan Mann and his colleagues. What they did is insightful, innovative, and important. I wish I had thought of it. My point is that it is really hard to get a handle on the abundance of mammals on any late Ice Age landscape, even those whose paleontology is well known.

The Great Basin isn't one of those places. As I noted in chapter 4, we have only 97 dates for the extinct late Ice Age mammals of the entire Great Basin. Not only are these dates spread across 12 genera, but a good number of them cannot be considered trustworthy. Because of this, we currently have no way of tracking the changing relative abundances of Ice Age mammals through time in the Great Basin. Even if we could, we would still have no idea how many of those animals were on the landscape at any given time.

Perhaps, though, we can use a different approach for the extinct Ice Age mammals of arid western North America. Perhaps the plants of this region know something we don't.

THE GHOSTS OF HERBIVORY: GRASSES, MEGAFAUNAL FRUITS, AND THE MEANING OF THORNS

The Implications of Great Basin Grasses

Over 30 years ago, two insightful botanists, Richard Mack and John Thompson, pointed out that the grasses of the Great Basin are so prone to overgrazing that their abundance declined dramatically as soon as exotic livestock were introduced to this region. They focused their analysis on such species as bluebunch wheatgrass (*Pseudoroegneria spicata*), squirreltail (*Elymus elymoides*), Sandberg bluegrass (*Poa secunda*), and Idaho fescue (*Festuca idahoensis*). To these, they could have added one of the Great Basin's most common grasses, Great Basin wildrye (*Leymus cinereus*).[23]

Mack and Thompson also observed that, in addition to the abundant grazing-intolerant grasses in the Great Basin, this region also lacks the abundant dung beetles (in particular, the genus *Onthophagus*) found in areas that evolved in the presence of significant numbers of large herbivores.[24] From this, they inferred that such large creatures must have been scarce in the Ice Age Great Basin. As they graciously noted, they were not the first to draw this inference from the nature of the vegetation of the Great Basin and immediately adjacent areas.[25] They were, however, the first to do it in such a powerful way. What comes next in this chapter simply follows in their footsteps.

Megafaunal Fruits

Anyone who spends a healthy part of his or her life reading the scientific literature knows that a lot of what gets published is trivial, a problem that has increased as the number of scientific journals has proliferated. Those journals need to fill their pages, and the more pages they need to fill, the more trivia gets published.

On the other hand, those of us who pursue science as a career also know that some deeply insightful, and at times game-changing, contributions have a hard time making it into print. In the scientific realm in which modern plants and Pleistocene mammals crosscut one another, the prime example is a paper written by ecologists Dan Janzen and Paul Martin. Before I tell you what that paper was about, and the implications it has for our understanding of Great Basin Ice Age mammals, I will tell you two things about its youthful history.

First, the paper was submitted to the journal *Science*, one of the most prestigious scientific journals in the world. *Science* rejected it, the editor implying that it sounded too much like a Hollywood script to appear in such a serious journal, according to Paul Martin. Somehow, Janzen convinced the journal to reconsider, and the paper appeared in *Science* in 1982.[26]

Second, the paper that appeared just before the Janzen and Martin contribution was very important for people who like to breathe. It analyzed the "size distribution of fine particles from coal combustion."[27] According to Google Scholar, that paper has been cited 241 times at the time of this writing. That is a lot. The Janzen and Martin paper, however, has been cited 512 times. That is a lot more.

The title of their initially rejected paper is "Neotropical Anachronisms: The Fruits the Gomphotheres Ate" (see chapter 3 to refresh your memory on gomphotheres). In retrospect, their argument seems obvious. That is often the case for me when people smarter than I am point something out that I wish I had been able to think of but didn't.

Janzen and Martin observed that many trees and tall shrubs in the lowland forests of Central America have fruits and seeds with traits that seem poorly matched with any living vertebrate seed dispersers. Among other things, these characteristics included large size, nutrient-rich pulp, seeds that would allow successful passage through the innards of a large herbivore, and fruits that did not open on their own to display or release their seeds (a characteristic known as indehiscence). They referred to this distinctive set of fruit and seed attributes as the "megafaunal dispersal syndrome"[28] and argued that the only possible primary dispersers for these seeds were now-extinct Pleistocene mammals—gomphotheres, for instance. As a result, the fruits seemed anachronistic.

This remarkable idea was at first controversial but has now been supported and refined.[29] There is even direct evidence that large, now-extinct Ice Age mammals distributed the seeds of such plants.[30] Although the introduction of the hypothesis focused on the Neotropics, it has been extended to many parts of the world, including the Sonoran and Mojave Deserts just south of the Great Basin (fig. 2.7).[31]

Obvious candidates for Sonoran and Mojave plants whose seeds may have been distributed primarily by large Pleistocene herbivores include, among others, buffalo gourd (*Cucurbita foetidissima*), fingerleaf gourd (*Cucurbita digitata*), coyote gourd (*Cucurbita palmata*), melon loco (*Apodanthera undulata*), jumping cholla (*Cylindropuntia fulgida*), honey mesquite (*Prosopis glandulosa*), screwbean mesquite (*Prosopis pubescens*), banana yucca, Joshua tree, and Mojave yucca (*Yucca schidigera*).

All these species have large, indehiscent fruits (the gourds, melon, cactus, and yuccas), or, if dehiscent, have substantial fruits that are held above the browsing range of today's large mammalian herbivores, along with a history of range expansion after the introduction of Eurasian livestock (the mesquites).

In no case, either in the arid southwestern United States or in other parts of the Americas, are large Pleistocene herbivores suggested to have been the only dispersers of the seeds of plants that form part of Janzen and Martin's megafaunal dispersal syndrome. After all, as botanist Henry Howe has emphasized, the extinctions occurred 10,000 or more years ago, yet the plants still exist. As Howe noted, these plants must and do have other means of dispersing their seeds.[32] It is now well established that these dispersers are vertebrates, water, and perhaps even wind.[33] These mechanisms, however, seem to be secondary to what was once the primary means of seed dispersal for these plants: the large, now extinct, Pleistocene herbivores.

The Joshua Tree Debate

The Joshua tree (*Yucca brevifolia*) is the best studied of these plants in the American Southwest. Ever since the 1930s, when Laudermilk and Munz published their analyses of Shasta ground sloth dung from Gypsum, Muav, and

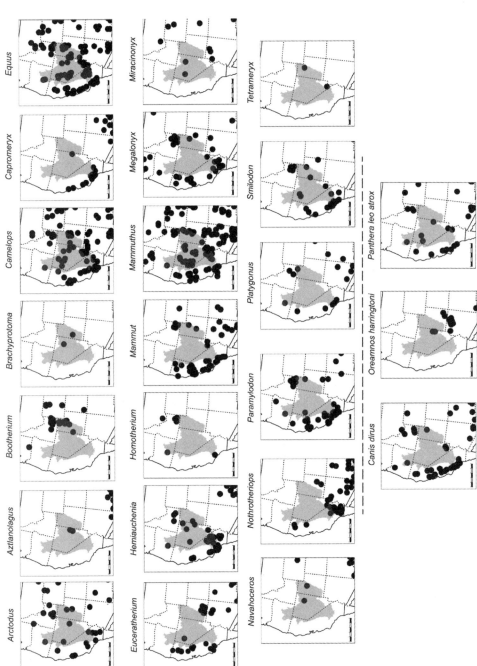

FIGURE 6.8. The distributions of the extinct late Pleistocene mammals of the Great Basin. Those below the dashed line are of extinct species from genera that still exist in North America; the hydrographic Great Basin is shown in gray. Scale bar = 300 miles. For sources, see "Sources for the Great Basin Ice Age Mammal Distribution Maps" in the reference list.

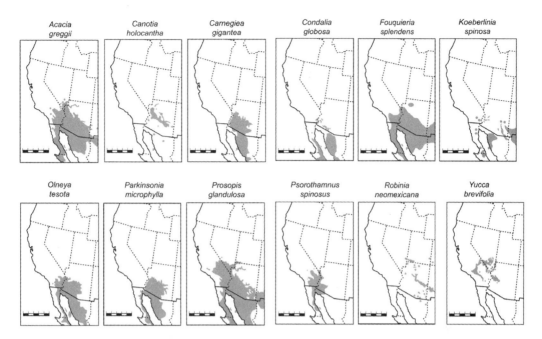

FIGURE 6.9. The distributions of selected mechanically defended plants over 6 feet tall in the warm deserts of western North America. Scale bar = 300 miles. For sources, see "Sources for Plant Distribution Maps" in the reference list.

Rampart Caves (chapters 3 and 5), it has been known that this animal ate Joshua tree leaves and fruits. It has been inferred from this that Shasta ground sloths were a primary disperser of Joshua tree seeds.[34] In fact, the northernmost Shasta ground sloths (*Nothrotheriops*) known from the Great Basin (fig. 6.8) coincide almost perfectly with the northernmost Joshua trees, both Pleistocene and Holocene (fig. 6.9).

Joshua trees, however, are still very much with us, and one reason is clear: seed dispersers still exist in the form of seed-eating rodents. Zoologists Stephen Vander Wall and Ben Waitman and their colleagues established that white-tailed antelope ground squirrels (*Ammospermophilus leucurus*) and Merriam's kangaroo rats (*Dipodomys merriami*) collect and cache Joshua tree seeds, and that subsequent pilfering and reburial by other rodents can disperse seeds farther than the original burial episode.[35] Obviously, Joshua trees are not dependent on now-extinct mammals for their continued existence.

Both of these research teams noted that Joshua tree seeds have thin coats and would likely be crushed during passage through a ground sloth's digestive tract. They also noted that many Joshua tree fruits may have been above the heights that ground sloths could reach, which is certainly true, and that the modern analog for the American Pleistocene horse, the introduced European *Equus caballus*, does not seem to be an effective disperser of Joshua tree seeds. Rodents, they concluded, are, and have been, this species' prime seed disperser.

This research makes it clear just how important small mammals are in helping Joshua trees reproduce today. However, it does not mean that they played the same primary role during the Ice Age. It is true that the Joshua tree seeds Laudermilk and Munz observed in sloth dung had been split. However, they also found stem fragments of saltbush (*Atriplex*) and jointfir (*Ephedra*) up to 1.5 inches long. Hunks of plants this size suggest that sloths could disperse viable Joshua tree seeds.[36] Just as important, there were other, now-extinct large herbivores that could have distributed those seeds during the Pleistocene (fig. 6.8). And even if the dietary preferences of introduced European horses can be validly extended to the caballine horses of the American Pleistocene, this was not the only North American Pleistocene horse lineage, as I discussed in chapter 3.

Biologist Christopher Smith and his colleagues have taken a very different approach to questioning the role that large Ice Age mammals may have played in distributing Joshua tree seeds.[37] To understand their critique requires knowing something about the research that led them to question this possibility.

Joshua trees in the eastern and western parts of their range differ in many ways, including the width and length of the leaves, the form of the flowers, and the height at which the lowest branch diverges from the tree's trunk. This last characteristic is the easiest to see from a distance. The lowest branch of western Joshua trees diverges more than 3 feet from the base of the tree (actually, a meter, but 3 feet is close enough). These trees are assigned to the subspecies *Yucca brevifolia brevifolia* (fig. 2.9). To the east, that lowest branch diverges less than 3 feet from the base, producing more shrub-like individuals assigned to *Yucca brevifolia jaegeriana*.[38] The two varieties are known to occur together in only one place—southern Nevada's Tikaboo Valley, just west of the Pahranagat Lakes (fig. 2.1). Genetic analysis suggests that these populations may have diverged from one another some five million years ago or so.[39] Even though that is a long time, the two forms of Joshua trees are still able to share genetic material, with more genetic material flowing from the eastern to the western subspecies than in the other direction.[40]

All species of the genus *Yucca* depend on moths for pollination.[41] Joshua trees depend on two closely related species of these moths: *Tegeticula synthetica* and *Tegeticula antithetica*.[42] These moths lay their eggs in Joshua tree

flowers and then pollinate those flowers, thus helping ensure that their cater-
pillars will have Joshua tree seeds to eat once they emerge. Genetic studies
suggest that the divergence of these two species of moths occurred a little
over one million years ago, much later than the divergence of the eastern and
western varieties of Joshua trees.[43] A lot has happened since then. The larger
of these two species of yucca moths, *Tegeticula synthetica*, focuses its attention
on the western variety of Joshua tree, while *Tegeticula antithetica* busies itself
with the eastern variety. The moths are known to meet each other only in the
same place that the two varieties of Joshua trees meet each other—in Tikaboo
Valley.[44]

Learning all of this required detailed information on a daunting variety
of topics, from the genetics and anatomy of plants and moths to the climatic
requirements of Joshua trees.[45] Some of that information led Smith and his
team to think that large, now-extinct Pleistocene mammals were not important dispersers of Joshua tree seeds. An analysis of Joshua tree genetic material
from across a broad expanse of their current range led them to argue that this
species did not undergo a dramatic reduction in population size after the Ice
Age ended. Instead, they argue that the vast expanse of territory lost in the
south—for instance in the Sonoran Desert—was replaced by territory gained
in the north. In addition, the climate models they built suggested to them that
all climatically appropriate habitats available for Joshua trees are now occupied by them.[46] Paleoecologist Ken Cole and his colleagues reached a similar
climate-based conclusion using a very different approach.[47]

Smith and his coworkers put these two things together to question the
possibility that large mammals played an important role in helping Joshua
trees move across the landscape. If, they observed, the extinction of large
mammals cost Joshua trees their primary dispersers, the abundance of these
trees should have declined, but their genetic evidence suggests that it did not.
If Joshua trees are now limited in distribution by the lack of their dispersers,
they should not occupy all climatically suitable habitats, but their climate
models suggest that they do.[48]

Neither of these arguments is compelling. As Cole and his colleagues have
pointed out, the evidence left by ancient woodrats shows no significant northward expansion of these trees since the end of the Ice Age—that is, since the
extinction of large Pleistocene herbivores.[49] Geneticist Tyler Starr and his
team note that there may have been "some slight expansion" to the north,[50]
which is certainly possible. However, there is no getting around the fact that
the history of Joshua trees during the last 10,000 years or so is one of territory
lost, not territory replaced.

How might we reconcile the genetic data suggesting that Joshua trees did not undergo a significant reduction in numbers after the Ice Age ended with the fossil record showing that these plants lost a vast amount of their territory during that same period? The answer might be simple. The genetic data are telling us about past population numbers, not past distributions. The fossil data are telling us about past distributions, not past population numbers. Perhaps what has been detected here is a change in population density, with Joshua trees, now dependent on the small-scale abilities of rodents to disperse their seeds,[51] more tightly packed in a fraction of the range they occupied during the late Ice Age.[52] The genetic and fossil data can easily be reconciled by realizing that they are telling us about distinctly different things. Once they are reconciled, a possible role for large Ice Age herbivores remains intact.

The climate models are not problematic for those of us who suspect that those herbivores played a significant role in distributing Joshua tree seeds. After all, both Smith and his colleagues, on the one hand, and Cole and his coworkers, on the other, have generated climate models suggesting that Joshua trees now live where they can. Nonetheless, the former team questions the role of Ice Age herbivores in dispersing Joshua tree seeds, and the latter suspects that they did exactly that. It is also true that many climate models built on the basis of the current distribution of organisms are likely to end up suggesting that where those organisms now live is where they can live—even if they might be able to thrive somewhere else. Certainly, the fit between those models and modern distributions tells us little about how they attained their past distributions, or about any animals that might have dispersed their seeds.

No matter what distributed Joshua tree seeds in the past, the yuccas as a group present another distributional oddity worth mentioning. The many species of this genus in North America are found from Central America to, just barely, southern Canada, with most of those species in the western half of North America.[53] In the Mojave and Sonoran Deserts, the multiple species of this genus, and the closely related genus *Hesperoyucca*, include forms with both indehiscent fruits (those that don't fall apart on their own and so form part of Janzen and Martin's megafaunal dispersal syndrome) and dehiscent ones (those that fall apart without help). The indehiscent species include Joshua tree, Mojave yucca, banana yucca, and chaparral yucca (*Hesperoyucca whipplei*). The dehiscent species include soaptree yucca (*Yucca elata*) and Utah yucca (*Yucca utahensis*).[54] The only species of either genus found in the floristic Great Basin, however, is Spanish bayonet (*Yucca harrimaniae*), a low-growing yucca with fruits that open on their own.[55]

If we accept that there is a diverse variety of plants in the Mojave and Sonoran Deserts whose primary dispersers during the Pleistocene were large, now-extinct herbivores, then it is striking that there is not a single species of such plants in the floristic Great Basin, not a single species with fruits that were built to be dispersed by large mammalian herbivores. This is entirely consistent with the argument that Mack and Thompson made on the basis of the nature of Great Basin grasses: that large mammals were not abundant on the ancestral Great Basin landscape. And, as we are about to see, there is an even bigger pattern here.

The Deeper Meaning of Thorns

About a decade ago, Paul Martin and I were on a field trip together in the Sonoran Desert near Tucson, Arizona. He saw me, probably incorrectly, as something of an expert on the Great Basin. I saw him, certainly correctly, as one of the world's experts on the Sonoran Desert. As we wandered around looking at, and talking about, plants, Paul asked me a question that took me by surprise. "Why," he asked, "are the plants of the southwestern deserts thorny but Great Basin plants oily?"

This was classic Paul Martin shorthand for what was actually a much more complex question. Plants throughout the deserts of arid western North America defend themselves through some combination of chemicals (Paul's "oils") and such things as spines, thorns, prickles, and spiny leaves. What Paul was actually asking was why mechanical defenses (his thorns) seemed so much more common in the warm deserts than in the Great Basin to the north. I had no answer at the time and neither did he. I have thought about this question ever since and present a possible answer here.

Unlike mammals, plants can't run away from their enemies. As a result, they have evolved an extraordinary suite of mechanisms to defend themselves from the herbivores, both vertebrate and invertebrate, that want to eat them. These mechanisms can take many forms, but the ones important to us here are either chemical or structural.

Plants have an astonishing ability to turn themselves into biochemical manufacturing facilities. They produce compounds that support growth, reproduction, and development, often referred to as primary metabolites, and compounds meant to do other things, including defense against herbivores. These are generally referred to as secondary metabolites, or "bioactive specialized compounds," though their functions are as essential to the growth and reproduction of the plant as the primary version. There are over 200,000 of these compounds known. While only a small fraction of them seem to be involved in warding off herbivores (especially insect herbivores), these compounds are extremely important to the well-being of the plants that deploy them. We are all familiar with some of them: nicotine, caffeine, capsaicin (think chili peppers), and morphine are all compounds produced by plants to deter herbivores. So are strychnine and cocaine.[56]

The presence of such chemicals triggers a remarkable battle between plants and the animals that would eat them. For example, creosote bush leaves are covered with a resin that contains a complex set of chemicals meant to deter herbivores. Desert woodrats (*Neotoma lepida*) that live where creosote bush is found have enzymes and gut microbes that can detoxify this resin, allowing the animals to feed on those leaves. Those that live in the botanical Great Basin to the north, where the plant is not found, lack these detoxifying mechanisms and so do not fare well when fed on a diet heavy in the same leaves their southern relatives in part depend on.[57]

Structural defenses are quite different. Rather than requiring the construction of a biochemical warfare system, these defenses require constructing plant parts that impede the ability of a herbivore to reach what it wants to eat. For an insect, these can be toughened or hairy leaves. For a mammal, it is spinescence—the presence of thorns, spines, prickles, and spiny-toothed leaves.[58]

Spinescence serves as a mechanical defense against vertebrate herbivores.[59] Among the cacti, spines play a significant role in thermoregulation, but even here their primary function appears to be the deterrence of vertebrate herbivores.[60] In fact, the colorful spines on cacti, agaves, and other plants may serve as a visible warning to herbivores—leave me alone or you will get hurt—much as the coloration in coral snakes warns us to stay away from them.[61] There is a reason my colleague Constance Millar refers to these as "ouchy plants."

Anyone who has tended roses, or backed into the spines of a jumping cholla, can sympathize with a mammal that bites into an ouchy plant. When there are no mammals to do that, plants don't produce armaments of this sort. The acacias provide a great example. In the American Southwest, the catclaw acacia (*Acacia greggii*) evolved in the presence of animals that wanted to eat it and has wicked, curved spines (fig. 6.10). The very name "acacia," in fact, makes its way back to the Greek word *akis*, which refers to a sharp point. In Hawaii, however, the endemic *Acacia koa*, used by Native Hawaiians to make their remarkable canoes, evolved in the absence of mammalian herbivores. It has no spines at all (fig. 6.11).[62] In some plants, spines lengthen in the presence of mammalian herbivores and shorten in their absence.[63] Similarly, the greater

FIGURE 6.10. The cat claws of catclaw acacia (*Acacia greggii*). Photograph by David A. Charlet.

the amount of large mammal browsing on acacias in the African savanna, the higher the abundance of thornier versions of those plants.[64]

Some plant defense mechanisms can be treated as either structural or chemical. The calcium oxalate crystals found in agave and yucca plants provide an example; this is the compound found in the dung of the Gypsum Cave Shasta ground sloths that ate those plants (chapter 5). Some botanists treat this as a structural defense; others, as a chemical one.[65] It doesn't matter, of course. What matters is what such things do—they deter herbivores.

Some plants rely on structural defenses against a particular predator or set of predators, some rely on chemical defenses against those enemies, and some rely on both.[66] Some of these defenses, called constitutive, are built into the plant; that is, they are there all the time. Others, called inducible defenses, can be deployed in response to attacks by herbivores, as with the lengthened spines in the acacias I mentioned above. Many plants can even

FIGURE 6.11. The spineless koa (*Acacia koa*) of Hawaii. Photograph by Heidi A. Lennstrom.

produce volatile compounds that attract the predators of whatever insect is attacking them.[67] But no matter how they do it, there are so many herbivores out there that plants must defend themselves against them somehow.

Because of Janzen and Martin's arguments about anachronistic fruits in the American tropics, and Connie Barlow's exploration of those arguments in her splendid book on such anachronisms,[68] I can't walk past the avocado display in the grocery store and not think about the large mammals that once dispersed the seeds of this plant.[69] But Janzen and Martin did more than talk about anachronistic fruits. They also observed that the thorns on tall plants that range from the Mojave Desert into Central America now seem anachronistic as well. That thorniness would have had a clear defensive function during the Pleistocene, prior to the extinction of such a broad range of large herbivorous mammals. Today, it often seems to have no function at all.

This brings us back to the question that Paul Martin asked me, and for which neither of us then had an answer. We both knew that when we passed from the southern floristic Great Basin into the Mojave and then the Sonoran Desert (fig. 2.7), we also passed from a region in which woody plants over

6 feet tall with thorns, spines, prickles, and spiny leaves are rare, into a region in which such plants are abundant. This is so obvious that it is hard not to be struck—"stuck" might be a better word—by this distinct difference. Why, Paul was asking, does this happen? Why do we pass from a region in which tall plants are defended primarily chemically, to one in which they are defended both chemically and mechanically?

A possible answer might already seem clear. This happens because we pass from a region in which large herbivorous mammals did not play an important role in the evolutionary history of these plants (the Great Basin) to one in which they did (the areas to the south).

It is easy to show how substantial this difference is. First, though, I have to explain why I defined a "tall" woody plant as one that can attain a height of over 6 feet.[70]

During the last 10,000 years—that is, since the end of the Pleistocene—the western North American deserts have supported six species of "large" mammalian herbivores: bison (shoulder height ca. 5.9 feet), elk (ca. 4.9 feet), bighorn sheep (ca. 3.6 feet), mule deer (ca. 3.3 feet), pronghorn (ca. 3.3 feet), and collared peccary (ca. 1.6 feet), though this last species is a very recent arrival in the southwestern United States. Bison don't really count because they are grazers, but whether they are included or not, the tallest of these recent herbivores of the arid West could feed on plants that reached no more than about 6 feet tall. That figure provides me with a definition of tall plants: they are, for my purposes, plants that reach more than 6 feet in height. Armed plants taller than this are defending themselves against mammalian herbivores that no longer exist.

Tall, mechanically defended plants are extremely common on the Mojave and Sonoran Deserts just to the south of the floristic Great Basin. In appendix 3, I list 99 of them, distributed across 21 plant families. Figure 6.9 shows the distribution of a small subset of them. In stark contrast, there are only six species of such plants with more than a token distribution in the floristic Great Basin, each belonging to a different family: Fremont's mahonia, Torrey's saltbush, silver buffaloberry, snow currant, Woods' rose, and water jacket (table 6.1). The distributions of all six are shown in figure 6.12.

Mapping the northern distribution of the tall, armed woody plants of the North American deserts shows that these plants drop dramatically in abundance as we cross from the Mojave Desert north into the floristic Great Basin. Since these plants all use potentially dangerous structures to defend themselves against large herbivorous mammals, I refer to this northern edge as the Mechanical Defense Line and have plotted it in figure 6.13.

TABLE 6.1. The tall, armed and dangerous plants of the Great Basin

Species	Common name	Family	Maximum height (in feet)
Mahonia fremontii	Fremont's mahonia	Berberidaceae	15
Atriplex torreyi	Torrey's saltbush	Chenopodiaceae	10
Shepherdia argentea	Silver buffaloberry	Elaeagnaceae	13
Ribes niveum	Snow currant	Grossulariaceae	10
Rosa woodsii	Woods' rose	Rosaceae	10
Lycium andersonii	Water jacket	Solanaceae	10

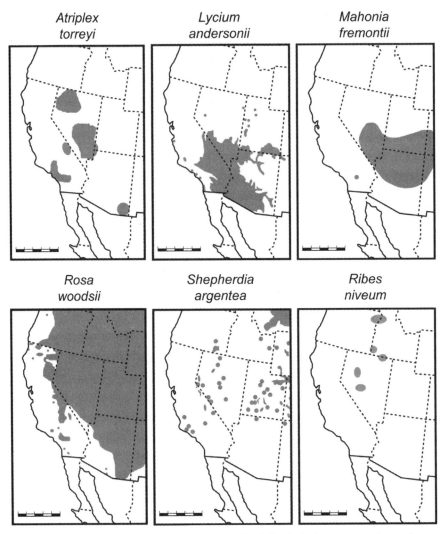

FIGURE 6.12. The distributions of mechanically defended plants over 6 feet tall in the floristic Great Basin. The distributions of *Rosa woodsii* and *Lycium andersonii* are truncated at the US–Mexico border. The isolated population of *Atriplex torreyi* in southeastern Arizona is treated as a separate species, *Atriplex griffithsii*, by USDA NRCS (2014), but as *A. torreyi* var. *griffithsii* by the *Flora of North America* (Welsh 2003). Scale bar = 300 miles. For sources, see "Sources for Plant Distribution Maps" in the reference list.

FIGURE 6.13. The Mechanical Defense Line.

THE CREATION OF THE MECHANICAL DEFENSE LINE

The Mechanical Defense Line marks the modern northern limit of abundant, tall, mechanically defended plants in the Intermountain West, species that grow to heights above the reach of any mammalian herbivore in this region during the last 10,000 years. It also marks the northern limit of plants bearing megafaunal fruits, plants whose primary seed dispersers had become extinct by about 10,000 years ago. This distributional match is not likely to be coincidental. What, then, created this demarcation?

An obvious answer is that the current position of the Mechanical Defense Line is due entirely to climate. We need look no further than creosote bush (fig. 6.14) to see that this must be the case.

Today, this species is the most abundant perennial plant in the warm deserts of North America.[71] Its northern distributional edge marks the boundary between the Mojave Desert and the floristic Great Basin[72] at the same time as

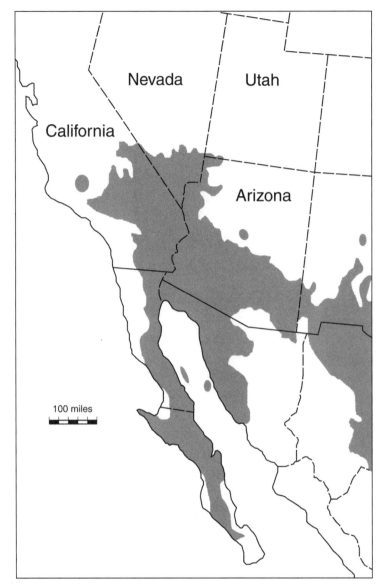

FIGURE 6.14. The distribution of creosote bush (*Larrea tridentata*) in western North America (after Canella 2003).

it coincides with the northern reach of the Mechanical Defense Line (compare figs. 6.13 and 6.14).

There is no mystery as to the cause of the northern limits of creosote bush, a species that is intolerant of prolonged bouts of subfreezing temperatures.[73] The floristic Great Basin winters are now too cold for it, a situation that may not last under conditions of global warming.[74] Just as important, creosote bush does not appear to have been in the Mojave Desert during the Ice Age.

It began to spread across this region during the early Holocene and did not colonize much of its current range until the middle Holocene or even later.[75] The current boundary between the floristic Great Basin and the Mojave Desert is quite recent and reflects climates that have been in place for only a few thousand years.

Insofar as we have paleobotanical records for them, the recent history of creosote bush is mirrored by that of many of the plants whose northern distributional edge lies at or near the Mechanical Defense Line. The earliest record for saguaro in the Sonoran Desert falls at 9,200 years ago, for blue paloverde (*Parkinsonia florida*) at 8,700 years ago. Ocotillo (*Fouquieria splendens*), desert ironwood (*Olneya tesota*), crown of thorns (*Koeberlinia spinosa*), and yellow paloverde (*Parkinsonia microphylla*) all arrive toward the current northern edge of their distribution during the middle or late Holocene, apparently migrating northward, one species at a time, from south of the US-Mexico border.[76]

However, other species that help define the position of the Mechanical Defense Line were very much in place during the late Pleistocene. Catclaw acacia is present in the Grand Canyon today and was there, to judge from the fossil record, before 36,000 years ago.[77]

The same can be said for the late Ice Age presence of Joshua tree at the northern edge of its general distribution in both California and Nevada. Records for this species in the Alabama Hills in Owens Valley, California, range from 31,450 to 13,350 years ago. Pleistocene records from the Sheep Range, the Specter Range, and the Amargosa Desert, Nevada, range from 30,470 to 10,010 years ago.[78]

In theory, we can turn this around and ask how long the mechanically defended plants of the floristic Great Basin (table 6.1) have been in this region. Unfortunately, we don't have much of a Great Basin Ice Age record for these plants. We do know that Woods' rose was here in the late Pleistocene,[79] and it is likely that the buffaloberry known from a number of late Ice Age sites along the western edge of the Lake Bonneville Basin is silver buffaloberry even though it has not been identified to the species level.[80] To my knowledge, there are no Pleistocene records for Fremont's mahonia, Torrey's saltbush, snow currant, or water jacket from the floristic Great Basin. This probably reflects the state of our knowledge in this realm, but it is possible that these plants were not here during the Ice Age.

It is no surprise that the armed and dangerous plants of the arid West have their own individual histories. The same is very much true for different species of mammals during the Pleistocene and continuing to today (chapter 7).[81] Nonetheless, nearly all these plants are found no farther north than the Mechanical Defense Line. The only substantial exception is Woods' rose (fig. 6.12), which has an extremely broad North American distribution.

The reason for this seems clear. Many, and perhaps all, of the species that reach, or come close to, the Mechanical Defense Line have northern distributions that are limited by extended bouts of deep winter cold. These species include catclaw acacia, saguaro, ocotillo, desert ironwood, yellow paloverde, and Joshua tree.[82] Since the northern distribution of creosote bush is limited by the same factor, it is no accident that the Mechanical Defense Line coincides with the southern boundary of the floristic Great Basin.

However, there must be more to it than this, because the Mechanical Defense Line is utterly ignored by plants that were within the reach of mammalian herbivores that did not become extinct—the ones that dine no higher than 6 feet. That becomes clear if we look at the distribution of armed and unarmed native shrubby plants that grow to no more than 6 feet in height. To do this, I thought about researching every species of shrubby plant in the Sonoran, Mojave, and floristic Great Basin Deserts to determine how tall it grows and whether it is armed. Since that would have taken until the start of the next Ice Age, I let other people do it for me. To see whether "short" armed plants ignore the Mechanical Defense Line, I used the compilations of shrubs in two important books on the plants of the Intermountain West. The first of these was Lyman Benson and Robert Darrow's famous *Trees and Shrubs of the Southwestern Deserts*. The second was Hugh Mozingo's valuable *Shrubs of the Great Basin*.[83] The first source provided me with 180 species of short shrubs from the Sonoran and Mojave Deserts; the second, with 74 species of short shrubs from the floristic Great Basin. The full lists, and associated information, are provided in appendixes 4 and 5.

The boiled-down results are shown in table 6.2. In the Mojave and Sonoran Deserts, 21.1 percent of these plants are armed and dangerous. In the floristic Great Basin, that number is 24.3 percent. In chapter 4, I used a simple statistical test, known as chi-square, to determine whether a similar difference was so small that it could be expected to occur by chance. If we apply the same test to the data in table 6.2, we discover that the difference in the distribution of armed and unarmed short plants north and south of the Mechanical Defense Line is so minor that it could be expected to occur by chance over half the time, not much different from the results we expect from flipping a coin. That is, short armed and unarmed plants ignore the Mechanical Defense Line. The same is not true for the tall armed and unarmed plants in my sample, just as

TABLE 6.2. Number of armed and unarmed shrub species less than 6 feet maximum height in the warm (Mojave and Sonoran) deserts and the floristic Great Basin

Region	Unarmed	Armed	Total
Great Basin	56 (75.7%)	18 (24.3%)	74
Warm deserts	142 (78.9 %)	38 (21.1%)	180

References: Warm desert sample from Benson and Darrow (1981; see appendix 4); floristic Great Basin sample from Mozingo (1987; see appendix 5).

TABLE 6.3. Number of armed and unarmed woody plant species greater than 6 feet maximum height in the warm (Mojave and Sonoran) deserts and the floristic Great Basin

Region	Unarmed	Armed	Total
Great Basin	19 (82.6 %)	4 (17.4%)	23
Warm deserts	57 (59.4%)	39 (40.6%)	96

Reference: Warm desert sample from Benson and Darrow (1981; see appendix 4); floristic Great Basin sample from Mozingo (1987; see appendix 5).

we would expect from what I said earlier. In my samples, only 17.4 percent of the tall shrubs of the floristic Great Basin are armed, compared to 40.6 percent of the tall shrubs to the south (table 6.3). This difference could be expected to occur by chance only 4 percent of the time.

Why would climate matter to a plant that was 6.5 feet tall, but not to one that was 5.5 feet tall? Maybe somebody can think of a reason, but I can't. To me, the obvious explanation is that plants both north and south of the Mechanical Defense Line had to worry about mammalian herbivores that could reach up and eat plant parts that were about 6 feet high, but only those south of the border had to worry about mammalian herbivores that could reach higher than this.

In short, the warm deserts of southwestern North America are remarkably rich in tall, mechanically defended plants. In contrast, the floristic Great Basin nearly lacks them even though it does not lack tall shrubs. The floristic Great Basin, the Mojave Desert, and the Sonoran Desert are all rich in short woody plants, but those plants defend themselves mechanically in equal proportions. The easiest way to explain this is by concluding that large mammalian herbivores were never particularly abundant in the floristic Great Basin. This is the same conclusion that Mack and Thompson drew from the nature of Great Basin grasses, and that I drew from the distribution of megafaunal fruits.

FIGURE 6.15. The short, spiky needles of singleleaf pinyon.

CONIFERS UP IN ARMS?

Some of my botanist friends have pointed out to me that my arguments about the meaning of the distribution of armed and dangerous plants in the arid West ignore the widespread conifers of the Great Basin. To them, some of these conifers seem as mechanically defended against herbivores as some of the other plants I have discussed.

In particular, they observe that both singleleaf pinyon and some tall junipers have prickly parts that either may or clearly do help ward off mammalian herbivores. Pinyons and junipers usually grow to about 20 feet in height, though western junipers can be over 60 feet tall, singleleaf pinyons over 40 feet, and Utah junipers over 20 feet.[84] Pinyon-juniper woodland now covers over 17 million acres of the Great Basin. If my botanist friends are right that these plants are mechanically armed, there must be something wrong with my argument.

I think, though, that they are wrong where it matters most. Singleleaf pinyon has short, stout, pointed needles (fig. 6.15); in young trees, they can take on something of the appearance of pincushions. In many ways, they would seem to provide an effective deterrent to hungry herbivores. Even if they do, though, this doesn't matter for the argument I am making here. As I discussed

FIGURE 6.16. The scaly foliage of a typical adult western juniper.

in chapter 2, pinyon pine was found only on the very southern edge of the floristic Great Basin during the Pleistocene and didn't begin to move northward until after the Ice Age ended. Those 17 million acres that pinyon-juniper covers in today's Great Basin were covered by something else until very recently, and those possibly defensive needles can't be telling us anything about the abundance of Ice Age mammals in this region.

The junipers are a different issue, since tall junipers were in the floristic Great Basin during the Ice Age. Young junipers have prickly, needlelike foliage, rather than the scaly foliage of most adult trees. It is easy to believe that this prickly foliage is meant to discourage mammalian herbivores. Occasionally, tall adult junipers can have this leaf shape as well,[85] and it appears that this needlelike morphology can be developed by adult junipers after they are attacked by mammalian herbivores (see figs. 6.16 and 6.17).[86]

This is yet another fascinating example of the myriad ways in which plants defend themselves. As is the case with pinyon pines, however, I don't think it matters for my arguments here, though for a very different reason.

One of the most annoying trees in my yard is a healthy, tall European holly (*Ilex aquifolium*) that was there well before we moved in. Every year, I have to

FIGURE 6.17. The needlelike foliage of an atypical adult western juniper (Devils Postpile National Monument, eastern Sierra Nevada, California).

remove holly sprouts, which, if given the chance and enough time, might well take over my garden. I have to be careful in doing that because the leaves of these sprouts have sharp spines, a clear defense against herbivores. The tall parental tree also has spiny leaves toward its base, but the frequency of defended leaves decreases the higher you go on the tree. As far as I can tell, the leaves 20 feet up, toward the top of the tree, have no spines at all. That is, the leaves of the holly kids and the lower leaves of the holly parent are very well defended, but the upper parental leaves have no spines at all.

This is a very well-known phenomenon in trees in general.[87] Juveniles, otherwise vulnerable targets for mammalian herbivores, defend themselves, but once their foliage is above the range of any possible herbivore, they leave the defenses behind. It is no surprise, then, that juvenile junipers do this, and the fact that they do it tells us nothing about now-extinct Ice Age plant predators.

Neither does the fact that juniper trees above 6 feet tall may retain their juvenile defensive foliage, or can develop it in response to being eaten by a mammal. The latter is a classic induced defense of the sort that I discussed above. Once induced, the defense can remain, at least in other organisms, even when the threat is gone.[88] Connie Millar's tall, defended junipers may have retained their juvenile defenses into adulthood, or they may have developed them in response to being munched on by a larger herbivore. Either way, the vast majority of tall junipers in the Great Basin have scaly and mechanically undefended leaves. This phenomenon may be telling us about the histories of those rare adult trees that have the defensive foliage that normally marks juvenile junipers, but it is not likely to be telling us anything about the extinct Ice Age mammals of the Great Basin.

SUBALPINE CONIFER HISTORY

But wait, there's more! As I discussed in chapter 2, during the latest Pleistocene, such subalpine conifers as bristlecone pine, limber pine, Engelmann spruce, and common juniper (*Juniperus communis*) extended to extremely low elevations in much of the floristic Great Basin, down to the bases of mountain ranges and onto appropriate bedrock outcrops in valley bottoms.[89]

As we saw in chapters 3 and 4, we know that a series of large, now-extinct mammals utilized the high elevations of at least some part of the Great Basin toward the end of the Ice Age. The Huntington Mammoth Site on the Wasatch Plateau yielded the remains of mammoth and giant bear at 8,990 feet, with the mammoth dated to 11,200 years ago and the bear to 10,870 years ago. The nearby Mastodon Sinkhole site, talked about in chapter 3, provided mastodon and horse at an elevation of 9,780 feet, with the mastodon dated to about 10,800 years ago. The Huntington mammoth had been eating the needles and twigs of subalpine fir (chapter 3). Although we don't know what the Mastodon Sinkhole mastodon had been eating before it died, we do know that the surrounding vegetation included healthy amounts of Engelmann spruce. And mastodons, of course, ate spruce.

But even though we know that at least some high elevations in the Great Basin supported large Pleistocene mammals, this does not change the fact that today, subalpine coniferous woodland does not support substantial numbers of a diverse set of large mammals, and it is not likely to have done so during the latest Pleistocene. It was precisely this woodland that extended to low elevations in many parts of the Great Basin during the latest Pleistocene.

THE IMPLICATIONS OF PLEISTOCENE LAKES

As I also talked about in chapter 2, the Great Basin saw the earth's greatest development of pluvial lakes during the late Pleistocene (fig. 2.13). When this system was at its maximum during the late Pleistocene, approximately 45,500 square miles of the lower elevations of the hydrographic Great Basin were covered by water, compared to about 3,900 square miles today. Given that the hydrographic Great Basin spans roughly 200,000 square miles, some 23 percent of this area was then covered by water. Much the same is true if we restrict our interest to the floristic Great Basin, which covers about 196,100 square miles, roughly 21 percent of which was covered by about 41,700 square miles of lake water during the late Pleistocene.[90]

That is, at the same time as subalpine conifers were descending to very low, valley edge, elevations during the late Pleistocene, about 20 percent of the Great Basin was covered by water, all of which was in the lowlands. So there was water, surrounding strips of land, and subalpine conifers above: this would not have left huge expanses of territory for teeming numbers of large herbivores. When the lakes began to fall, and subalpine conifers to rise, now-extinct mammals would have had a challenge in repopulating the Great Basin from elsewhere. The hydrographic Great Basin is bounded by the Sierra Nevada and southern Cascades on the west, and by a series of formidable high-elevation barriers to the east. Easy recolonization could have occurred from the south or north but would have been restricted by massive topographic features in other directions.

SLIM PICKINS FOR PLEISTOCENE HUNTERS

Earlier in this chapter, I suggested that the plants of arid western North America may be telling us something about the abundance of large Pleistocene mammals on the Great Basin landscape. In doing this, I have followed in the footsteps of botanists Richard Mack and John Thompson, who pointed out that the abundance of grazing-intolerant grasses in the Great Basin suggests that large herbivores could not have been common here during the Ice Age. We have now seen that tall, mechanically defended plants are extremely rare in the floristic Great Basin, even though short armed plants are common. Plants whose primary seed dispersers were likely to have been large vertebrate herbivores are absent from the floristic Great Basin. Subalpine conifers descended to very low elevations during the late Ice Age, producing habitat that is not likely to have supported large numbers of various kinds of now-extinct large mammals. Some 20 percent of the Great Basin, either hydrographic (ca. 23 percent) or botanical (ca. 21 percent), was covered by water when pluvial lakes were at their maximum during the Pleistocene. Go far from those lakes into the mountains, and you would, in many areas, have encountered low-elevation subalpine conifers. Taken together, this suggests that large Pleistocene mammals were never abundant in the floristic Great Basin, and that they were particularly rare here toward the very end of the Ice Age.

If this inference is correct, the implications for understanding the adaptations of the early peoples of the Great Basin are fairly clear. Although the absence of strong associations between artifacts and the remains of extinct Pleistocene mammals in this region may be due to the extreme scarcity of known buried archaeological sites of the right age, there is no compelling reason to think that many such sites will ever be found. In suggesting this, I am not at the same time suggesting that the Mojave and northern Sonoran Deserts teemed with such mammals. As I noted earlier, many, though not all, of the tall, armed plants that lie south of the Mechanical Defense Line appear to have resided south of the US-Mexico border during the late Pleistocene. In fact, not only are there no kill associations between artifacts and extinct mammals in the floristic Great Basin, there are also none in the Mojave or northern Sonoran Deserts. The only site of this sort in the Sonoran Desert is El Fin del Mundo. This site, as I discussed in chapter 3, provided Clovis fluted points from the same deposits that yielded the remains of two gomphotheres (figs. 3.58 and 3.62). El Fin del Mundo is in the Sonoran Desert of northwestern Sonora, Mexico—south, obviously, of the border between the United States and Mexico.

The famous Clovis kill sites of the southwestern United States are from the San Pedro River drainage of southeastern Arizona. Here, at least four sites— Escapule, Lehner, Murray Springs, and Naco—have yielded Clovis fluted points and the remains of mammoths in such close association as to leave no doubt that these sites were created by people who were exploiting those animals.[91] This area is in neither the Sonoran nor the Chihuahuan Desert (figs. 2.7 and 3.62). Instead, it is in the desert grasslands that fall between the two. It is also in an area that, at the time, supported grasses that would have provided rich summer forage for large grazing herbivores, including mammoths. That combination would have provided productive hunting for people armed with an effective, Clovis, tool kit.[92]

The ghosts of herbivory are everywhere, perhaps even in the chemical defenses of plants that no longer need them,[93] but the ghosts of large mammal herbivory are curiously rare in the floristic Great Basin. For this reason, and the others I have discussed, it seems likely that large, now-extinct mammals were never common in the floristic Great Basin and that they were particularly uncommon toward the very end of the Ice Age. Since people arrived here toward that very end, I doubt we will ever find many sites showing that people hunted or scavenged such beasts in the Great Basin.

7

Clovis, Comets, and Climate: Explaining the Extinctions

As the Ice Age came to an end, the Americas lost 77 genera of mammals. Most of the animals involved were large, weighing over 100 pounds, and some can only be described as huge—a sloth as tall as a giraffe and as bulky as an elephant about sums it up. As impressive as these losses were, we probably still don't know the full scope of the extinctions that occurred in the Americas at this time. After all, the notoungulate *Mixotoxodon* was not reported from North America until 2013, the same year that brought us knowledge of a completely new kind of extinct South American ground sloth (*Megistonyx*; see chapter 3).

The Great Basin was not immune to any of this, having lost 20 genera of mammals, along with all those birds that I discussed, sometime toward the end of the Pleistocene (tables 3.2 and 3.4). While it is unlikely that future research in this region is going to reveal an extinct genus of mammal previously unknown to science, we may well learn that some extinct mammals that have yet to be found in the Great Basin did, in fact, live here. The tapir (fig. 3.31) is the most obvious candidate, and it is not unreasonable to hold out hope for the dhole (fig. 3.14).

The real challenge, however, is not adding new animals to a list, as interesting as that might be to many of us. Instead, it is trying to figure out why these extinctions occurred in the first place.

Attempts to explain the losses of Ice Age mammals began as soon as those losses were first established, around 1800, and have continued unabated ever since. Even so, there is still no agreement about the cause or causes of them.[1] This is not for lack of interest or lack of trying. Scientists from many disciplines have recognized that understanding the reasons for these extinctions is important not just for understanding the past, but perhaps also for understanding the future of our planet and life on it. As a result, these scientists have generated a vast literature. Nonetheless, we may not be that much closer to understanding why these losses occurred than were the scientists who struggled with the same issues during the late nineteenth century, even though we know a lot more about the extinctions now than they did then.

Today, three prime movers are called on to explain why the Americas saw such massive extinctions as the Ice Age came to an end. The first of these blames human hunting and is routinely referred to as the Overkill Hypothesis. The second argues that a late Pleistocene extraterrestrial impact event caused it all. The third blames climate change. The best that can be said of any of these explanations is that one may be less unconvincing than the others. There is a fourth approach as well, which attempts to lay the blame on a combination of human hunting and climate change; I discuss it toward the end of the chapter.

By the time we have surveyed these three possibilities, you may agree with Charles Darwin's thoughts on this very same issue:

> Whenever we can precisely say why this species is more abundant in individuals than that; why this species and not another can be naturalized in a given country; then, and not till then, we may justly feel surprise why we cannot account for the extinction of this particular species or group of species.[2]

PLEISTOCENE OVERKILL
AND THE CLOVIS PHENOMENON

It was Paul Martin who most fully developed the argument that the arrival of human hunters caused the extinction of the full suite of late Ice Age American mammals. In most cases, he argued, those extinctions were directly driven by human hunting. In this view, efficient human hunters crossed the Bering Land Bridge and moved south of glacial ice, arriving sometime around 11,000 years ago. These new arrivals—the people we now call Clovis (chapter 6)—encountered a rich array of large mammals naive to sophisticated human hunters. Fed by easy prey, Clovis people expanded southward in an ever-expanding front, leaving extinct populations, and ultimately extinct genera, of mammals behind them. By the time the descendants of these people reached the southern tip of South America, they had caused otherwise inexplicable damage to the mammalian fauna of the Americas. The giant sloths, horses, camels, peccaries, pronghorns, giant beavers, even Aztlán rabbits—each and every one fell prey to human hunters.

Martin did not think that all of the extinctions had occurred this way. The carnivores, including the sabertooth cat, the scimitar cat, the American cheetah, and the dire wolf, were lost because the early hunters eliminated the herbivores on which they depended for their existence. The others—the herbivores—were, quite literally, hunted to death.[3]

That is the overkill argument. It would be hard to overemphasize how popular it is. You find it in ecology textbooks, in books written for popular audiences, in newspapers, in magazines, and on television.[4] I wouldn't be surprised to hear that kids learn about it in school.

At its heart, Martin's argument is built on four major statements about the past.

First, he pointed out that the prehistoric human colonization of islands was routinely followed by the extinction of the vertebrates that lived on those islands.

Second, he argued that the archaeological complex known as Clovis, which dates to a few centuries on either side of 11,000 years ago, was probably created by the first people to have entered the Americas, and that these people were impressively good at hunting large mammals.

Third, he argued that Clovis peoples preyed on a diverse variety of large mammals, the same mammals that they ended up driving to extinction.

Finally, he argued that the North American extinctions all occurred at or near 11,000 years ago, about the same time that the Clovis phenomenon shows up in the archaeological record.

From this, Martin concluded what I outlined above: Clovis hunters drove the herbivores to extinction, and the human-driven extinction of the herbivores caused the extinction of the carnivores that depended on them.

Because this is such a visible argument, both in the popular media and in scientific literature, I will try to convince you that it has very little going for it. In fact, the more you know about the late Ice Age archaeology and paleontology of North America, the less impressive it becomes. There is a reason that paleontologists and archaeologists whose expertise includes late Ice Age North America tend to discount the overkill position, as Martin himself recognized.[5]

I begin with the argument that Clovis peoples were the first to enter North America, about 11,000 years ago. We have already seen that this is not true. The Monte Verde site, in southern Chile, dates to 12,500 years ago. Paisley Cave 5, which I discussed in chapter 6, has a series of ancient human coprolites of Native American origin, the oldest of which dates to 12,400 years ago. Only archaeologists at the outermost fringes of the discipline think that Clovis represents the earliest peoples of the Americas.

Here, though, we run into something of a problem. Martin routinely asserted that Clovis peoples were the first to enter the Americas, and he continued to do so even after Monte Verde was excavated, reported in exquisite detail, and vetted by other professionals.[6] In fact, Martin made it clear that he didn't think much of those who even tried to look for pre-Clovis people in the Americas, comparing that attempt to "the eternal search for Bigfoot" and referring to it as "less than serious science."[7] Martin was no fan of any pre-Clovis site in the Americas because such a site might damage the very argument on which he staked so much of his career—overkill.

He actually didn't need to do this since he also pointed out that "whether or not prehistoric people were in North America earlier, 11,000 [years ago] is the time of unmistakable appearance of Paleo-Indian hunters using distinctive projectile points."[8] This, it turns out, is as true now as it was 30 years ago, when he said it. The earliest American sites with impressive weaponry associated with the remains of extinct mammals are, in fact, Clovis sites.

So let's move on to another of his key assertions, that Clovis people preyed on a diverse variety of the now-extinct herbivores. As I discussed in chapter 3, evidence for this is rare (fig. 3.62). There are 13 sites with Clovis-aged artifacts tightly associated with the remains of mammoth and mastodon, one that has them with the remains of camel and horse, and there is El Fin del Mundo, which has Clovis points associated with the gomphothere *Cuvieronius* (fig. 3.59).[9]

We also know that Clovis diets changed across space and that they included a broad variety of much smaller things that did not become extinct.[10] There is absolutely no evidence, not even a hint, that Clovis people hunted a diverse variety of now-extinct large mammals. No sloths in Clovis sites, no giant beavers, no glyptodonts, no capybaras, no mountain deer, no extinct peccaries or pronghorns, no anything except for the elephant-like mammals and the scattered horses and camels.

For decades, archaeologists saw Clovis people as "big-game hunters," people who made their living by preying on large, and now-extinct, mammals. Martin's overkill argument was hatched when this view was accepted by most archaeologists who dealt with this time period. I am not aware of any archaeologist who believes this anymore,[11] yet the overkill argument requires it. If people did not hunt these animals, and hunt them in great number, they could not have caused their extinction.

In a 1968 *Star Trek* episode, "Spectre of the Gun," the *Star Trek* crew created a tranquilizer in an attempt to avoid being on the losing side in a showdown at the OK Corral. The tranquilizer did not work, even though it had to. To

Spock, the implication was clear: "If the tranquilizer does not function, which is clearly impossible, then a radical alteration of our thought patterns must be in order."[12] To me, the lack of evidence, or even a hazy hint of evidence, that Clovis people were heavily invested in the hunting of large mammals indicates that the overkill hypothesis does not function and that a radical alteration of the thought patterns of those who think it does is likewise in order.

Then there is the assumption that all the extinctions occurred around 11,000 years ago. This assumption provides the critical chronological correlation with Clovis, since it makes both things happen at the same time. It is a simple step from that correlation to concluding that the latter must have caused the former.

For those interested in how science works, this one is really interesting. Until the mid-1960s, it was widely believed that the extinctions had been scattered through time and that they had continued until well after 10,000 years ago.[13] All this began to change in 1964, when the eminent geologist C. Vance Haynes demonstrated that Clovis sites dated to between 11,000 and 11,500 years ago, and that mammoth populations may have undergone a "dramatic decline"[14] after 11,000 years ago.

That demonstration had a huge impact on Martin. Just a few years earlier, he had noted that "the Pleistocene megafauna survived all climatic change until the arrival of early Man,"[15] but he also thought that at least some of the animals had survived beyond 10,000 years ago. Now, armed with Haynes's careful conclusions, he thought differently. Not only did Clovis date to around 11,000 years ago, but all the extinctions dated to that time as well. "The main wave of extinction," he said, "appears to have occurred around 11,000 [years ago]."[16] He did not reach this conclusion because radiocarbon dating actually supported it. At the time, there were 34 extinct genera of North American mammals on his list. Of those, he could show that only 13 had made it to 11,000 years ago or so.[17] Instead, he concluded it because it fit his belief that early human hunters—Clovis people—caused the extinctions. That is, he derived his chronology for the extinctions from his explanation of those extinctions. Then, he used that chronology as support for his explanation.

We are not, and should not be, accustomed to this kind of logic—it is called circular—from scientists, but Martin specialized in it. For instance, he also rejected any hint of pre-Clovis peoples in the Americas because "eminently huntable animals such as ground sloths continued to occupy their dung caves in North and South America as long as they did."[18] That is as circular as it gets, but this is the kind of logic that has propelled the overkill argument to its highly visible position.[19] If you don't read the argument carefully, it actually makes sense. Once it is carefully read, a radical alteration of thought patterns becomes essential.

Martin wasn't the only one to infer the timing of the extinctions from his favored explanation of them. Those who disagreed with Martin and thought that the cause was to be found in the substantial climate change that occurred around 11,000 years ago did the same thing. In both cases, scientists were comfortable assuming that all the extinctions occurred around 11,000 years ago because that is what their explanations required.[20]

Once virtually everyone assumed that the extinctions occurred at about 11,000 years ago, the extinctions themselves became an event, something that had occurred rapidly. The speed of the extinctions became an essential aspect of the process that had to be explained. Even today we can only show that 17 of the 37 genera lasted until about 11,000 years ago (chapter 4), but many continue to believe that all the extinctions occurred at this time. Scientists aren't supposed to assume critical aspects of what it is they are trying to explain based on the explanations they believe in the first place. Those attempting to explain Pleistocene extinctions, however, have done just that. This kind of thing drives me nuts.

In short, three of the four premises of Martin's overkill argument have serious problems. Clovis wasn't first, there is no evidence that Clovis peoples hunted the full range of the now-extinct herbivores, and even though many seem happy to assume that all the extinctions occurred around 11,000 years ago, we are far from being able to show that.

That leaves Martin's observation about islands, that the prehistoric human colonization of these landforms was soon followed by the extinction of many of the vertebrates that lived there. This is critical to Martin's argument because, if true, it shows that prehistoric peoples caused vertebrate extinctions, and caused them very close to their time of arrival.

Unlike the other three critical legs of Martin's overkill argument, this one is decidedly correct. We can pick pretty much any island that was occupied prehistorically to see this process in action, but the process is so well known from New Zealand that it has almost become the exemplar for this phenomenon.

In one of the most astonishing of human accomplishments, people managed to colonize New Zealand around 750 years ago, about the same time that they reached Hawaii and Easter Island.[21] At the time of their arrival, New Zealand supported nine species of the huge flightless birds known as moas, the largest of which weighed more than 500 pounds.[22] Within a few hundred years, all were extinct.

Everyone agrees that moas were driven to extinction by the impacts of human colonists. Everyone agrees that human hunting had something to do with this, since there are well over 100 archaeological sites that contain the remains of these birds.[23] Moas were not alone in their fate. At least 25 other species of birds, along with other organisms, were lost after people arrived.[24] In all instances, human arrival appears to have been to blame.

Such things happened on island after island, in Oceania and elsewhere.[25] To Martin, such losses establish that people can cause extinctions to occur. They establish that, to use the words of Charles Lyell, one of the nineteenth century's finest geologists, "we wield the sword of extermination as we advance."[26]

The extinctions that rapidly followed prehistoric human colonization in island settings, however, do not tell us what formed the cutting edge of the exterminating sword. In the case of moas, human hunting was clearly involved. The archaeology tells us that. But Martin agreed that human impacts on islands were of multiple sorts, all of which may have played a role in driving vertebrate extinction.[27] In New Zealand, for instance, people set the landscape on fire soon after, and perhaps as soon as, they arrived.[28] They also introduced dogs and Pacific rats (*Rattus exulans*). There is every reason to believe that it was not just human hunting that caused these extinctions, but instead the multiple impacts of all these things. In a depressing number of instances, we can say that prehistoric human colonization of an island caused vertebrate extinction, but in no case can we say with certainty that it was hunting and hunting alone that did this. It is not surprising that people can have this impact on islands. Islands are tightly bounded areas that can undergo rapid environmental change. The vertebrates on islands have small populations to begin with, which cannot be replenished from elsewhere. If vertebrates are there long enough, they may lose whatever coping mechanisms they have for dealing with novel predators, diseases, and competitors.[29]

Continents are totally different from islands in these all-important ways. As a result, there is no reason to think that the fate of vertebrates on islands after human arrival tells us anything about what happens when people first arrive on continents. Nonetheless, this is exactly the inference Martin wished us to draw from the fate of so many vertebrates on so many islands in prehistoric times.

Finally, let's go back to the Great Basin. This 200,000-square-mile area supported 20 genera of now-extinct late Ice Age mammals. Not a single site suggests that people interacted with them, but they became just as extinct here as they did elsewhere.

Once looked at in this fashion, Martin's famous overkill hypothesis, so widely visible in the popular media, has almost nothing going for it. Remember this the next time you run across it in the newspaper, on television, in the popular literature, or anywhere else. Remember it if your child comes home from school and tells you that the ancestors of Native Americans killed off all the mammoths. Remember it when you read in *National Geographic* that American Ice Age horses may have become extinct because "they were killed for food by humans."[30] The argument is intuitively appealing because we all know the kinds of impacts we can have, and have had, on the environment. But while self-flagellation might make us feel good on occasion, the overkill whip has no real sting to it.

A KILLER COMET?

In 1981, geoscientist G. Robert Brakenridge suggested that the explosion of a star in the Vela constellation about 11,000 years ago might have had significant impacts on our planet, including an increase in nitrogen dioxide in the atmosphere and a decrease in ozone. This, in turn, may have caused an increase in ultraviolet light, and a decrease in visible light, received at the earth's surface. He also suggested that the increased nitrogen that was transferred from the atmosphere to the surface might have led to the creation of algal blooms, much as the use of nitrogen-enriched fertilizers can do today, and that those algal blooms might have led to the creation of particular kinds of sediments at the end of the Ice Age, sediments that might serve as a marker for this event.[31]

Having said all that, Brakenridge cautiously noted that the age of the Vela supernova roughly coincided with two other well-known aspects of the late Ice Age. First, it coincided with the onset of the Younger Dryas, that intense cold snap dating from about 10,000 to 10,100 years ago that I discussed in chapter 2. Second, it coincided with the appearance of black mats in the southwestern United States and elsewhere.

Black mats are widespread in arid western North America. They first became well known because they cap the famous Clovis sites of southeastern Arizona—Escapule, Lehner, Naco, and Murray Springs (figs. 3.62 and 6.2). At Lehner, the base of the black mat has been particularly well dated at about 10,800 years ago; Clovis artifacts, and the associated mammoths, lie immediately beneath it.

Although the most famous black mats are the ones that formed in some areas at the very end of the Ice Age, they are not confined to that period. Much older versions are known in the arid West, and they continued forming here during much of the past 10,000 years, with a conspicuous gap between about 6,000 and 3,000 years ago.[32]

Geologist Jay Quade and his colleagues have shown that the black mats of arid western North America are organically enriched deposits that form only under certain conditions. It has to be wet enough to promote spring discharge and high water tables.[33] That is why the arid western black mat record shows that big gap I mentioned—this was a time of greatly increased warmth and aridity in this region. Once the moisture returned, so did the black mats.[34]

Brakenridge had a different explanation for the late Ice Age black mats of the American Southwest and elsewhere. He suggested they were formed by the algal blooms ultimately fostered by the Vela supernova. Brakenridge returned to this argument in 2011, suggesting that the Vela supernova might have played a role in the extinction of North America's late Ice Age mammals.[35] As hard as he tried, though, his arguments did not gain much direct traction. I say "direct traction" because the core of Brakenridge's arguments is very similar to that of some that have now attracted a huge amount of attention.

In 2001, nuclear physicist Richard Firestone and archaeologist William Topping published an astonishing paper in a popular newsletter targeted at readers interested in the early peoples of the Americas. The title set the tone for what the authors were about to argue: "Terrestrial Evidence of a Nuclear Catastrophe in Paleoindian Times."[36]

In 1981, Brakenridge had been the model of a cautious and thoughtful scientist. In 2001, Firestone and Topping were the models of something quite different. They argued that at about 41,000 years ago, the earth was flooded with radiation that originated from a supernova, and that shock waves from that event hit the earth again at about 33,000 and 12,500 calendar years ago. This "catastrophic nuclear irradiation"[37] was focused on the Great Lakes region, where it may have heated the atmosphere to over 1,000°C (1,830°F) and melted significant amounts of glacial ice.

That, though, was just the start. Among other things, the incoming radiation altered the radiocarbon clock dramatically. Organic material that dated to the latest Ice Age—11,000 years ago, for instance—might be as much as 40,000 years older than that. That somehow made sense to Firestone and Topping, since they saw similarities between the artifacts left behind by the early peoples of the Americas and the Mousterian cultural tradition in Europe, a tradition tightly associated with Neandertals. Add 30,000 years to dates for Clovis, and sure enough, you are back into Neandertal times, but Neandertals never made it even remotely close to the Bering Land Bridge.

According to Firestone and Topping, the radiation that bathed the earth did not simply reset the radiocarbon clock. It also caused the extinction of late Pleistocene mammals in North America. Not everything it did was bad, though. By causing fortunate genetic changes in plants, this otherwise cat-astrophic event allowed people to domesticate those plants. That popcorn you had last weekend? Thank Firestone and Topping's catastrophe, since they claimed that corn "probably evolved by macro-mutation at that time."[38]

There are enough similarities between Brakenridge's 1981 arguments and what Firestone and Topping claimed 20 years later to think that they had taken the cautious geoscientist's arguments and warped them into something radi-cally different. The warping did not make a positive impression on the profes-sional radiocarbon community. Two of that community's most knowledge-able members, John Southon and Erv Taylor, responded in detail to Firestone and Topping's claims about the reset radiocarbon clock, concluding that their arguments were "at best, highly problematical and, at worst, difficult to take seriously."[39] To many of us, that conclusion seemed politely restrained.

To my knowledge, Firestone never repeated his bizarre claims about a reset radiocarbon clock, but he did not let go of his arguments concerning the possible impacts of an extraterrestrial event on late Ice Age earth history. In 2006, he and two coauthors offered up a book-length and much-altered version of the arguments he had first made in 2001. In *The Cycle of Cosmic Catastrophes*, they argued that around 41,000 years ago, a nearby supernova exploded close enough to earth to cause vertebrate extinctions in Australia and southeastern Asia, including the near extinction of many of the peoples of this latter area. Some 8,000 years later, the first shock wave from this explosion reached earth, with the second reaching it around 16,000 years ago. Then came the real biggie. Around 13,000 calendar (or 11,000 radiocarbon) years ago, the Northern Hemisphere was pummeled by a series of comet-like objects. Those objects caused continent-wide firestorms, massive tsunamis that devastated coastal human populations, the melting of massive amounts of glacial ice, the late Ice Age mammal extinctions, "the destruction of most vegetation in the Northern Hemisphere," and the Younger Dryas cold snap.[40] As if that were not enough, vast quantities of frozen methane were released from beneath the ocean and caught fire. "Here and there miles-long tongues of orange-blue flames flickered and danced across the sea surface. For weeks, the sea burned or boiled with escaping methane."[41]

If all this sounds nuts, that's because it is. The same press that published *Cosmic Catastrophes* has also published books claiming that Neandertals had telepathic abilities and were the driving force behind the lost civilization of Atlantis (Colin Wilson's *Atlantis and the Kingdom of the Neanderthals*), and that we all descend in part from aliens (Zecharia Sitchin's *Gods, Demigods, and Human Ancestry: The Evidence of Ancient DNA*). We are known by the company we keep.

It is, though, a truism of science that it does not matter where intriguing ideas come from, and Firestone, along with his *Cosmic Catastrophe* coauthor Allen West, convinced some very good scientists that some of the evidence for an impact event at 11,000 radiocarbon years ago might actually be worth pursuing. In 2007, they and 24 coauthors reworked their older and often bizarre arguments, adding a series of new and more compelling analyses and publishing their results in the prestigious *Proceedings of the National Academy of Sciences.*[42]

In that paper, they argued that at about 10,900 years ago, one or more extraterrestrial objects, most likely from a fragmenting comet that was more than 13 feet across, exploded somewhere over glacial ice in the eastern part of northern North America. That explosion resulted in a shock wave and fireball that devastated everything it encountered. Wildfires consumed forests and grasslands on a continental scale. Those fires destroyed the food source for North America's large Pleistocene mammals and produced charcoal, soot, toxic fumes, and ash that spewed into the air and blocked incoming sunlight. The impact of the fireball on the atmosphere depleted it of significant parts of its ozone shield, allowing greatly increased amounts of ultraviolet light to reach the surface and further harm the organisms living there. Brakenridge had suggested the very same thing years before, although he had a different cause—the Vela supernova—in mind. Huge sections of the eastern, Laurentide, ice mass melted, tossing water vapor into the atmosphere and dumping masses of cold, fresh water into the Atlantic Ocean. As a result, the oceanic circulation system that allows warm waters to flow to the northern Atlantic from the south shut down, in turn forcing dramatic changes in the earth's climate.

To Firestone and his colleagues, the results of all this included the formation of the late Ice Age black mats, the onset of the Younger Dryas cold snap, the extinction of many of North America's Ice Age mammals, and the demise of Clovis. Minus the cosmic catastrophes that occurred as early as 41,000 years ago, these impacts parallel what Firestone and his colleagues argued in their 2006 book. The *Proceedings of the National Academy of Sciences*, however, tends to avoid wacko science, and it did so here as well. Firestone's team of accomplished scientists amassed an impressive array of evidence to support their argument that the earth had been the scene of a major extraterrestrial impact about 10,900 years ago.

That evidence included the presence, in geological deposits that seemed to be of the right age, of

1. tiny magnetic spherules and magnetic grains thought to have been produced as part of the impact;

2. elevated levels of iridium (elevated levels of this element were key to the discovery that a massive meteor strike led to the extinction of dinosaurs some 66 million years ago);
3. elevated levels of charcoal, soot, and chemical compounds produced by wildfires;
4. tiny carbon spheres produced by burning;
5. distinctive, cage-like carbon structures known as fullerenes, said to be associated with meteorites and to contain extraterrestrial helium;
6. glass-like carbon that seemed to have been formed by melting; and
7. tiny diamonds—nanodiamonds—which they suggested are rare on earth but are known from meteorites.

Firestone's team argued that these markers are routinely found just beneath black mats that began to form as the Younger Dryas took hold, the same kind of black mat that covers Clovis artifacts at such sites as Lehner and Murray Springs. These mats, Firestone and his colleagues claimed, are composed of decaying plant and animal material and ash, all resulting from the fires ignited by the initial impact event or events. Black mat experts disagree.[43]

These arguments set off what can justly be described as a scientific, and public media, firestorm of their own. By my count, over 80 papers on this topic have now been published. One set of those papers convincingly demolishes the Younger Dryas Impact Hypothesis, as it now tends to be called. The other set vigorously defends it. Some of the original markers—for instance, the magnetic grains and fullerenes—have been dropped or significantly downplayed. Others—for instance, a particular form of tiny diamond known as lonsdaleite and a form of silica-rich glass known as lechatelierite—have been added.[44]

Through all this, some things have become abundantly clear. First, there is no evidence for continent-wide wildfires at the time called for by the impact hypothesis. In fact, on a global scale, the onset of the Younger Dryas was associated with a significant decrease in fire frequency, exactly the opposite of what the impact argument predicts.[45] In addition, recent analyses have shown that few of the deposits said to contain impact markers are securely dated to 10,900 years ago, and a good number of them clearly do not date to this time.[46] It may even be that the Younger Dryas–like cold snap might not have been unique to the end of the last glacial period.[47] If this were the case, then a unique event would not be needed to explain it.

The impact hypothesis debate is certainly going to continue for quite some time. From our perspective, however, this debate is important for only one reason: the argument that the suggested impact was the cause of late Ice Age extinctions in North America. If this were the case, a debate that has gone on for well over a century would be over and those of us who have struggled with this issue for our entire careers would be pleased to find something else to struggle with. Unfortunately, even if a substantial extraterrestrial object collided with the earth about 10,900 years ago—and that remains a huge "if"—it did not necessarily lead to the demise of all the animals we have talked about here. There are many, many reasons to feel that way.

The prime mechanism that the impact group suggests was the cause of the extinctions is those continent-wide wildfires. As I have mentioned, however, there is no evidence at all that these actually occurred.

Perhaps it was ozone depletion that led to the demise of these animals, even though the various research teams that see the impact event as the cause of the extinctions do not seem to call on this as the mechanism. However, physicist Adrian Melott and his colleagues have estimated that an airburst produced by an object as big as the one hypothesized by Firestone and his colleagues in that 2007 paper would produce about a 30 percent global decrease in the atmosphere's ozone content—some ten times the average global depletion the atmosphere has gone through recently.[48] This, in turn, would dramatically increase the amount of ultraviolet light reaching the earth. Could it be that this led to the demise of all those animals?

The answer here is "no" as well. By claiming that the hypothesized catastrophic extraterrestrial event caused the North American extinctions, the proponents of this explanation neglect the fact that whatever was happening in North America was happening in South America as well (chapter 3). The extinctions cannot be said to be particularly well dated in North America, but they are even less so to the south. Whether they are well dated or not, South America underwent massive extinctions—54 genera—as the Ice Age came to an end. If the hypothesized impact caused extinctions, it must have caused them from the southern end of Tierra del Fuego to the northernmost edge of Canada.

Perhaps atmospheric mixing saw to it that ozone was depleted on a global scale, thus affecting animals in both North and South America. That might provide the needed mechanism, even though it would not explain why such animals as bison and caribou survived.

Global ozone depletion at killing levels, though, wouldn't work either. Although Africa saw its share of late Ice Age losses at both the genus and species levels, those losses were far less pronounced than they were in the Americas.[49] Though we certainly have a lot to learn about late Ice Age Africa, as things stand now it seems that the Great Basin lost more genera of mammals during the late Ice Age (20 genera) than did Africa as a whole (7 genera). Globally depleted ozone cannot explain this, nor can it explain why so many African mammals were able to survive.

Finally, there is the intriguing situation presented by Europe and northern Asia. Along with so much of the rest of the world, these areas were not immune to the loss of large mammals as the Ice Age came to an end. Here, however, these losses are far better dated than they are anywhere else. As a result, and as I discussed in earlier chapters, we know that they were staggered in space and time.[50] Woolly mammoths, for instance, disappeared from much of Europe soon after 12,000 years ago or so but lasted until 10,000 years ago in Estonia, and until about 9,700 years ago on the mainland of far northern Siberia. They also lasted until 5,700 years ago on Saint Paul Island in the Bering Strait and until about 3,700 years ago on Wrangel Island, located in the Arctic Ocean north of the Siberian mainland.[51]

None of this is consistent with late Ice Age extinctions having been caused by ozone depletion resulting from an extraterrestrial impact. Nor is any of it consistent with an extraterrestrial impact having played any role in causing late Ice Age extinctions anywhere. How could an impact have caused mammoth extinctions in Alaska around 11,000 years ago[52] but spared those mammoths living on Wrangel and Saint Paul Islands, both about 300 miles away?

As if that were not enough, a significant increase in global ozone levels would have had a significant impact on shallow-water organisms and on organisms that live near the top of the water column, including phytoplankton. That, in turn, would have reduced primary food production in the oceans. Even the comparatively minor decrease in ozone that has happened during the past few decades appears to have wreaked havoc with amphibians.[53] The paleontological record does not show any of these predictable impacts at the end of the Ice Age. There are no known extinctions of large oceanic vertebrates and no wholesale losses of amphibians or fishes. While killer whales (or orcas, *Orcinus orca*) seem to have undergone a significant population decline across much of their realm during the late Pleistocene, that decline appears to have been during the last glacial maximum, thousands of years prior to the suggested impact event.[54] Even though they are at the top of the oceanic food chain, killer whales survived the killer impact just fine.

There is no question that extraterrestrial impacts have caused extinctions in the past. Dinosaurs might provide the most famous example, but an impact in eastern Argentina seems to have caused significant local extinctions about 3.3 million years ago.[55] Such things happen. But whether an extraterrestrial impact occurred in North America about 11,000 years ago or not, it played no role in the losses we have been discussing here.

CLIMATE

A few years ago, paleoecologist Jacquelyn Gill and her colleagues took a close look at tiny bits of late Ice Age history extracted from three lakes and marshes in Indiana and New York. Those tiny bits included charcoal, pollen, and the minute spores of the fungus known as *Sporormiella*.[56]

To a paleoecologist, *Sporormiella* is a wondrous thing because it thrives on herbivore dung. The more dung there is on the landscape, the happier *Sporormiella* becomes. If you listen carefully while walking through a cow pasture, you can sometimes hear the contented chirpings of happy dung fungus. The more herbivores there are on the landscape, the more *Sporormiella* spores get produced, and the more of them end up in such places as lakes and ponds. Since the spores preserve well, they can be used to track the abundance of herbivores on the landscape through time. There are some tricks to doing this, but the method works well and has been used in many parts of the world, with results that reach back into the Ice Age.[57]

Gill and her colleagues showed that the decline in *Sporormiella* spores in their samples began as early as about 12,500 years ago and continued to decline from there. By 10,000 years ago, the spores were virtually gone. We have no idea which herbivores produced the dung behind all this, but we can be certain that whatever they were, they were relatively common 12,500 years ago and pretty much gone a few thousand years later.

The meaning of this from our perspective seems fairly clear. The declines in the number of herbivores that were occurring near these lakes and marshes did not occur all at once. Rather than being an event, these declines were a process that spanned at least several thousand years. Assuming that the animals involved were part of the now-extinct late Ice Age fauna, this does not match the requirements of the Clovis-driven overkill hypothesis or those of the Younger Dryas impact argument.

If Clovis hunters and an extraterrestrial impact are unlikely to have driven North (and South) America's Pleistocene mammals to extinction, then what is left? A hypervirulent disease has been suggested as a possibility,[58] and there certainly is ample evidence that diseases can have significant impacts on entire species of mammals.[59] We also know that a variety of late Pleistocene mammals, including mastodon and helmeted muskox, had to deal with infectious tuberculosis. There is, however, no evidence that a significant number of the now-extinct North American mammals that made it to the late Pleistocene suffered in this way. Not even those who have provided the evidence for infectious diseases in late Ice Age mammals see it as the cause of the extinctions of those mammals.[60]

That leaves the obvious choice. The extinctions that we know occurred at the very end of the Ice Age—those 17 genera of mammals that I discussed in chapter 4—were lost during a time of major climate change. Perhaps, then, it was climate change that brought them to their knees.

Many of us who have spent a significant part of our careers studying these animals suspect that this was the case. As a result, a number of attempts have been made to show precisely how climate change could have led to the extinctions that impacted North America as the Ice Age came to an end.

These tend to be of two sorts. First, there have been bold attempts to show how late Ice Age climate change could have driven many kinds of North American mammals to extinction all at once. I will refer to these as "continental" explanations, with the continent being North America. Second, there have been bold attempts to show how climate change at the end of the Pleistocene could have driven a single species, and the genus to which it belonged, to extinction. I will refer to these as "single-species" explanations. We have learned a lot from the continental approach, though it has not shed much direct light on why the extinctions occurred. The single-species approach, on the other hand, is proving to be impressively productive.

There are a number of continental explanations for the North American late Ice Age extinctions, but the most important was developed by paleontologists Russ Graham and Ernie Lundelius and their colleagues in a series of important papers.[61] They argued that while the late Pleistocene saw significantly lower temperatures, temperature differences between summer and winter were likely much less than they became after the Ice Age ended. This, they argued, was because the extremely cold air masses that now flow into the central United States from the Arctic were then blocked by the massive ice that covered so much of Canada and parts of the northern United States during the late Pleistocene. Once that massive ice had melted enough, Arctic air could again enter the heart of what is now the United States, producing far colder winters. Average annual temperatures rose at the same time as seasonal swings in temperature from winter to summer increased dramatically.

Graham and Lundelius argued that these increases in seasonal temperature swings led to huge changes in the plant communities of North America, with more complex late Ice Age plant assemblages replaced by the structurally simpler ones that have been with us ever since. It was this reorganization of plant communities, coupled with the drastic swings in temperature from winter to summer and back again, that led to the extinctions.

Graham and Lundelius were driven to make this argument for the same reason that a similar argument was made during the nineteenth century in Europe.[62] Animals that today do not, and quite clearly cannot, live together were very much doing so during the late Ice Age. For instance, the plains pocket gopher (*Geomys bursarius*) and the taiga vole (*Microtus xanthognathus*) do not occur remotely close to one another today. The former lives in the plains of the central United States and far southern Canada; the latter lives in far northwestern North America. During the late Pleistocene, however, they were living on the same spot at the same time in Tennessee.

Sets of small animals like this are so consistently found in the North American paleontological record that they get a name. They are called "nonanalog faunas," with "nonanalog" referring to the lack of a modern analog for them. Their very existence implies that the climatic conditions under which they lived have no modern counterpart. And if that were the case, it would also follow that the large mammals that lived alongside them also lived in climatic conditions that no longer exist. Those nonanalog faunas disappeared as the Pleistocene ended. In fact, they seem to have disappeared at about the same time as some of the larger mammals disappeared.

To Graham and Lundelius, only one explanation seemed likely. As the Ice Age came to an end, colder winters and warmer summers forced these pairs to split up. The same climate changes that caused these splits, they argued, caused the extinction of larger mammals.

Although this argument had been made before, the version Graham and Lundelius presented was far more refined, and far more powerful, than any of the earlier versions. They had a mechanism to account for changing seasonal swings in temperatures—the presence, and then absence, of the massive glaciers in the north. They also had empirical evidence that could be best explained by those changes—the nonanalog pairs. They were even able to provide radiocarbon dates that documented that the members of these pairs were living at the same place at the same time.[63]

This impressive body of work accomplished some very important things that I will talk about shortly. One thing it did not accomplish, however, was explaining all of the extinctions that occurred at the end of the North American Ice Age. To take an obvious example, the late Ice Age faunas of California west of the Sierra Nevada saw huge losses, superbly documented at such sites as Rancho La Brea (chapter 3). The mechanism that Graham and Lundelius called on to cause the climate change so critical to their argument—the melting of glacial ice in the center of the continent—could not have played this role here. While their explanation might help us understand what was happening to some mammals in some parts of North America, it cannot account for the full run of extinctions.

What, then, did this work accomplish to merit the space I have given it here?

First, Graham and Lundelius focused their attention not on the large, charismatic mammals that had previously drawn so much attention, but on such small animals as chipmunks, gophers, and voles. Others had paid attention to these animals before,[64] but Graham and Lundelius drove home the fact that what was happening to the large mammals could not be adequately understood without understanding the fate of the small ones. Faunal change at the end of the Ice Age formed a continuum, with many big animals going extinct at the same time as small animals were moving across the landscape.

This, in fact, is exactly what one would predict if climate change were calling the shots. Paleontologists have long pointed out that under conditions of climate change, big mammals with slow reproductive rates can be expected to be far more vulnerable to extinction than small ones with fast reproductive rates.[65] It is not just paleontologists who have noted this. Through the analysis of a large sample of contemporary mammal species, biologists Christy McCain and Sarah King have shown that during very recent times, the largest responses to climate change have come from the largest mammals. Such species as caribou and polar bears, they showed, are 27 times more likely to respond to climate change than are shrews and mice. These large-bodied forms, they concluded, are at the most risk of extinction under conditions of climate change in the future.[66]

That big mammals are in general more at risk in times of significant climate change does not in itself tell us much about the causes of the extinctions. The same is true if human hunting were the cause. However, only climate change can account for the fact that at the same time as at least 17 of the large mammal extinctions were occurring in North America, small mammals were undergoing dramatic changes in their ranges. It might also help explain why the Great Basin today supports only one species of large scavenging bird, even though it supported six, and perhaps seven, of them during the Ice Age (chapter 3).

BUILDING INDIVIDUAL SPECIES HISTORIES

The second major contribution that Graham and Lundelius's work made in this realm was to emphasize that the assemblages of plants and animals that we see around us today are not composed of species that are tightly bound together by their adaptations to particular climates and to each other. Instead, this work helped confirm the truth of the view championed by botanist Henry Gleason much earlier in the century.

Gleason argued that "every species of plant is a law unto itself, the distri-bution of which in space depends on its individual peculiarities of migration and environmental requirements."[67] In this view, if conditions changed, the plant species in a biological community would respond not as a group—for instance, by moving across the landscape as a coherent assemblage—but each would instead respond in its own way. The associations of plants that we see today, he concluded, did not have to be that way but instead represented "merely a coincidence." This view he famously referred to as the "individual-istic concept of the plant association."[68]

Gleason's view, now referred to as "Gleasonian individualism," was radi-cally different from what was then the standard view of a plant community, which saw such communities as organic wholes. Gleason's view was also radically correct. That his view captures the way the world works started to become clear in the 1950s, as a result of studies of modern assemblages of plants. It became even clearer when those working with more ancient plant communities—for instance, those from the Ice Age—observed that the members of those communities did not migrate across space together under conditions of climate change, but instead went their own individual ways. By the 1970s, it had become clear that Gleason's view did, in fact, apply to the plant world.

Gleason was a botanist, his arguments meant to apply to the botanical world. Russ Graham and Ernie Lundelius, on the other hand, are specialists in the deeper histories of mammals. Their work played a major role in showing that Gleasonian individualism applies to mammals as well as it does to plants. This was a huge accomplishment. It is now well recognized that understanding the past and future of any set of organisms requires that the history of each species in that set be deciphered on its own. Gleason's view rules the day.

The implications of this for understanding late Ice Age extinctions in North America, or anywhere else, are clear. Rather than trying to explain them all at once, we need to focus on individual species. We need to know not only when they became extinct, but also their population histories as ex-tinction neared. We need to know the precise nature of climate change as that was happening, and we need to know, insofar as it can be known, the nature of their interactions with people during those times. We need, in short, single-species histories and single-species explanations.

We already have such histories for a wide variety of European and north-ern Eurasian mammals—such things as woolly rhinoceros and the giant deer or "Irish Elk." For many of these animals, we know that the losses occurred in different places at different times, we know that many of the animals had long been human prey, and we often, though not always, know quite a bit about the nature of the landscapes and climates as the time of extinction approached. As a result, no one argues that human hunting was the driving cause of these losses. A recent, impressive study that combined an analysis of ancient DNA, climate change, and archaeology concluded that climate change seemed to be the prime driver in these extinctions, with the possible exception of decreased populations of Eurasian bison and horses. These two animals, however, survived the end of the Pleistocene. In fact, they survive today.[69]

The only truly comparable analyses we have for North America come from the far north. For Alaska, thanks to the efforts of such scientists as Dale Guthrie, Dan Mann, and their colleagues, we have detailed radiocarbon chronologies for a variety of late Pleistocene mammals, including horse (gone by about 10,500 years ago) and mammoth (gone by about 11,500 years ago).[70] There is no compelling evidence for human predation on these animals in Alaska, but there is evidence, provided by the analysis of ancient plant DNA preserved in permafrost, of substantial vegetational change as these animals were reaching extinction.[71] That, in turn, matches evidence, provided by Guthrie, that Alaskan horses were becoming smaller as their extinction neared.[72] On Alaska's North Slope, Mann and his colleagues have shown, the extinctions coincided with changes in temperature, vegetation, and moisture regimes that were inimical to such animals as mammoth and horse but well suited to moose and caribou. Given the lack of evidence that people interacted with the animals that became extinct, all of these authors conclude that climate change was the prime driver of the losses.

It also appears that people could not have been involved in the extinction of the mastodon in Alaska and the Yukon. Paleontologist Grant Zazula and his colleagues obtained over 50 radiocarbon dates for three dozen mastodon bones and teeth from this region and found that virtually all of the animals had died long before 30,000 years ago.[73] No matter what caused the extinction of the mastodon far to the south, the far northern populations must have succumbed to something else: there were no people in far northern North America at the time these animals were lost (chapter 6).

Paleontologist Eric Scott has tried something similar for western North America, including the southern Great Basin. Even though he was not armed with an impressive array of radiocarbon dates, his bold work emphasized that the composition of late Ice Age faunas in the American West changed through time, and that unless those changes and their timing are understood, we are not likely to be able to understand why the extinctions occurred.[74] Even more recently, archaeologist Daniel Gilmour and his colleagues have begun to build a detailed species history for the extinct late Ice Age mammals of Oregon's Willamette Valley, an important first for this region.[75] Whenever studies of this sort are done, the twin lessons are the same: First, we have no chance of understanding late Ice Age extinctions unless we attempt to understand the detailed history of each species on its own. Second, when such individual species histories have been built, they routinely point to climate change as the prime cause.

If, it might be asked, climate change toward the end of the last glacial episode caused the North American extinctions, why weren't the ends of earlier glacial episodes during the Pleistocene also accompanied by extinctions? This is one of the most frequent, and long-standing, criticisms lobbed at those who suspect that climate was behind it all.[76]

It is also a criticism made by those who assume that all the losses occurred at the same time, and that what explains one of them should explain all of them. It is this very set of assumptions that led to so little progress in understanding late Ice Age extinctions during the past two centuries.

Lacking detailed information on the later Pleistocene histories of each of the animals that was lost, and having no idea of the key environmental variables that may have played a role in causing any given extinction, we have no idea what to look for at the end of earlier glacial episodes. If we don't know what to look for, it is easy to conclude that the ends of prior glacial episodes were just like the one that happened at the very end of the Ice Age. Once you conclude that, it is far easier to conclude that climate could not have caused the losses than it is to conclude that it is only your own assumptions that make you think this. Science is a tricky enterprise.

A FOND FAREWELL

As things stand now, there is no compelling reason to think that people hunted many of North America's late Pleistocene herbivores to extinction. There are also no compelling climate-based explanations for the extinction of those animals south of glacial ice.

Because of this, there has been a recent tendency to conclude that multiple factors working together led to these losses, that some combination of climate change and human activities caused the extinctions.[77] This approach might work well were we able to combine the strengths that each of these explanations might have separately. However, because such strengths are hard to come by, many of these arguments accomplish little more than combining the weaknesses of two very different approaches to understanding the past.

Others, however, are far more interesting and may show the way to a more productive explanatory future. Zoologist Graham Prescott and his colleagues, for instance, used a sophisticated mathematical model to argue that the rapid cooling, great changes in temperature, and low average temperatures of the latest Pleistocene may have set it apart from earlier Ice Age times. This, they suggest, might have played a role in bunching extinctions at the end of the Ice Age, and folding in the human arrival in the Americas may have simply made things worse.[78]

This may well have been the case, but Prescott and his colleagues were interested in Ice Age extinctions on a global scale. As a result, they drew their information on climate from an ice core in Antarctica. Since they wanted to extend their study deep into the Pleistocene, this was the only approach they could have used. But if we are interested in knowing what happened to mammoths in the Great Basin, ground sloths in the Southwest, or mastodons in the Great Lakes region, a very different approach is needed.

The only way out of this explanatory morass is to tackle the extinct animals one at a time, building local individual species histories as I have talked about in this chapter. It worked in Europe, and there is no reason to think that it will not work in North America. It will take a lot of effort, but until that has been done, we are not likely to see any deeper understanding of these losses than we have now. Even Paul Martin, the author of the overkill hypothesis, might have agreed with this:

> Most fossils are proof of nothing more than an organism's existence and death. In very few cases can we infer an unambiguous cause of individual mortality, much less of its extinction as a species. This makes it difficult to extract definitive answers from the fossil record.[79]

Just as Darwin predicted over 150 years ago, difficult is exactly what explaining American Ice Age extinctions has proved to be.

The Relationship between Radiocarbon (^{14}C) and Calendar Years for 10,000 to 25,000 Radiocarbon Years Ago

^{14}C years	Calendar years	^{14}C years	Calendar years	^{14}C years	Calendar years	^{14}C years	Calendar years
10,000	11,478	14,100	17,139	18,200	21,730	22,300	26,844
10,100	11,700	14,200	17,274	18,300	21,844	22,400	27,121
10,200	11,902	14,300	17,386	18,400	21,967	22,500	27,281
10,300	12,095	14,400	17,520	18,500	22,141	22,600	27,321
10,400	12,273	14,500	17,654	18,600	22,239	22,700	27,411
10,500	12,466	14,600	17,769	18,700	22,316	22,800	27,640
10,600	12,559	14,700	17,882	18,800	22,387	22,900	27,777
10,700	12,622	14,800	18,000	18,900	22,477	23,000	27,867
10,800	12,678	14,900	18,279	19,000	22,590	23,100	27,950
10,900	12,767	15,000	18,250	19,100	22,758	23,200	28,045
11,000	12,867	15,100	18,243	19,200	22,885	23,300	28,141
11,100	12,993	15,200	18,497	19,300	23,010	23,400	28,214
11,200	13,107	15,300	18,588	19,400	23,113	23,500	28,270
11,300	13,197	15,400	18,640	19,500	23,317	23,600	28,342
11,400	13,271	15,500	18,707	19,600	23,452	23,700	28,439
11,500	13,352	15,600	18,762	19,700	23,567	23,800	28,543
11,600	13,434	15,700	18,829	19,800	23,663	23,900	28,663
11,700	13,551	15,800	18,950	19,900	23,761	24,000	28,804
11,800	13,652	15,900	19,113	20,000	23,888	24,100	28,920
11,900	13,759	16,000	19,153	20,100	24,036	24,200	29,023
12,000	13,856	16,100	19,255	20,200	24,122	24,300	29,164
12,100	13,944	16,200	19,371	20,300	24,228	24,400	29,309
12,200	14,050	16,300	19,454	20,400	24,343	24,500	29,389
12,300	14,224	16,400	19,513	20,500	24,453	24,600	29,449
12,400	14,451	16,500	19,606	20,600	24,596	24,700	29,514
12,500	14,657	16,600	19,737	20,700	24,691	24,800	29,595
12,600	14,878	16,700	19,835	20,800	24,790	24,900	29,727
12,700	15,053	16,800	19,980	20,900	24,916	25,000	29,883

¹⁴C years	Calendar years	¹⁴C years	Calendar years	¹⁴C years	Calendar years	¹⁴C years	Calendar years
12,800	15,211	16,900	20,097	21,000	25,042		
12,900	15,401	17,000	20,206	21,100	25,170		
13,000	15,617	17,100	20,298	21,200	25,299		
13,100	15,869	17,200	20,394	21,300	25,422		
13,200	16,115	17,300	20,518	21,400	25,611		
13,300	16,376	17,400	20,739	21,500	25,750		
13,400	16,574	17,500	20,826	21,600	25,898		
13,500	16,683	17,600	20,993	21,700	26,028		
13,600	16,763	17,700	21,174	21,800	26,127		
13,700	16,836	17,800	21,301	21,900	26,287		
13,800	16,901	17,900	21,382	22,000	26,442		
13,900	16,964	18,000	21,454	22,100	26,505		
14,000	17,043	18,100	21,561	22,200	26,640		

Note: Conversions were done with CALIB Radiocarbon Calibration Version 6.0 (http://calib.qub.ac.uk/calib/), using a standard deviation of 50 years and the IntCal09 calibration dataset. The calendar year is the median age provided by CALIB for each radiocarbon age. Calibrations for the last 10,000 years are in Grayson (2011).

Common and Scientific Names of Plants
Discussed in the Text

Common name	Scientific name	Common name	Scientific name
Acacia	*Acacia*	Koa	*Acacia koa*
Agave	*Agave*	Limber pine	*Pinus flexilis*
Banana yucca	*Yucca baccata*	Melon loco	*Apodanthera undulata*
Big sagebrush	*Artemisia tridentata*	Mojave yucca	*Yucca schidigera*
Bluebunch wheatgrass	*Pseudoroegneria spicata*	Mormon tea	*Ephedra viridis*
Blue paloverde	*Parkinsonia florida*	Nevada jointfir	*Ephedra nevadensis*
Boxelder	*Acer negundo*	Ocotillo	*Fouquieria splendens*
Bristlecone pine	*Pinus longaeva*	Pine	*Pinus*
Buffalo gourd	*Cucurbita foetidissima*	Saguaro	*Carnegiea gigantea*
Catclaw acacia	*Acacia greggii*	Saltbush	*Atriplex*
Chaparral yucca	*Hesperoyucca whipplei*	Sandberg bluegrass	*Poa secunda*
Common juniper	*Juniperus communis*	Screwbean mesquite	*Prosopis pubescens*
Common reed	*Phragmites australis*	Singleleaf pinyon	*Pinus monophylla*
Coyote gourd	*Cucurbita palmata*	Soaptree yucca	*Yucca elata*
Creosote bush	*Larrea tridentata*	Spanish bayonet	*Yucca harrimaniae*
Crown of thorns	*Koeberlinia spinosa*	Squirreltail	*Elymus elymoides*
Desert globemallow	*Sphaeralcea ambigua*	Subalpine fir	*Abies lasiocarpa*
Desert holly	*Atriplex hymenelytra*	Texas mulberry	*Morus microphylla*
Desert ironwood	*Olneya tesota*	Twoneedle pinyon	*Pinus edulis*
Engelmann spruce	*Picea engelmannii*	Utah agave	*Agave utahensis*
Fingerleaf gourd	*Cucurbita digitata*	Utah juniper	*Juniperus osteosperma*
Fourwing saltbush	*Atriplex canescens*	Utah yucca	*Yucca utahensis*
Fremont cottonwood	*Populus fremontii*	Water birch	*Betula occidentalis*
Great Basin wildrye	*Leymus cinereus*	Western juniper	*Juniperus occidentalis*
Honey mesquite	*Prosopis glandulosa*	Whitebark pine	*Pinus albicaulis*
Idaho fescue	*Festuca idahoensis*	White bursage	*Ambrosia dumosa*
Jointfir	*Ephedra*	White fir	*Abies concolor*
Joshua tree	*Yucca brevifolia*	Willow	*Salix*
Jumping cholla	*Cylindropuntia fulgida*	Yellow paloverde	*Parkinsonia microphylla*

Scientific name	Common name	Scientific name	Common name
Abies concolor	White fir	*Juniperus osteosperma*	Utah juniper
Abies lasiocarpa	Subalpine fir	*Koeberlinia spinosa*	Crown of thorns
Acacia	Acacia	*Larrea tridentata*	Creosote bush
Acacia greggii	Catclaw acacia	*Leymus cinereus*	Great Basin wildrye
Acacia koa	Koa	*Morus microphylla*	Texas mulberry
Acer negundo	Boxelder	*Olneya tesota*	Desert ironwood
Agave	Agave	*Parkinsonia florida*	Blue paloverde
Agave utahensis	Utah agave	*Parkinsonia microphylla*	Yellow paloverde
Ambrosia dumosa	White bursage	*Phragmites australis*	Common reed
Apodanthera undulata	Melon loco	*Picea engelmannii*	Engelmann spruce
Artemisia tridentata	Big sagebrush	*Pinus*	Pine
Atriplex	Saltbush	*Pinus albicaulis*	Whitebark pine
Atriplex canescens	Fourwing saltbush	*Pinus edulis*	Twoneedle pinyon
Atriplex hymenelytra	Desert holly	*Pinus flexilis*	Limber pine
Betula occidentalis	Water birch	*Pinus longaeva*	Bristlecone pine
Carnegiea gigantea	Saguaro	*Pinus monophylla*	Singleleaf pinyon
Cucurbita digitata	Fingerleaf gourd	*Poa secunda*	Sandberg bluegrass
Cucurbita foetidissima	Buffalo gourd	*Populus fremontii*	Fremont cottonwood
Cucurbita palmata	Coyote gourd	*Prosopis glandulosa*	Honey mesquite
Cylindropuntia fulgida	Jumping cholla	*Prosopis pubescens*	Screwbean mesquite
Elymus elymoides	Squirreltail	*Pseudoroegneria spicata*	Bluebunch wheatgrass
Ephedra	Jointfir	*Salix*	Willow
Ephedra nevadensis	Nevada jointfir	*Sphaeralcea ambigua*	Desert globemallow
Ephedra viridis	Mormon tea	*Yucca baccata*	Banana yucca
Festuca idahoensis	Idaho fescue	*Yucca brevifolia*	Joshua tree
Fouquieria splendens	Ocotillo	*Yucca elata*	Soaptree yucca
Hesperoyucca whipplei	Chaparral yucca	*Yucca harrimaniae*	Spanish bayonet
Juniperus communis	Common juniper	*Yucca schidigera*	Mojave yucca
Juniperus occidentalis	Western juniper	*Yucca utahensis*	Utah yucca

Tall (> 6 Feet) Mechanically Defended Plants of the Mojave and Sonoran Deserts (Mx: Species found in the Sonoran Desert of Mexico)

Achatocarpaceae
Phaulothamnus spinescens (Mx)
Agavaceae
Yucca schidigera
Yucca ×schottii
Yucca valida (Mx)
Arecaceae
Washingtonia filifera
Asteraceae
Chloracantha spinosa
Bombacaceae
Ceiba acuminata (Mx)
Cactaceae
Carnegiea gigantea
Cylindropuntia acanthocarpa
Cylindropuntia arbuscula
Cylindropuntia bigelovii
Cylindropuntia cholla (Mx)
Cylindropuntia echinocarpa
Cylindropuntia ×fosbergii
Cylindropuntia fulgida
Cylindropuntia munzii
Cylindropuntia prolifera
Ferocactus cylindraceus
Ferocactus emoryi
Ferocactus wislizeni
Myrtillocactus cochal (Mx)
Pachycereus pecten-aboriginum (Mx)
Pachycereus pringlei (Mx)

Pachycereus schottii
Stenocereus gummosus (Mx)
Stenocereus thurberi
Capparaceae
Koeberlinia spinosa
Celastraceae
Canotia holacantha
Euphorbiaceae
Adelia virgata (Mx)
Fabaceae
Acacia brandegeana (Mx)
Acacia cochliacantha (Mx)
Acacia constricta
Acacia farnesiana
Acacia greggii
Acacia kelloggiana (Mx)
Acacia millefolia
Acacia occidentalis (Mx)
Acacia pacensis (Mx)
Acacia peninsularis (Mx)
Acacia pennatula (Mx)
Acacia pringlei (Mx)
Cercidium praecox (Mx)
Ebenopsis confinis (Mx)
Erythrina flabelliformis
Haematoxylon brasiletto (Mx)
Havardia mexicana (Mx)
Havardia sonorae (Mx)
Mimosa distachya

Mimosa grahamii
Mimosa margaritae (Mx)
Mimosa palmeri (Mx)
Olneya tesota
Parkinsonia aculeata
Parkinsonia florida
Parkinsonia microphylla
Pithecellobium leucospermum (Mx)
Prosopidastrum mexicanum (Mx)
Prosopis glandulosa
Prosopis palmeri (Mx)
Prosopis pubescens
Prosopis velutina
Psorothamnus spinosus
Robinia neomexicana
Fouquieriaceae
Fouquieria burragei (Mx)
Fouquieria diguettii (Mx)
Fouquieria macdougallii (Mx)
Fouquieria splendens
Polemoniaceae
Acanthogilia gloriosa (Mx)
Rhamnaceae
Colubrina californica
Condalia brandegeei (Mx)
Condalia globosa
Condalia warnockii
Ziziphus amole (Mx)
Ziziphus obtusifolia
Ziziphus parryi
Rosaceae

Prunus fremontii
Rubiaceae
Randia echinocarpa (Mx)
Randia megacarpa (Mx)
Randia obcordata (Mx)
Randia sonorensis (Mx)
Randia thurberi (Mx)
Rutaceae
Zanthoxylum arborescens (Mx)
Zanthoxylum sonorense (Mx)
Sapotaceae
Sideroxylon peninsulare (Mx)
Simaroubaceae
Castela emoryi
Castela peninsularis (Mx)
Castela polyandra (Mx)
Solanaceae
Lycium andersonii
Lycium berlandieri
Lycium brevipes
Lycium exsertum
Lycium fremontii
Lycium macrodon
Lycium parishii
Lycium torreyi
Solanum hindsianum
Theophrastaceae
Jacquinia macrocarpa (Mx)
Ulmaceae
Celtis ehrenbergiana
Celtis iguanaea (Mx)

References: Shreve and Wiggins 1964; Wiggins 1980; Benson and Darrow 1981; R. Turner, Bowers, and Burgess 1995; Baldwin et al. 2002, 2012.

APPENDIX 4

Maximum Height and Armature
of Sonoran and Mojave Desert Shrubs

Species	Armed?	Maximum height (m)	Species	Armed?	Maximum height (m)
Acanthaceae			*Ambrosia ilicifolia*	Yes	1.0
Anisacanthus thurberi	No	2.0	*Amphipappus fremontii*	No	0.6
Justicia californica	No	2.0	*Baccharis brachyphylla*	No	2.0
Justicia candicans	No	1.0	*Baccharis emoryi*	No	3.0
Anacardiaceae			*Baccharis pteronioides*	No	2.0
Rhus aromatica	No	2.5	*Baccharis salicifolia*	No	4.0
Rhus kearneyi	No	5.5	*Baccharis sarothroides*	No	3.0
Rhus ovata	No	4.0	*Baccharis sergiloides*	No	2.0
Toxicodendron diversilobum	No	2.5	*Barkleyanthus salicifolius*	No	1.5
			Bebbia juncea	No	1.0
Toxicodendron rydbergii	No	0.3	*Brickellia atractyloides*	No	0.4
Apocynaceae			*Brickellia baccharidea*	No	1.0
Haplophyton crooksii	No	0.6	*Brickellia californica*	No	1.5
Asclepiadaceae			*Brickellia coulteri*	No	1.0
Asclepias albicans	No	4.0	*Brickellia floribunda*	No	1.5
Asclepias subulata	No	3.0	*Brickellia frutescens*	No	0.4
Asteraceae			*Brickellia incana*	No	1.3
Acamptopappus shockleyi	No	0.5	*Brickellia knappiana*	No	0.3
Acamptopappus sphaerocephalus	No	0.5	*Brickellia longifolia*	No	3.0
			Brickellia microphylla	No	0.6
Adenophyllum cooperi	No	0.5	*Brickellia oblongifolia*	No	0.5
Adenophyllum porophylloides	No	0.6	*Brickellia venosa*	No	0.7
			Chrysothamnus albidus	No	1.0
Ambrosia ambrosioides	No	1.0	*Chrysothamnus pulchellus*	No	0.5
Ambrosia cordifolia	No	1.0			
Ambrosia deltoidea	No	0.6	*Encelia farinosa*	No	0.6
Ambrosia dumosa	No	0.5	*Encelia frutescens*	No	1.0
Ambrosia eriocentra	No	1.0	*Ericameria brachylepis*	No	2.0

Species	Armed?	Maximum height (m)
Ericameria cooperi	No	1.0
Ericameria cuneata	No	1.0
Ericameria laricifolia	No	0.8
Ericameria linearifolia	No	1.0
Ericameria nauseosa	No	2.0
Ericameria palmeri	No	2.0
Ericameria paniculata	No	2.0
Ericameria pinifolia	No	3.0
Ericameria teretifolia	No	1.0
Gutierrezia californica	No	0.5
Gutierrezia microcephala	No	0.7
Gutierrezia sarothrae	No	0.4
Gutierrezia serotina	No	0.6
Gymnosperma glutinosum	No	1.0
Hazardia brickellioides	Yes	0.8
Hecastocleis shockleyi	Yes	0.8
Hymenoclea monogyra	No	4.0
Hymenoclea salsola	No	1.0
Isocoma acradenia	No	1.0
Isocoma pluriflora	No	1.0
Isocoma tenuisecta	No	1.0
Lepidospartum latisquamum	No	2.0
Lepidospartum squamatum	No	2.0
Machaeranthera carnosa	No	1.0
Oxytenia acerosa	No	2.0
Parthenium incanum	No	0.6
Peucephyllum schottii	No	4.0
Pleurocoronis pluriseta	No	1.0
Pluchea sericea	No	3.0
Porophyllum gracile	No	1.0
Psilostrophe cooperi	No	0.5
Senecio flaccidus	No	2.0
Tetradymia axillaris	Yes	1.0
Tetradymia comosa	Yes	1.0
Tetradymia glabrata	No	1.0
Tetradymia stenolepis	Yes	1.0
Thymophylla acerosa	No	0.2
Trixis californica	No	0.8

Species	Armed?	Maximum height (m)
Viguiera deltoidea	No	0.5
Viguiera reticulata	No	1.0
Xylorhiza cognata	No	1.0
Xylorhiza orcuttii	No	1.0
Xylorhiza tortifolia	No	1.0
Zinnia acerosa	No	0.3
Berberidaceae		
Berberis harrisoniana	Yes	1.0
Mahonia fremontii	Yes	5.0
Mahonia haematocarpa	Yes	2.0
Mahonia nevinii	Yes	4.0
Mahonia trifoliolata	Yes	2.5
Bignoniaceae		
Chilopsis linearis	No	8.0
Tecoma stans	No	2.0
Brassicaceae		
Lepidium fremontii	No	1.0
Buddlejaceae		
Buddleja sessiliflora	No	6.0
Buddleja utahensis	No	0.5
Burseraceae		
Bursera fagaroides	No	5.0
Bursera microphylla	No	5.0
Capparaceae		
Atamisquea emarginata	No	3.0
Cleome isomeris	No	1.5
Koeberlinia spinosa	Yes	3.5
Caprifoliaceae		
Sambucus caerulea	No	12.0
Celastraceae		
Canotia holacantha	Yes	6.0
Mortonia scabrella	No	1.0
Chenopodiaceae		
Allenrolfea occidentalis	No	1.0
Atriplex acanthocarpa	No	1.0
Atriplex canescens	No	2.0
Atriplex confertifolia	Yes	0.8
Atriplex hymenelytra	Yes	1.0
Atriplex lentiformis	Yes	3.0
Atriplex parryi	Yes	0.3
Atriplex polycarpa	No	1.0

Species	Armed?	Maximum height (m)	Species	Armed?	Maximum height (m)
Atriplex spinifera	Yes	0.8	*Dalea formosa*	No	1.0
Grayia spinosa	Yes	1.0	*Dalea pulchra*	No	0.8
Krascheninnikovia lanata	No	0.8	*Dalea versicolor*	No	1.2
Sarcobatus vermiculatus	Yes	1.5	*Erythrina flabelliformis*	Yes	8.0
Suaeda moquinii	No	1.0	*Eysenhardtia orthocarpa*	No	6.0
Suaeda suffrutescens	No	1.0	*Hoffmannseggia microphylla*	No	2.0
Crossosomataceae					
Crossosoma bigelovii	No	2.0	*Lysiloma divaricatum*	No	3.0
Glossopetalon spinescens	Yes	2.0	*Mimosa aculeaticarpa*	Yes	2.0
Cupressaceae			*Mimosa distachya*	Yes	3.0
Juniperus monosperma	No	5.0	*Mimosa dysocarpa*	Yes	2.0
Ephedraceae			*Mimosa grahamii*	Yes	0.6
Ephedra californica	No	1.2	*Parkinsonia aculeata*	Yes	15.0
Ephedra nevadensis	No	1.3	*Parkinsonia microphylla*	Yes	8.0
Ephedra torreyana	No	1.0	*Prosopis glandulosa*	Yes	17.0
Ephedra trifurca	Yes	2.0	*Prosopis juliflora*	Yes	17.0
Ephedra viridis	No	1.5	*Prosopis pubescens*	Yes	10.0
Euphorbiaceae			*Prosopis velutina*	Yes	17.0
Acalypha californica	No	1.0	*Psorothamnus arborescens*	Yes	1.0
Argythamnia brandegeei	No	2.0	*Psorothamnus emoryi*	No	1.0
Bernardia myricifolia	No	2.0	*Psorothamnus fremontii*	Yes	2.0
Croton ciliatoglandulifer	No	1.0	*Psorothamnus polydenius*	Yes	2.5
Croton sonorae	No	2.0	*Psorothamnus schottii*	Yes	2.1
Euphorbia misera	No	1.5	*Psorothamnus spinosus*	Yes	8.0
Jatropha cardiophylla	No	1.0	*Senna armata*	Yes	0.6
Jatropha cinerea	No	6.0	*Sophora arizonica*	No	2.0
Jatropha cuneata	No	2.5	*Zapoteca formosa*	No	4.0
Manihot angustiloba	No	2.0	**Fagaceae**		
Manihot davisiae	No	3.0	*Quercus turbinella*	No	7.0
Sebastiania bilocularis	No	6.0	**Fouquieriaceae**		
Tetracoccus hallii	No	2.0	*Fouquieria splendens*	Yes	9.0
Tetracoccus ilicifolius	Yes	1.5	**Garryaceae**		
Fabaceae			*Garrya flavescens*	No	4.0
Acacia angustissima	No	1.0	**Grossulariaceae**		
Acacia constricta	Yes	3.0	*Ribes quercetorum*	Yes	1.0
Acacia farnesiana	Yes	10.0	**Hydrophyllaceae**		
Acacia greggii	Yes	7.0	*Eriodictyon angustifolium*	No	1.0
Acacia millefolia	No	3.0	*Eriodictyon crassifolium*	No	3.0
Calliandra eriophylla	No	1.0	*Eriodictyon trichocalyx*	No	1.5
Cercis occidentalis	No	7.0	**Krameriaceae**		
Coursetia glandulosa	No	2.0	*Krameria erecta*	No	0.5

Species	Armed?	Maximum height (m)	Species	Armed?	Maximum height (m)
Krameria grayi	Yes	0.5	*Eriogonum fasciculatum*	No	3.0
Lamiaceae			*Eriogonum heermannii*	No	2.0
Hyptis emoryi	No	3.0	*Eriogonum plumatella*	No	0.6
Poliomintha incana	No	1.0	*Eriogonum wrightii*	No	1.0
Salazaria mexicana	Yes	1.0	**Rhamnaceae**		
Salvia apiana	No	3.0	*Colubrina texensis*	No	2.0
Salvia dorrii	Yes	1.0	*Condalia correllii*	Yes	3.0
Salvia eremostachya	No	1.0	*Condalia ericoides*	Yes	1.0
Salvia funerea	Yes	1.5	*Condalia globosa*	Yes	5.0
Salvia greatae	Yes	1.0	*Condalia warnockii*	Yes	3.0
Salvia mohavensis	No	1.0	*Ziziphus obtusifolia*	Yes	3.0
Salvia vaseyi	No	1.0	*Ziziphus parryi*	Yes	3.0
Loasaceae			**Rosaceae**		
Petalonyx nitidus	No	0.4	*Amelanchier alnifolia*	No	6.0
Petalonyx parryi	No	1.5	*Cercocarpus montanus*	No	8.0
Petalonyx thurberi	No	1.0	*Coleogyne ramosissima*	Yes	1.5
Malvaceae			*Fallugia paradoxa*	No	2.0
Abutilon berlandieri	No	2.0	*Prunus andersonii*	Yes	2.0
Abutilon incanum	No	4.0	*Prunus fasciculata*	Yes	3.0
Abutilon palmeri	No	2.0	*Prunus fremontii*	Yes	5.0
Abutilon thurberi	No	0.6	*Purshia glandulosa*	No	6.0
Gossypium thurberi	No	2.0	*Purshia mexicana*	No	3.5
Hibiscus coulteri	No	1.0	*Purshia ×subintegra*	No	10.0
Hibiscus denudatus	No	1.0	*Rosa woodsii*	Yes	3.0
Horsfordia alata	No	4.0	*Vauquelinia californica*	No	8.0
Horsfordia newberryi	No	3.0	**Rubiaceae**		
Malvastrum bicuspidatum	No	2.0	*Cephalanthus occidentalis*	No	8.0
Oleaceae			**Rutaceae**		
Forestiera acuminata	No	4.0	*Thamnosma montana*	No	0.8
Forestiera shrevei	No	4.0	**Salicaceae**		
Fraxinus anomala	No	7.0	*Salix exigua*	No	7.0
Fraxinus gooddingii	No	8.0	*Salix lasiolepis*	No	15.0
Menodora scabra	No	0.8	*Salix sessifolia*	No	10.0
Menodora spinescens	Yes	1.0	**Sapindaceae**		
Polemoniaceae			*Dodonaea viscosa*	No	3.0
Linanthus pungens	No	1.0	*Sapindus saponaria*	No	5.0
Polygalaceae			**Sapotaceae**		
Polygala acanthoclada	Yes	0.9	*Sideroxylon lanuginosum*	Yes	4.0
Polygonaceae			**Scrophulariaceae**		
Eriogonum deserticola	No	1.5	*Diplacus aridus*	Yes	0.5
			Keckiella antirrhinoides	No	1.5

Species	Armed?	Maximum height (m)
Simaroubaceae		
Castela emoryi	Yes	4.0
Simmondsiaceae		
Simmondsia chinensis	No	2.0
Solanaceae		
Lycium andersonii	Yes	3.0
Lycium berlandieri	Yes	2.5
Lycium brevipes	Yes	3.0
Lycium californicum	Yes	1.5
Lycium cooperi	Yes	2.0
Lycium exsertum	Yes	3.0
Lycium fremontii	Yes	3.0
Lycium macrodon	Yes	3.0
Lycium pallidum	Yes	2.0
Lycium parishii	Yes	3.0
Lycium torreyi	Yes	3.0
Nicotiana glauca	No	6.4
Sterculiaceae		
Ayenia compacta	No	0.5
Ayenia microphylla	No	0.6
Hermannia pauciflora	No	0.3
Waltheria indica	No	2.0
Ulmaceae		
Celtis ehrenbergiana	No	3.0
Verbenaceae		
Aloysia wrightii	No	2.0
Lantana urticoides	No	2.0
Zygophyllaceae		
Fagonia laevis	No	0.6
Larrea tridentata	No	3.0

References: Sonoran and Mojave Desert shrub sample: Benson and Darrow 1981.
Plant heights: Shreve and Wiggins 1964; Benson and Darrow 1981; R. Turner, Bowers, and Burgess 1995; Felger, Johnson, and Wilson 2001; Welsh et al. 2008; Baldwin et al. (2012).
Plant names: USDA NRCS 2014.

APPENDIX 5

Maximum Height and Armature
of Great Basin Shrubs

Species	Armed?	Maximum height (m)
Aceraceae		
Acer glabrum	No	8.0
Anacardiaceae		
Rhus trilobata	No	2.5
Toxicodendron diversilobum	No	4.0
Toxicodendron rydbergii	No	0.3
Asteraceae		
Artemisia arbuscula	No	0.5
Artemisia cana	No	1.5
Artemisia nova	No	0.5
Artemisia tridentata	No	3.0
Brickellia californica	No	1.0
Brickellia microphylla	No	0.7
Brickellia oblongifolia	No	0.6
Chrysothamnus albidus	No	1.5
Chrysothamnus viscidiflorus	No	1.0
Ericameria bloomeri	No	0.9
Ericameria nana	No	0.5
Ericameria nauseosa	No	2.0
Ericameria parryi	No	0.6
Ericameria suffruticosa	No	0.4
Gutierrezia sarothrae	No	0.6
Hymenoclea salsola	No	2.0
Picrothamnus desertorum	Yes	0.8
Tetradymia axillaris	Yes	1.2

Species	Armed?	Maximum height (m)
Tetradymia canescens	No	0.9
Tetradymia glabrata	No	1.2
Tetradymia nuttallii	Yes	1.2
Tetradymia spinosa	Yes	1.2
Tetradymia tetrameres	No	2.0
Berberidaceae		
Mahonia repens	Yes	1.2
Betulaceae		
Alnus incana	No	10
Brassicaceae		
Lepidium fremontii	No	0.8
Caprifoliaceae		
Lonicera conjugialis	No	1.8
Lonicera involucrata	No	2.0
Lonicera utahensis	No	2.0
Sambucus nigra	No	8.0
Sambucus racemosa	No	2.0
Symphoricarpos acutus	No	1.0
Symphoricarpos longiflorus	Yes	1.0
Symphoricarpos oreophilus	No	1.5
Symphoricarpos rotundifolius	No	1.8
Chenopodiaceae		
Allenrolfea occidentalis	No	1.4
Atriplex bonnevillensis	No	0.8
Atriplex canescens	No	2.0

Species	Armed?	Maximum height (m)	Species	Armed?	Maximum height (m)
Atriplex confertifolia	Yes	0.8	**Polemoniaceae**		
Atriplex torreyi	Yes	3.0	*Linanthus pungens*	No	0.3
Atriplex tridentata	No	0.8	**Polygonaceae**		
Bassia americana	No	0.5	*Eriogonum heermannii*	No	2.0
Grayia spinosa	Yes	1.5	*Eriogonum lobbii*	No	2.0
Krascheninnikovia lanata	No	1.0	*Eriogonum microthecum*	No	1.5
Sarcobatus baileyi	Yes	1.0	*Eriogonum nummulare*	No	2.0
Sarcobatus vermiculatus	Yes	2.1	*Eriogonum sphaerocephalum*	No	0.4
Suaeda moquinii	No	1.5			
Cornaceae			*Eriogonum wrightii*	No	1.0
Cornus sericea	No	4.0	**Rhamnaceae**		
Crossosomataceae			*Ceanothus cordulatus*	Yes	< 1.5
Glossopetalon spinescens	Yes	1.0	*Ceanothus velutinus*	No	1.5
Cupressaceae			*Frangula rubra*	No	< 2.0
Juniperus communis	No	< 1.0	**Rosaceae**		
Elaeagnaceae			*Amelanchier alnifolia*	No	5.0
Shepherdia argentea	Yes	4.0	*Amelanchier pallida*	No	4.0
Shepherdia canadensis	No	2.0	*Amelanchier pumila*	No	3.0
Ephedraceae			*Amelanchier utahensis*	No	4.0
Ephedra nevadensis	No	1.0	*Cercocarpus intricatus*	No	2.0
Ephedra viridis	No	1.0	*Cercocarpus montanus*	No	4.0
Ericaceae			*Chamaebatiaria millefolium*	No	2.0
Arctostaphylos nevadensis	No	0.5			
			Coleogyne ramosissima	Yes	2.0
Arctostaphylos patula	No	3.0	*Holodiscus discolor*	No	1.5
Vaccinium uliginosum	No	0.6	*Physocarpus alternans*	No	1.2
Fabaceae			*Prunus andersonii*	No	< 3.0
Psorothamnus kingii	No	0.4	*Prunus emarginata*	No	4.0
Psorothamnus polydenius	Yes	1.5	*Purshia stansburiana*	No	3.5
Fagaceae			*Purshia tridentata*	No	2.0
Chrysolepis sempervirens	No	2.5	*Rosa woodsii*	Yes	3.0
Grossulariaceae			**Salicaceae**		
Ribes aureum	No	3.0	*Salix exigua*	No	8.0
Ribes cereum	No	2.0	**Solanaceae**		
Ribes velutinum	Yes	2.0	*Lycium andersonii*	Yes	3.0
Lamiaceae			*Lycium cooperi*	Yes	2.0
Salvia dorrii	Yes	0.8	*Lycium shockleyi*	Yes	0.6

References: Floristic Great Basin shrub sample: Mozingo 1987.
Plant heights: Welsh et al. 2008; Baldwin et al. 2012; relevant volumes of *Intermountain Flora.*

NOTES

Chapter 1: A Sloth in Prison

1. "Footprints of monster men" 1882; this story was taken from the August 8 edition of the *San Francisco Chronicle*.
2. "Mysterious tracks in stone" 1882; this story was taken from the August 4 edition of the *San Francisco Call*.
3. Bancroft 1890
4. Gibbes 1882:7
5. Ibid.
6. Ibid.
7. Leviton and Aldrich 1997
8. Ibid.; Orsi 2005. In 1884, Scupham also helped arrange California's exhibit at the World's Industrial and Cotton Centennial Exposition in New Orleans.
9. Gibbes 1882
10. Ibid.; Leviton and Aldrich 1997
11. "Footprints of monster men" 1882
12. Harkness 1882:2–3
13. Harkness 1882
14. Cope 1883e:70
15. "Fossil human foot-prints" 1883:137–38
16. Le Conte 1882:6
17. Ibid.:8; emphasis in original
18. Ibid.
19. For example, "Fossil footprints of Nevada" 1882; "Carson footprints" 1883
20. Cope 1883c
21. For example, "Recent discoveries of fossil footprints" 1882
22. Cope 1882
23. Gibbes 1882:6
24. Leviton and Aldrich 1997
25. Stephens 1982
26. Le Conte 1883
27. Grayson 1983
28. Argyll 1883:579
29. Cope 1883c
30. "Nevada footprints" 1883
31. Leviton and Aldrich 1997

32. "Dr. Harkness" 1883; "Carson footprints" 1883; "Carson footprints: Second paper" 1883
33. G. Davidson 1883
34. Leviton and Aldrich 1997
35. G. Davidson 1883:7
36. Ibid.
37. Blake 1884
38. Leviton and Aldrich 1997
39. Tylor 1885:154
40. Bancroft 1890:314
41. Yerington 1895:8–9
42. Ibid.:28
43. "Carson footprints: An amateur professional" 1883
44. For example, Twain 1885; Emerson 1980; Marché 1986; Berkove 2006
45. Twain (1872) 1981; Bancroft 1890
46. Stock 1917
47. Stock 1920
48. Stock 1925
49. Stock 1936b:27
50. H. McDonald 2007
51. Glen Whorton, former director of the Nevada Department of Corrections, personal communication, 2014
52. Skrupa 2014
53. The last execution in the Nevada State Prison was in 2006. Death Penalty Information Center, http://www.deathpenaltyinfo.org/node/5741#NV.
54. Nevada Legislature 2014

Chapter 2: The Great Basin Now and Then

1. Grayson 2011
2. Frémont 1845:175
3. Ibid.:275
4. Jackson 1970
5. Oregon State University 2015
6. Preuss 1958
7. Frémont and Preuss 1848
8. Jackson 1970
9. Manly 1894; L. Johnson and Johnson 1987; Landon 1999
10. Grayson 2011
11. Cronquist et al. 1972
12. Thompson and Anderson 2000
13. Lachniet et al. 2014
14. Kleman et al. 2010; Lachniet et al. 2014
15. England et al. 2006
16. Grayson 2011
17. Joughin, Smith, and Medley 2014; Rignot et al. 2014
18. Hu et al. 2010
19. Grayson 2011

20. Ibid.

21. See, for instance, Nelson and Madsen (1978) and K. Adams and Wesnousky (1998).

22. As I have mentioned, calendar years and radiocarbon years are not the same thing (see chapter 4). In radiocarbon years, the Younger Dryas fell between about 11,000 and 10,100 years ago. In calendar years, however, it fell between about 12,900 and 11,700 years ago and thus lasted about 1,200 "real" years (S. Rasmussen et al. 2006; Steffensen et al. 2008).

23. There is an immense literature on the Younger Dryas; for starters, see S. Rasmussen et al. 2006; Broecker et al. 2010; Steffensen et al. 2008; Condron and Winsor 2012; Lane et al. 2013; and A. Moreno 2014.

24. Bridgham et al. 2013

25. O. Davis 2002

26. Maher 1964

27. Woolfenden 1996, 2003; Mensing 2001

28. Mensing et al. 2004

29. Louderback and Rhode 2009

30. Betancourt, Van Devender, and Martin 1990; Rhode 2001

31. Rhode 2003; Rhode and Louderback 2007

32. West, Tausch, and Tueller 1998

33. Jennings and Elliott-Fisk 1993; Grayson 2011; Cole et al. 2013

34. Lanner 1984; Zouhar 2001

35. Bradley and Fleishman 2008

36. Rhode and Madsen 1995; Rhode 2000; Wigand and Rhode 2002

37. Wells 1983; Thompson 1984, 1992; Rhode and Madsen 1995; Rhode 2000; Grayson 2011

38. Wigand and Mehringer 1985; Wigand and Nowak 1992; Grayson 2000a, 2000b; Wigand and Rhode 2002

39. For very different views of the abundances of large mammals during the early Holocene in the Great Basin, see Broughton et al. (2008) and Grayson (2011:236–48).

40. Grayson 2011; Cole et al. 2013

41. Pavlik 2008

42. Cole at al. 2013

43. International Commission on Stratigraphy, http://stratigraphy.org/

44. Sanders 2002; Sanders, Weems, and Albright 2009; E. Scott 2010

Chapter 3: A Zoologically Impoverished World

1. Wallace 1876:149

2. Wallace 1876:150

3. Wroe et al. 2013; Faith 2014

4. Pope (1733–1734) 1963, epistle 1:8

5. Lovejoy 1964:184

6. T. Jefferson 1799:255–56; for a discussion of this paper, see Thomson (2008, 2011).

7. Parkinson 1804:469–70

8. Grayson 1983

9. Ray and Sanders 1984; G. Jefferson 1989a; Sanders 2002; Graham 2003

10. Weinstock et al. 2005

11. Ferrusquía-Villafranca et al. 2010

12. E. Scott et al. 2010

13. P. Martin 1967

14. Fariña, Vizcaíno, and De Iuliis (2013) provide an excellent discussion of the approaches to, and the pitfalls of, estimating the weights of extinct Pleistocene mammals.

15. G. Morgan 2008

16. A. Gardner 2007

17. I have placed all the glyptodonts in a single family; not all agree (Porpino, Fernicola, and Bergqvist 2010).

18. A. Gardner 2007

19. R. Nowak 1999

20. Sources for the body weights given in the text are provided in table 3.1.

21. Krause et al. 2008; Figueirido and Soibelzon 2010

22. Rincón, Prevosti, and Parra 2011

23. Fariña 1996

24. Larivière 2001

25. Ortiz, Pardiñas, and Steppan 2000; Pardiñas, D'Elía, and Ortiz 2002

26. Garrott and Oli 2013

27. Alberdi and Prado 1998; Weinstock et al. 2005; Orlando et al. 2009

28. Alberdi and Prado 1995; Forasiepi, Martinelli, and Blanco 2007

29. For instance, Guérin and Faure 1999; Fariña, Vizcaíno, and De Iuliis 2013

30. For instance, C. Scherer et al. 2007

31. Geist 1998; Lord 2007; Merino and Rossi 2010

32. Merino and Rossi 2010

33. Menégaz and Ortiz Jaureguizar 1995; Cione et al. 1999; Alcaraz and Zurita 2004

34. Welker et al. 2015

35. Webb 2006; G. Morgan 2008; Woodburne 2010; Montes et al. 2012, 2015; Coates and Stallard 2013; Hoorn and Flantua 2015; Bacon et al. 2015

36. G. Simpson 1980

37. Bond 1999

38. Bond et al. 2001; Geist 1998

39. Bond, Cerdeño, and López 1995; Bond 1999

40. Bond, Cerdeño, and López 1995; Bond 1999; McFadden 2005

41. Rodríguez-de la Rosa, Guzmán-Gutiérrez, and Ortega-Hurtado de Mendoza 2011; Lundelius et al. 2013

42. Arroyo-Cabrales et al. 2007; on the possibility that mammoths or mastodons made it to South America, see Fariña, Vizcaíno, and De Iuliis (2013:200–201).

43. H. McDonald and Perea 2002; H. McDonald and De Iuliis 2008; Corona, Perea, and McDonald 2013

44. Pujos 2000; Miño-Boilini and Carlini 2009; Saint-André et al. 2010

45. McKenna and Bell 1997; Sedor et al. 2004

46. Shoshani 1996; Shoshani and Tassy 2005; Lucas 2008; Lucas and Alvarado 2010

47. For instance, Ferretti 2008

48. Alberdi and Prado 1995; Prado et al. 2001; Gutiérrez et al. 2005; Encina and Alberdi 2011; Alberdi et al. 2012; Prado et al. 2012

49. Asevedo et al. 2012; Mothé et al. 2012; Mothé, Avilla, and Cozzuol 2013; Mothé and Avilla 2015

50. Mothé et al. 2012
51. Dillehay 1997; Frassinetti and Alberdi 2000, 2005; Encina and Alberdi 2011
52. Hartwig 1995; Guedes and Salles 2005; Halenar 2011; Halenar and Rosenberger 2013
53. Cartelle and Hartwig 1996; Guedes and Salles 2005; Halenar and Rosenberger 2013
54. Bond 1999; Guérin and Faure 1999, 2004; C. Scherer, Pitana, and Ribeiro 2009
55. To create these maps, I began with Faunmap II (Graham and Lundelius 2010), which focuses on the United States. I then added sites that are not currently in that database, including those from Mexico north of the Tropic of Cancer and from Canada. The references used to do that are so lengthy that they have not been included here.
56. Unless otherwise indicated, body weights for extinct North American mammals come from F. Smith et al. (2003).
57. Gillette and Ray 1981; Gillette and Whisler 1986
58. Cartelle and De Iuliis 2006; Fariña, Vizcaíno, and De Iuliis 2013
59. H. McDonald and Lundelius 2009
60. G. Jefferson 1987; G. Jefferson 2003
61. Mawby 1967
62. Gillette, McDonald, and Hayden 1999
63. Lull 1930; Adam 1999
64. Laudermilk and Munz 1938; P. Martin, Sabels, and Shutler 1961; Hansen 1978
65. Laudermilk and Munz 1934; Poinar et al. 1998; Hofreiter et al. 2000; F. Gill et al. 2009
66. Eames 1930
67. Thompson et al. 1980
68. Spaulding and Martin 1979
69. Green 2009
70. Eames 1930; Lull 1930
71. G. Jefferson 1990; C. Davis and Smith 1981; R. Reynolds et al. 1991
72. Mawby 1967
73. H. McDonald and Pelikan 2006
74. Saunders 1977
75. F. Moreno 1899b:144; see also F. Moreno (1899a, 1902). For a more detailed, and slightly different, version of Moreno's discovery, see Fariña, Vizcaíno, and De Iuliis (2013). Moreno himself gave the date as both 1897 (F. Moreno 1899b, 1902) and 1898 (1899a).
76. F. Moreno 1899b:145
77. Ibid.
78. Vizcaíno et al. 2001; Fariña, Vizcaíno, and De Iuliis 2013
79. Ameghino 1898:324
80. For example, Nordenskjold 1899; Woodward 1900
81. Hesketh Prichard 1902:xiii
82. Ibid.
83. Oren 2001:3
84. W. Miller 1976
85. Stock 1925; Elftman 1931

86. Goldstein et al. 2008
87. Durbin et al. 2008
88. Ripoll et al. 2010
89. Arroyo-Cabrales and Polaco 2003; Arroyo-Cabrales and Carranza Castañeda 2009; Ferrusquía-Villafranca et al. 2010
90. E. Anderson 1996, 2004
91. Hockett and Dillingham 2004
92. Heaton 1985
93. Christiansen 1999; Sorkin 2006; Figueirido et al. 2010
94. Madsen 2000a
95. Emslie and Czaplewski 1985
96. Schubert and Kaufmann 2003
97. Figueirido et al. 2010
98. Kurtén 1967a, 1976b; Emslie and Czaplewski 1985; Matheus 1995, 2001
99. Sorkin 2006; Figueirido, Palmqvist, and Pérez-Claros 2009; Figueirido et al. 2010
100. Soibelzon and Schubert 2011
101. Richards, Churcher, and Turnbull 1996
102. O'Keefe, Fet, and Harris 2009
103. L. Miller 1910a:4
104. Barnett et al. 2005
105. Christiansen and Harris 2005
106. Meachen-Samuels and Van Valkenburg 2010; Meachen-Samuels 2012
107. Antón 2013:196
108. L. Martin, Babiarz, and Naples 2011
109. Merriam and Stock 1932; Akersten 1985; Anyonge 1996; Wheeler 2011
110. Feranec 2004, 2008
111. Akersten 1985; Antón et al. 2004; McHenry et al. 2007; Wheeler 2011
112. Meachen-Samuels and Van Valkenburgh 2010; L. Martin, Babiarz, and Naples 2011; Wheeler 2011
113. DeSantis et al. 2012
114. Coltrain et al. 2004; Christiansen and Harris 2005
115. McHorse, Orcutt, and Davis 2012
116. Binder and Van Valkenburgh 2010
117. Carbone et al. 2009; Van Valkenburgh et al. 2009
118. McCall, Naples, and Martin 2003
119. Van Valkenburgh, Teaford, and Walker 1990; L. Martin, Babiarz, and Naples 2011
120. Heaton 1985
121. Stout 1986; Livingston 1991; Land 1992; G. Jefferson, McDonald, and Livingston 2004
122. Fortsch 1978; G. Jefferson 1989b
123. San Bernardino County Museum 2012
124. L. Martin, Babiarz, and Naples 2011
125. Rawn-Schatzinger 1992; L. Martin, Naples, and Babiarz 2011; Antón 2013
126. Anyonge 1993; Marean and Erhardt 1995; F. Smith et al. 2003; Haas 2005
127. Evans 1961; Meade 1961; Graham 1976b; Rawn-Schatzinger 1992; Marean and Erhardt 1995. Graham, Lundelius, and Meissner (2013) provide a very thorough review of all the work done at Friesenhahn Cave.

128. Marean and Erhardt 1995
129. Graham 1976b
130. Feranec 2008
131. L. Martin, Babiarz, and Naples 2011
132. Churcher 1966; L. Martin, Naples, and Babiarz 2011
133. Van Valkenburgh 1988, 2009
134. Rawn-Schatzinger 1992; Marean and Erhardt 1995; Wheeler 2011; for a recent, detailed discussion of *Homotherium*, see Antón et al. (2014).
135. Kitchen 1974; Byers 1997; O'Gara 2004
136. Byers 1997:12
137. Byers 1997; A. Turner and Antón 1997
138. D. Adams 1979; Krausman and Morales 2005
139. Ibid.
140. A. Russell and Bryant 2001
141. D. Adams 1979; Van Valkenburgh, Grady, and Kurtén 1990
142. Van Valkenburgh, Grady, and Kurtén 1990
143. Ibid.
144. A. Turner and Antón 1997
145. Barnett et al. 2005
146. G. Simpson 1941; Kurtén 1976a; Ray 1967
147. Krausman and Morales 2005
148. Orr 1969
149. Hockett and Dillingham 2004
150. Deméré 2007
151. S. Jenkins and Busher 1979; P. Reynolds 2002; Hopkins 2008
152. Parmalee and Graham 2002; McDonald and Bryson 2010
153. Mead 1987; Hafner 1994; Hafner and Sullivan 1995; Mead and Grady 1996; Grayson 2005; Millar and Westfall 2010; Calkins et al. 2012
154. B. Russell and Harris 1986; Harris 2014
155. Jass 2007
156. R. Guthrie 2001; Elias and Crocker 2008; Stuart and Lister 2012
157. Hulbert 1995; Guérin and Patou-Mathis 1996
158. Cozzuol et al. 2013
159. Ray and Sanders 1984; G. Jefferson 1989a; Hulbert 1995; Sanders 2002; Graham 2003
160. Winans 1989; Azzaroli 1998
161. Weinstock et al. 2005; Orlando et al. 2008, 2009
162. Hulbert, Morgan, and Kerner 2009
163. Czaplewski 2012
164. Lundelius 1960; Nye 2007
165. Beebe 1980; Zazula and Froese 2013
166. Hillenius 1992
167. Lundelius 1960; Guilday, Hamilton, and McCrady 1971; Finch, Whitmore, and Sims 1972; Wetzel et al. 1975; Nye 2007
168. van Roosmalen et al. 2007
169. Lewis 1970
170. Finch, Whitmore, and Sims 1972

171. Munson 1991
172. Wetzel et al. 1975; Mayer and Wetzel 1986
173. Jim Patton, personal communication, March 8, 2006; quoted with his permission
174. Cozzuol et al. 2013; van Roosmalen et al. 2007
175. H. McDonald 2002
176. Pickford, Morales, and Soria 1995; Titov and Logvynenko 2006; Sen 2010
177. Honey et al. 1998; Heintzman et al. 2015
178. Heintzman et al. 2015
179. Webb 1974; Webb and Stehli 1995; Meachen 2003, 2005; Meachen and Hallman 2002
180. Feranec 2003; Kohn, McKay, and Knight 2005; Hoppe and Koch 2006; Semprebon and Rivals 2010
181. Graham 1992; Webb and Stehli 1995
182. Graham 1992; Kohn, McKay, and Knight 2005
183. Harrison 1979; Honey et al. 1998; Meachen 2003
184. Kohn, McKay, and Knight 2005
185. Dompierre and Churcher 1996; Semprebon and Rivals 2010; Feranec and MacFadden 2000; Feranec 2003
186. Webb and Stehli 1995
187. Webb 1965
188. Akersten, Foppe, and Jefferson 1988; Dompierre and Churcher 1996; Semprebon and Rivals 2010
189. Heintzman et al. 2015
190. Vetter, Lachniet, and Rowland 2007
191. R. Guthrie 2006; Meiri et al. 2014
192. Burns 2010
193. J. McDonald, Ray, and Grady 1996
194. Bedetti, Palombo, and Sardella 2001; García and Arsuaga 2003; Banks et al. 2008
195. Grayson and Delpech 2005
196. Churcher and Pinsof 1987; Farlow and McClain 1996; Geist 1998
197. Breda 2005, 2008
198. Pinsof 1995; Geist 1998; Schubert et al. 2004; Breda and Marchetti 2005; Breda 2008; C. Long and Yahnke 2011
199. H. Kahlke 1990; Lister 1993; Geist 1998; Harington 2011
200. For example, Churcher 1991; Churcher and Pinsof 1987; Breda and Marchetti 2005; Breda 2008
201. Breda and Marchetti 2005; Harington 2011
202. Schultz and Howard 1936
203. Kurtén 1975
204. Blackford 1995
205. Kurtén 1975; Kurtén and Anderson 1980; Blackford 1995; Geist 1998
206. Morejohn and Dailey (2004) claim that the remains assigned to *Navahoceros* actually belong to *Odocoileus*, the genus to which our mule deer belong. Because they provide no analyses to support this claim, and because they seem to have been unaware of the detailed analysis of the postcranial skeleton of *Navahoceros* provided by Blackford (1995), there is no reason to accept this claim.
207. Webb 1992, 2000

208. For a discussion of the guemal skeleton, see Flueck and Smith-Flueck (2011).

209. C. Gilbert, Ropiquet, and Hassanin 2006

210. Nelson and Madsen 1987; Blackford 1995

211. Hockett and Dillingham 2004

212. Byers 1997

213. Goss 1983

214. O'Gara 1990; O'Gara and Janis 2004

215. Skinner 1942; Webb 1973; E. B. Davis 2007

216. Roosevelt and Burden 1934

217. White 2008:370

218. Stock 1930b, 1932; Colbert and Chaffee 1939; Skinner 1942; Furlong 1943; Webb 1973; Janis and Manning 1998; E. B. Davis 2007

219. Skinner 1942; Furlong 1943; White and Morgan 2011

220. Geiser 1953

221. Lull 1921:167

222. Lull 1921; Dalquest 1974; Janis and Manning 1998

223. The horn cores that led Lull to describe Shuler's pronghorn can be seen on the Shuler Museum's website, http://smu.edu/shulermuseum/type_specimens /SMU60004.html.

224. Gillette and Miller 1999

225. Mawby 1967

226. White and Morgan 2011

227. Saysette 1999; Stock 1956

228. Geist 1998; Lord 2007; Merino and Rossi 2010

229. R. Nowak 1999

230. J. Morgan and Morgan 1995; Jiménez-Hidalgo, Carranza-Castañeda, and Montellano-Ballesteros 2004; White and Morgan 2011

231. White and Morgan 2011

232. Hernández Fernández and Vrba 2005

233. Leslie and Sharma 2009; Baskaran et al. 2011

234. G. Jefferson 1989b, 1991a, 1991b

235. Downs et al. 1959

236. S. Miller 1979

237. Sokolov 1974; Heptner, Nasimovich, and Bannikov 1988; Harington and Cinq-Mars 1995; Milner-Gulland et al. 2001; J. Young et al. 2010; J. Berger, Young, and Berger 2008

238. R. Kahlke 1991; García and Arsuaga 2003; Harington 1981

239. J. McDonald and Bartlett 1983; Crégut-Bonnoure 1984; Guérin and Patou-Mathis 1996; García and Arsuaga 2003; Campos, Willerslev, Sher, et al. 2010; Graham and Lundelius 2010

240. MacPhee et al. 2002; Boeskorov 2006; MacPhee and Greenwood 2007

241. Campos, Sher, et al. 2010

242. Sinclair 1904; Sinclair and Furlong 1904; Sinclair 1905; Stock and Furlong 1927

243. Campos, Sher, et al. 2010

244. Furlong 1905; Stock and Furlong 1927; Gazin 1933; Frick 1937

245. Knopf, Mead, and Anderson 2007

246. Mead, O'Rourke, and Foppe 1986

247. McGuire 1980; Hockett and Dillingham 2004

248. Hattori 1982; Dansie and Jerrems 2005

249. G. Morgan and Lucas 2005

250. Harington 1968; J. McDonald 1984; J. McDonald, Neusius, and Clay 1987; J. McDonald and Ray 1989; R. Guthrie 1992

251. R. Guthrie 1992

252. Nelson and Madsen 1978, 1980, 1987; Gillette and Miller 1999; W. Miller 2002

253. Nelson and Madsen 1980

254. Lundelius et al. 2013

255. Rodríguez-de la Rosa, Guzmán-Gutiérrez, and Ortega-Hurtado de Mendoza 2011; Arroyo-Cabrales et al. (2010) report a possible record from Tamaulipas in northern Mexico, but to my knowledge this has not been confirmed.

256. Sanders 2002

257. Lundelius 1972

258. Mead et al. 2006; Bright et al. 2010; Nunez et al. 2010

259. Sanchez et al. 2014; see also Meltzer 2014

260. Saunders 1996; Polaco et al. 2001; Arroyo-Cabrales et al. 2007; Lucas, Alvarado, et al. 2007

261. Saunders 1996:274

262. King and Saunders 1984; G. Haynes 1991; Saunders 1996; Hubbard, Fisher, and Kordulias 2000; Shoshani and Marchant 2001

263. Lepper et al. 1991; Koch, Hoppe, and Webb 1998; Hoppe and Koch 2006; Gobetz and Bozarth 2001; Green, Semprebon, and Solounias 2005; Teale and Miller 2012

264. Newsom and Mihlbachler 2006

265. Saunders 1977; King and Saunders 1984; Saunders 1996; G. Haynes 1991; Shoshani and Marchant 2001

266. Shoshani and Marchant 2001; Shoshani 2003

267. Poole 1994; Fisher 2009

268. Fisher 2008:280

269. Fisher 2008, 2009

270. Grayson and Meltzer 2002, 2015

271. Graham et al. 1981; Graham and Kay 1988

272. Shipman, Fisher, and Rose 1984

273. Gustafson and Manis 1984

274. Gilbow 1981; Grayson and Meltzer 2002

275. Waters et al. 2011

276. Poole 1994

277. Gustafson, Gilbow, and Daugherty 1979; Gustafson 1985

278. Meltzer 2012, personal communication

279. Gustafson and Manis (1984) note these skeletal pathologies in passing, but later discussions of the Manis mastodon do not mention them; they are evident on even casual inspection of the skeleton.

280. Spurr 1903:157; G. Jefferson, McDonald, and Livingston 2004

281. Nelson and Madsen 1987

282. W. Miller 1987; Madsen 2000a

283. Lucas, Alvarado, et al. 2007; Arroyo-Cabrales et al. 2007

284. G. Haynes 1991; Agenbroad and Mead 1996; Lister and Bahn 2007

285. Saunders et al. 2010; Pasenko and Schubert 2004
286. Shoshani and Tassy 2005
287. Mol et al. 2001
288. Vereshchagin and Baryshnikov 1982; Kubiak 1982; R. Guthrie 1990; G. Haynes 1991; Ukraintseva 2013; Kharlamova et al. 2014
289. White 1986
290. Römpler et al. 2006
291. K. L. Campbell et al. 2010; for a detailed discussion of the genetics of the woolly mammoth's adaptation to the cold, see Lynch et al. (2015).
292. Van Geel et al. 2008
293. Lister and Sher 2001; Lister et al. 2005
294. Barnes et al. 2007; Debruyne et al. 2008; M. Gilbert, Drautz, et al. 2008; Nyström et al. 2010
295. Nikolskiy, Sulerzhitsky, and Pitulko 2011
296. Vartanyan et al. 2008
297. Veltre et al. 2008
298. Enk et al. 2009; Nyström et al. 2010
299. Grayson and Meltzer 2002, 2015
300. Meltzer and Sturtevant 1983; Griffin et al. 1988
301. Purdy et al. 2011; MacFadden et al. 2012; Purdy 2012
302. Haynes 1991
303. Hansen 1980; O. Davis et al. 1984; Mead, Agenbroad, et al. 1986; Agenbroad and Mead 1996; Mead and Swift 2012
304. Koch et al. 1998; Hoppe and Koch 2006; Metcalfe et al. 2011
305. Hoppe 2004; Hoppe and Koch 2007; Hoppe et al. 1999; Hoppe and Koch 2006; Metcalfe et al. 2011
306. Agenbroad 2003, 2009
307. Herridge and Lister 2012; Simmons 1999; Muhs et al. 2015
308. Rick et al. 2012; Muhs et al. 2015
309. Roth 1996; Agenbroad 1999, 2009, 2012; Agenbroad et al. 2005; Rick et al. 2012
310. Enk et al. 2011
311. Saunders et al. 2010; Pasenko and Schubert 2004
312. Fisher 2001; Hoyle et al. 2004
313. Hall 1981; Taulman and Robbins 1996; A. Gardner 2007
314. Letts 2011
315. Klippel and Parmalee 1984; Schubert and Graham 2000; Carbot-Chanona 2010; Letts 2011; Shapiro, Graham, and Letts 2014
316. Turner and Antón 1997; Montellano-Ballesteros and Carbot-Chanona 2009; Stuart and Lister 2011
317. G. Simpson 1941; Christiansen and Harris 2009
318. Barnett et al. 2009
319. Graham, Farlow, and Vandike 1996
320. Meachen-Samuels and Binder 2010; Wheeler and Jefferson 2009. For somewhat different estimates of the weights of American lions, see Christiansen and Harris (2009).
321. Van Valkenburgh and Hertel 1998
322. Elftman 1931

323. Firby, Mawby, and Davis 1987
324. Merriam 1915; G. Simpson 1941
325. S. Miller 1979
326. Mawby 1967
327. G. Jefferson 1991b
328. E. McCain and Childs 2008
329. S. Miller 1979; G. Jefferson 1990, 1991a
330. There is a huge and complex literature on the genetic relationship between dogs and wolves. If you are interested, good places to start are Thalmann et al. (2013) and Freedman et al. (2014). On the archaeology of ancient dogs, see Morey (2014).
331. R. Nowak 1979
332. Dundas 1999; Prevosti and Rincón 2007; Prevosti, Tonni, and Bidegain 2009
333. Kurtén 1984; Anyonge and Roman 2006
334. Mech 1974
335. Van Valkenburgh and Sacco 2002; McHorse, Orcutt, and Davis 2012
336. Mech 1974; Anyonge and Roman 2006
337. Elftman 1931
338. W. Miller 1976
339. T. Allen 2012
340. Tsong 2010; "Federal judge dismisses" 2012
341. Lyman 1998
342. Utah Division of Wildlife Resources 2013a, 2013b
343. Utah Division of Wildlife Resources 2013b:4
344. Lori Hunsaker, Utah deputy state historic preservation officer, personal communication
345. Kelly 1995
346. Hall 1981
347. Campos, Willerslev, Mead, et al. 2010
348. Stock 1936a
349. Mead 1983; Mead, Martin, et al. 1986; Mead et al. 1987; Mead and Lawler 1994; Jass, Mead, and Logan 2000
350. Mead 1983; Mead, O'Rourke, and Foppe 1986; Mead and Lawler 1994
351. Emslie and Czaplewski 1985; Mead and Lawler 1994
352. Howard 1962, 1971b; Feduccia 1974; Steadman et al. 1994; Emslie 1985; D. Guthrie 2009; G. Morgan and Emslie 2010
353. Ornithologists, the scientists who study birds, capitalize the common names of their species; mammalogists, who study mammals, do not. I follow that convention here.
354. L. Miller 1910b; DeMay 1941a, 1941b; Emslie 1998
355. Howard 1955; G. Jefferson 1985, 2003
356. Hayward et al. 1976; Ryser 1985; Alcorn 1988
357. Ryser 1985; Alcorn 1988
358. Shufeldt 1892
359. Howard 1946
360. Howard 1955; G. Jefferson 1985
361. G. Jefferson 1985, 2003
362. Ross 1935; Howard 1936:35

363. Howard 1964a, 1964b; Bickart 1990; S. Olson and Rasmussen 2001; Steadman and Carranza-Castañeda 2006
364. Steadman and Mead 2010
365. Howard 1964a, 1964b
366. Meyer de Schauensee 1970
367. L. Miller 1910a
368. L. Miller and Howard 1937; Hertel 1995; Van Valkenburgh and Hertel 1998
369. Howard 1974; Emslie 1988
370. Howard 1962, 1974
371. Corona-M. 2002
372. Howard 1952
373. Ibid.:53; Emslie 1987b
374. Emslie 1990
375. L. Miller 1931; Emslie 1987a
376. Steadman and Miller 1987; Emslie 1990; G. Morgan and Emslie 2010
377. Emslie 1987a
378. Howard 1962, 1968
379. Hertel 1994, 1995; Fox-Dobbs et al. 2006
380. Steadman et al. 1994
381. Howard 1952
382. Howard 1962
383. L. Miller 1909; Howard 1962; Emslie 1998; K. E. Campbell 1980b; K. E. Campbell and Tonni 1983; Steadman et al. 1994; Chatterjee, Templin, and Campbell 2007
384. Mawby 1967; Emslie and Heaton 1987
385. Howard 1952, 1972; K. E. Campbell, Scott, and Springer 1999
386. K. E. Campbell and Tonni 1983; K. E. Campbell, Scott, and Springer 1999
387. Howard 1962; K. E. Campbell and Tonni 1983; Fox-Dobbs et al. 2006
388. Chatterjee, Templin, and Campbell 2007:12398
389. L. Miller 1916:109
390. Howard 1932; Zhang, Feduccia, and James 2012
391. Howard 1962, 1971b; Feduccia 1974; Steadman et al. 1994; Emslie 1985; D. Guthrie 2009; G. Morgan and Emslie 2010
392. L. Miller 1916; Howard 1932, 1962
393. Hertel 1995; Van Valkenburgh and Hertel 1998
394. Howard 1952, 1962; Hager 1972; Steadman et al. 1994, D. Guthrie 2009
395. L. Miller 1911a; Howard 1932; Global Raptor Information Network 2013
396. Howard 1946
397. Howard 1935
398. Hertel 1995; Van Valkenburgh and Hertel 1998
399. L. Miller 1911a
400. Hayward et al. 1976; Alcorn 1988
401. Corona-M. 2002; Howard 1932; Howard and Miller 1933; Emslie 1998; G. Morgan and Emslie 2010
402. James 2004
403. F. Gill and Donsker 2014
404. P. Martin and Steadman 1984; Van Valkenburgh and Hertel 1998
405. Chamberlain et al. 2005; Fox-Dobbs et al. 2006

406. Grayson 2006a; Grayson and Fisher 2009
407. Grayson 1977, 1993; K. E. Campbell and Tonni 1981; Olson 2007

Chapter 4: Dating an Ass

1. Lull 1930:347
2. Harrington 1933:171
3. Gilreath 2009
4. Arnold and Libby 1951; Harrington 1951:25
5. Broecker, Kulp, and Tucek 1956
6. Grayson (2011) provides a table showing these calibrations for the past 25,000 years.
7. Hattori 1982; Orr 1956, 1974; Stafford et al. 1987; Dansie and Jerrems 2005
8. Stafford et al. 1987
9. Dansie and Jerrems 2005
10. Schubert 2010
11. P. Martin 1958
12. P. Martin 1958, 1963
13. Dillehay 1997; M. Gilbert, Jenkins, et al. 2008
14. Vartanyan et al. 2008; Stuart et al. 2004
15. P. Martin 1967
16. FAUNMAP Working Group 1994; Graham and Lundelius 2010
17. Faith and Surovell 2009

Chapter 5: A Stable of Ground Sloths

1. D. Jenkins 2007; M. Gilbert, Jenkins, et al. 2008; D. Jenkins et al. 2012, 2014
2. Thomas 1985; Grayson 1985
3. Beck and Jones 2009; Cannon, Livingston, and Broughton 2009
4. Harrington and Simpson 1961; Wormington and Ellis 1967
5. For starters, see Quade (1986); Quade et al. (1998), Quade, Forester, and Whelan (2003), and the discussion in Grayson (2011).
6. National Park Service, "Tule Springs Fossil Beds National Monument," http://www.nps.gov/tusk/index.htm
7. For example, Springer et al. 2005, 2010
8. For instance, Cope 1878a; Condon 1902, 1910; McCornack 1914, 1920, 1928
9. Clark 1989; Elftman 1931; Oregon Blue Book 2014; Oregon State Archives 2014
10. Condon 1902:145
11. "Eighty acres" 1877
12. For example, Cope 1889a, Shufeldt 1892
13. "Wonderful fossil beds" 1877; Clark 1989; McCornack 1928
14. Condon 1902, 1910
15. Jackman and Long 1964
16. Clark 1989; Packard 1952
17. Clark 1989; McCornack 1928
18. Clark 1989
19. Ibid.
20. Sternberg 1903:91

21. Sternberg 1881, 1884
22. Sternberg 1881:601
23. Cope 1878b, 1878c
24. Cope 1878b:125
25. Ibid.:125–26
26. "Important notice" 1877; Conklin 1944
27. Sternberg 1881:601
28. Sternberg 1884:599
29. Osborn 1929:170
30. Sternberg 1898:93
31. T. Gill 1897
32. Cope 1889a:979; see also Osborn 1931
33. For example, Cope 1880, 1883a, 1883b, 1883d, 1889b, 1893, 1895
34. Cope 1883a:292, 1883d:70
35. Cope 1893:335
36. Cope 1895:599
37. T. Gill 1897
38. International Commission on Stratigraphy 2013
39. Grayson 1983, 1986
40. Cope 1885
41. Meltzer 2006, 2009, 2015
42. Sternberg 1898:82, 1903:90–93
43. For example, Sternberg 1903, 1909
44. Cope 1883b
45. For example, G. Gilbert 1890:394
46. "Wonderful fossil beds" 1877:609
47. Minor and Spencer 1977; Pinson 1997
48. Minor and Spencer 1977
49. Kooyman et al. 2012
50. Gibson and Spencer 1977
51. John Breyer, personal communication, 2013
52. Friedel 1993
53. For example, Cope 1878a, 1883b, 1889a
54. For example, Cope 1883b; Jordan 1907; Allison and Bond 1983
55. Shufeldt 1912
56. Cope 1878a
57. See, for instance, Shufeldt (1907) and J. Davidson (1997).
58. For example, Shufeldt 1891a, 1891b, 1891c, 1892, 1912, 1913a, 1913b, 1913c
59. Stein 1997
60. L. Miller 1911b, 1912
61. Howard 1946
62. Gabrielson and Jewett 1940; Grayson and Maser 1978; Marshall, Hunter, and Contreras 2006
63. For example, Howard 1964a; Jehl 1967; R. Storer 1989
64. Hargrave 2009
65. Avibase, http://avibase.bsc-eoc.org

66. For example, Merriam 1906, 1910; L. Miller 1910a, 1910b, 1911a; Bocheński and Campbell 2006; Friscia et al. 2008; K. E. Campbell and Bocheński 2010; McHorse, Orcutt, and Davis 2012
67. Cope 1878a, 1884, 1889b
68. For example, Matthew 1902; Sinclair 1904
69. Elftman 1931
70. Allison 1966
71. J. Martin 1996
72. Goodwin 1995; Mead et al. 2010
73. Cope 1883c
74. Matthew 1902:320
75. Sinclair 1904; Hay 1927
76. G. Gilbert 1890; Matthew 1902; Sinclair 1904; L. Miller 1911a, 1911b, 1912; Hay 1927; Elftman 1931
77. Merriam 1906
78. Merriam and Stock 1921
79. Hoffman et al. 1927
80. For example, DeMay 1941a, 1941b; Howard 1946
81. Hallett, Hills, and Clague 1997; Zdanowicz, Zielinski, and Germani 1999
82. Allison 1966
83. J. Martin et al. 2005
84. Ibid.
85. Retrum 2010
86. Hargrave 2009
87. Gilreath 2009
88. Mark Harrington 1933, 1934b
89. Mark Harrington 1940; Marie Harrington 1985
90. Nordenskjold 1899; Woodward 1900, 1902; Brown 1903; Mark Harrington 1934b
91. Lull 1929; see also Lull 1930
92. The permit was renewed the following year: see Nussbaum (1930, 1931).
93. Mark Harrington 1930; D. Simpson 1965
94. Mark Harrington 1931:233
95. Haskin and Sells 1931; CPI Inflation Calculator 2014
96. Mark Harrington 1932a; Haskin and Sells 1931
97. Mark Harrington 1940; Marie Harrington 1985
98. Ibid.
99. Mark Harrington 1933:14, 1940
100. Scherer 1930
101. Mark Harrington 1940
102. J. Scherer 1930; Mark Harrington 1933
103. Rowland and Tanke 2007
104. Mark Harrington 1940:59
105. Mark Harrington 1930:41
106. Ibid.:42
107. Merriam 1930:61
108. Meltzer, in press
109. Mark Harrington 1933:73

110. Mark Harrington 1940:66
111. Mark Harrington 1932b
112. Mark Harrington 1933:171, 1934b:65
113. Mark Harrington 1940
114. Stock 1930a, 1931; Mark Harrington 1933
115. For example, Brattstrom 1954; Dundas 1999
116. L. Miller 1931; Brattstrom 1954, 1958, 1961
117. Lull 1929:19
118. Eames 1930; Stock 1931
119. Mark Harrington 1933
120. Laudermilk 1931, 1933; Munz 1959
121. Stock 1933:330
122. Laudermilk and Munz 1934:36
123. Franceschi and Nakata 2005; Prychid and Rudall 1999
124. Salinas, Ogura, and Soffchi 2011
125. For example, Fowler 1986, 1995; Louderback, Pavlik, and Spurling 2013; Rhode 2002
126. Wendy Hodgson, personal communication, 2014
127. Laudermilk and Munz 1934:34
128. Gardner 1994; D. Ward, Spiegel, and Saltz 1997
129. Rappaport and Hochman 1988
130. Poinar et al. 1998; Hofreiter et al. 2000; J. Gill et al. 2009
131. Grayson 2011
132. Mark Harrington 1933
133. Emslie 1990
134. Mark Harrington 1933; "Human material" 1967; Berger and Libby 1967; Heizer and Berger 1970
135. Mark Harrington 1933:73
136. Heizer and Berger 1970:14
137. Gilreath 2009
138. Glowiak and Rowland 2008
139. Even though Gypsum Cave is on the National Register of Historic Places and is of spiritual importance to Native Americans in the area (Bengston 2009), it has not been treated well recently. As hard as the Bureau of Land Management has tried to protect this national treasure, the front of the cave has been spray-painted by vandals and the interior harmed by a variety of unauthorized activities, including those related to drug use. The BLM is in the process of cleaning up this important site and taking steps to ensure its protection (Renee Kolvet and Bryan Hockett, personal communications, 2014).
140. Bryan 1979; Tuohy 1979
141. Mark Harrington 1934a:166
142. J. Harrington 1943
143. Mark Harrington 1934a, 1934c, 1935, 1936
144. Stock 1936a:149
145. Howard 1935
146. Howard 1952
147. L. Miller 1909; Howard 1971a; T. Storer 1972; K. E. Campbell 1980a

148. Howard 1972; K. E. Campbell, Scott, and Springer 1999
149. Harrington 1934a
150. Bryan 1979; Thompson 1985; Bryan and Tuohy 1999; Goebel et al. 2011
151. See Grayson (2011) for a discussion of that archaeology.
152. Bryan 1979; Beck and Jones 2010
153. S. Miller 1979
154. Bell (2000) provides a great review of the use of voles to tell time.
155. Jass 2011
156. Jass 2007
157. Brattstrom 1958, 1976; Mead, Thompson, and Van Devender 1982
158. E. Anderson 1970, 1994; Grayson 1984c, 1985; Youngman and Schueler 1991; Meyers 2007
159. Bryan 1979, 1988
160. Bonnichsen et al. 2001
161. Mark Harrington 1928; Mark Harrington and Simpson 1961
162. Cressman 1939, 1940, 1942, 1946, 1966
163. Hockett and Dillingham 2004
164. McGuire 1980
165. Gruhn and Bryan 1981; McGuire 1982; Lyman 1987
166. Hockett and Dillingham 2004
167. McGuire 1980
168. Hockett and Dillingham 2004; Hockett and Jenkins 2013
169. W. Miller 1982; Heaton 1985, 1990; Emslie and Heaton 1987; Morrison and Dwyer 2012
170. Heaton 1985:346
171. Gillette and Madsen 1992, 1993; Madsen 2000a
172. Lewis 1990
173. D. B. Madsen, personal communication, 2013
174. Schubert and Wallace 2009
175. Madsen 2000a:100
176. Elias 1990, 1991, 1995, 1996
177. Schaedler et al. 1992
178. Tuohy 1979:8
179. Mead, Bell, and Murray 1992; Bell and Mead 1998
180. L. Davis, Macfarlan, and Henrickson 2012
181. For example, Madsen 2000b

Chapter 6: Extinct Mammals, Dangerous Plants, and the Early Peoples of the Great Basin

1. Waters and Stafford 2007; Sanchez et al. 2014. Grayson (2011) used a shorter time span (11,200 to 10,800 years ago), but El Fin del Mundo, discussed in chapter 3 (Sanchez et al. 2014), confirms other suggestions of an earlier beginning for Clovis (Ferring 2001).
2. D. Jenkins 2007; M. Gilbert, Jenkins, et al. 2008; D. Jenkins et al. 2012, 2014; Hockett and Jenkins 2012. There a few Paisley doubters (Sistiaga et al. 2014), but I am not among them.
3. Dillehay 1997

4. Meltzer 2009; Grayson 2011
5. Pitulko et al. 2004
6. Goebel et al. 2013
7. Stanford and Bradley 2012
8. Straus, Meltzer, and Goebel 2005; Meltzer 2009; O'Brien et al. 2014.
9. Wilke, Flenniken, and Ozbun 1991; Meltzer 2009
10. M. Rasmussen et al. 2014
11. Haury, Sayles, and Wasley 1959; C. Haynes and Huckell 2007
12. Grayson 1993, 2011; Beck and Jones 2009
13. Rhode 1987 and personal communication
14. Meltzer 2006, 2009
15. Basgall 1993; Jones and Beck 1999; Beck and Jones 2007, 2009; Grayson 2011
16. Hattori 2008; Beck and Jones 2013; Moss and Erlandson 2013
17. Connolly 1999
18. D. Jenkins et al. 2012, 2014; Beck and Jones 2013
19. McNeil et al. 2005, 2007; Lucas, Allen, et al. 2007; B. Allen, Love, and Meyers 2009
20. Louderback, Grayson, and Llobera 2011
21. Grayson 2000a, 2000b
22. Mann et al. 2013
23. Mack and Thompson 1982
24. Howden and Cartwright 1963
25. F. Larson 1940; Tisdale 1961
26. P. Martin 2000
27. McElroy et al. 1982
28. Janzen and Martin 1982:26
29. For example, Howe 1985; Janzen 1986; Barlow 2000; Guimarães, Galetti, and Jordano 2008; Zaya and Howe 2009; Jansen et al. 2012
30. Newsom and Mihlbachler 2006
31. P. Martin 1999, 2005; Lenz 2001
32. Howe 1985
33. Wallen and Ludwig 1978; Bregman 1988; Vander Wall, Kuhn, and Beck 2005; Jansen et al. 2012
34. Lenz 2001; Cole et al. 2011
35. Vander Wall et al. 2006; Waitman, Vander Wall, and Esque 2012
36. Laudermilk and Munz 1934
37. C. Smith et al. 2011
38. Lenz 2007; Godsoe et al. 2008; C. Smith et al. 2009; Starr et al. 2012
39. C. Smith et al. 2008
40. Starr et al. 2013; Yoder et al. 2013
41. Pellmyr 2003
42. Pellmyr and Segraves 2003; Godsoe et al. 2008; C. Smith et al. 2008, 2009
43. C. Smith et al. 2008
44. C. Smith et al. 2009; Godsoe et al. 2010
45. For example, Godsoe et al. 2009; C. Smith et al. 2011; Starr et al. 2013
46. C. Smith et al. 2011
47. Cole et al. 2011
48. C. Smith et al. 2011

49. Cole et al. 2011:145

50. Starr et al. 2013:447

51. Vander Wall et al. 2006; Waitman, Vander Wall, and Esque 2012

52. See Cole et al. (2011) on increasing densities of Joshua trees in the northern part of their current range.

53. Pellmyr et al. 2007

54. *Yucca utahensis* is variously treated as a separate species (*Flora of North America*, http://www.efloras.org/florataxon.aspx?flora_id=1&taxon_id=242102079) or as a subspecies of *Yucca elata* (USDA PLANTS, http://plants.usda.gov/core/profile?symbol=YUEL).

55. Kartesz 1988; Albee, Shultz, and Goodrich 1988

56. Hanley et al. 2007; Salminen and Karonen 2011; Mithöfer and Boland 2012; Fürstenberg-Hägg, Zagrobelny, and Bak 2013; Neilson et al. 2013

57. Magnanou, Malenke, and Dearing 2009; Kohl et al. 2014

58. Baldwin et al. 2002

59. Owen-Smith and Cooper 1987; Cooper and Owen-Smith 1986; J. Schmidt 1989; Grubb 1992; T. Young and Okello 1998; Hanley et al. 2007

60. Nobel 1978, 1980, 1983; Nobel and Loik 1999

61. Lev-Yadun 2001, 2003

62. Wagner, Herbst, and Sohmer 1990; Zeigler 2002

63. Abrahamson 1975; T. Young and Okello 1998; Karban 2010

64. Ford et al. 2014

65. Hanley et al. 2007; Mithöfer and Boland 2012

66. Koricheva, Nykänen, and Gianoli 2004

67. Dicke 1999; Agrawal 2001; Karban 2010; McCormick, Unsicker, and Gershenzon 2012; Mithöfer and Boland 2012; Fürstenberg-Hägg, Zagrobelny, and Bak 2013; Coley and Kursar 2014

68. Barlow 2000

69. The thick pulp of the avocado fruit is a result of its domestication, but the large size of its seed is not (Gama-Campillo and Gomez-Pampa 1992; Ben-Ya'acov, Solis-Marina, and Bufler 2003).

70. Actually, in what follows, I used 2 meters, not 6 feet, but the two are close enough that the difference does not matter.

71. S. Smith, Monson, and Anderson 1997

72. Cronquist et al. 1972

73. Pockman and Sperry 1997; Martínez-Vilalta and Pockman 2002; Medeiros and Pockman 2011

74. Shafer, Bartlein, and Thompson 2001; Weiss and Overpeck 2005

75. See the review in Grayson (2011).

76. Cole 1990; Van Devender 1990; McAuliffe and Van Devender 1998; Betancourt et al. 2001; Holmgren, Betancourt, and Rylander 2006; Holmgren, Norris, and Betancourt 2007; Cole et al. 2013; Holmgren et al. 2014

77. Hansen 1978; Mead and Phillips 1981

78. Spaulding 1985, 1994; Koehler and Anderson 1995; Cole et al. 2011

79. C. Nowak et al. 1994a, 1994b

80. Thompson 1984; Rhode 2000; David Rhode, personal communication

81. Graham 1985b; Cole 1990; Grayson 2007
82. McAuliffe and Van Devender 1998; Dimmit 2000; Holmgren, Norris, and Betancourt 2007; Cole et al. 2011
83. Benson and Darrow 1981; Mozingo 1987
84. Baldwin et al. 2012
85. Cronquist et al. 1972; Little 2006; R. Adams 2011
86. Bartel 1994; Constance Millar, personal communication
87. Crawley 1983; Kozlowski 1971
88. Tollrian and Harvell 1999
89. See the review in Grayson (2011).
90. Cronquist et al. 1972; lake area figures are calculated from table 5.2 in Grayson (2011).
91. C. Haynes and Huckell 2007
92. Holmgren, Betancourt, and Rylander 2006; Holmgren, Norris, and Betancourt 2007
93. Carmona, Lajeunesse, and Johnson 2011

Chapter 7: Clovis, Comets, and Climate

1. Grayson 1980, 1984a
2. Darwin 1859:322
3. P. Martin 1967, 1973, 1990, 2005; P. Martin and Steadman 1999
4. Lomolino, Riddle, and Brown 2006; Diamond 1992; P. Ward 2000
5. P. Martin and Burney 1999
6. Dillehay 1997; Meltzer 1997; Meltzer et al. 1997; P. Martin and Steadman 1999; P. Martin 2005
7. P. Martin 2005:133, 278
8. P. Martin 1984:363
9. Grayson and Meltzer 2015
10. Stanford 1991, 1999; Cannon and Meltzer 2004, 2008; Holliday and Meltzer 2010
11. Meltzer 2009
12. Spectre of the Gun, CBS Studios, http://www.chakoteya.net/startrek/56.htm
13. P. Martin 1958, 1963
14. C. Haynes 1964:1412
15. P. Martin 1963:65
16. P. Martin 1967:89
17. Grayson 1987, 1991, 2007
18. P. Martin 2005:136
19. Grayson 1984a, 2006b
20. Grayson 1991, 2001
21. Wilmshurst et al. 2011
22. Worthy and Holdaway 2002; Huynen et al. 2003; Olson and Turvey 2013
23. A. Anderson 1989
24. Worthy and Holdaway 2002; Tennyson and Martinson 2006
25. Grayson 2001
26. Lyell 1832:155
27. P. Martin 1984

28. McWethy et al. 2010
29. Steadman 1989
30. Quammen 2014:117
31. Brakenridge 1981
32. Quade et al. 1998
33. Quade 1986; Quade and Pratt 1989; Quade et al. 1998; Quade, Forester, and Whelan 2003; Pigati et al. 2014
34. Grayson 2011
35. Brakenridge 2011
36. Firestone and Topping 2001
37. Ibid.:9
38. Ibid.:14–15
39. Southon and Taylor 2002:17
40. Firestone, West, and Warwick-Smith 2006:148
41. Ibid.:144
42. Firestone et al. 2007
43. Firestone 2009; Firestone et al. 2010; Quade et al. 1998; C. Haynes 2008
44. For an introduction to this literature on both sides of the debate, see Holliday et al. (2014), Meltzer et al. (2014), and Kinzie et al. (2014).
45. Marlon et al. 2009; Daniau, Harrison, and Bartlein 2010
46. van Hoesel et al. 2014; Meltzer et al. 2014, but see also J. Kennett et al. 2015
47. D. Kennett et al. 2008; Broecker et al. 2010; Carlson 2010
48. Melott and Thomas 2009; Melott et al. 2010
49. Faith 2014; Faith et al. 2014
50. MacPhee and Greenwood 2007; Stuart and Lister 2007, 2012
51. Vartanyan et al. 2008; Kuzmin 2010
52. R. Guthrie 2006
53. Melott and Thomas 2009
54. Moura et al. 2014
55. Schulte 2010; Vizcaíno et al. 2004. For a fascinating general review of impact-related issues, see Barrientos and Masse (2014).
56. J. Gill et al. 2009
57. O. Davis 1987; Burney, Robinson, and Burney 2003; Robinson, Burney, and Burney 2005; O. Davis and Shafer 2006; Raper and Bush 2009; Parker and Williams 2011; Feranec et al. 2011; Wood et al. 2011; Wood and Wilmshurst 2012
58. MacPhee and Marx 1997; for South America, see Ferigolo 1999
59. MacPhee and Greenwood 2013
60. Rothschild 2003; Rothschild and Martin 2003, 2006; Rothschild and Laub 2006
61. Graham 1976a, 1985a, 1985b, 1990; Graham and Semken 1976; Graham and Lundelius 1984; Graham and Mead 1987; Lundelius et al. 1983; FAUNMAP Working Group 1996
62. Grayson 1984b
63. Stafford et al. 1999; Semken et al. 2010; Fulton et al. 2013
64. For example, Guilday 1967
65. Guilday 1967; Grayson 1991; Barnosky et al. 2004; Koch and Barnosky 2006
66. C. McCain and King 2014
67. Gleason 1926:26

68. Ibid.:16

69. For example, Grayson and Delpech 2003, 2008; Stuart et al. 2004; Vartanyan et al. 2008; Lorenzen et al. 2011; Stuart and Lister 2012, 2014

70. R. Guthrie 2006; Mann et al. 2013

71. Willerslev et al. 2014

72. R. Guthrie 2003

73. Nalawade-Chavan et al. 2014; Zazula et al. 2014

74. E. Scott 2010

75. Gilmour et al. 2015

76. P. Martin 1967, 2005; P. Martin and Steadman 1999; Barnosky et al. 2004; Koch and Barnosky 2006

77. Barnosky et al. 2004; Robinson, Burney, and Burney 2005; Koch and Barnosky 2006; Polyak et al. 2012

78. On the value of modeling in this realm in general, see Yule et al. (2014).

79. P. Martin 2005:159

REFERENCES

Abrahamson, W. G. 1975. Reproductive strategies in dewberries. *Ecology* 56:721–26.

Adam, P. J. 1999. *Choloepus didactylus. Mammalian Species* 621:1–8.

Adams, D. B. 1979. The cheetah: Native American. *Science* 205:1155–58.

Adams, K. D., and S. G. Wesnousky. 1998. Shoreline processes and the age of the Lake Lahontan highstand in the Jessup embayment, Nevada. *Bulletin of the Geological Society of America* 110:1318–32.

Adams, R. P. 2011. *Junipers of the World: The Genus* Juniperus. Trafford, Bloomington, IN.

Agenbroad, L. D. 1999. Pygmy mammoths *Mammuthus exilis* from Channel Islands National Park, California (USA). In *Mammoths and the Mammoth Fauna: Studies of an Extinct Ecosystem*, edited by G. Haynes, J. Klimowicz, and J. W. F. Reumer, 89–102. *Deinsea* 6.

———. 2003. New absolute dates and comparisons for California's *Mammuthus exilis*. In *Advances in Mammoth Research*, edited by J. W. F. Reumer, J. De Vos, and D. Mol, 1–16. *Deinsea* 9.

———. 2009. *Mammuthus exilis* from the California Channel Islands: Height, mass, and geologic age. In *Proceedings of the 7th California Islands Symposium*, edited by C. C. Damiani and D. K. Garcelon, 15–19. Institute for Wildlife Studies, Arcata, CA.

———. 2012. Giants and pygmies: Mammoths of Santa Rosa Island, California (USA). *Quaternary International* 255:2–8.

Agenbroad, L. D., J. R. Johnson, D. Morris, and T. W. Stafford Jr. 2005. Mammoths and humans as late Pleistocene contemporaries on Santa Rosa Island. In *Proceedings of the 6th California Islands Symposium*, edited by D. K. Garcelon and C. A. Schwemm, 3–7. Institute for Wildlife Studies, Arcata, CA.

Agenbroad, L. D., and J. I. Mead. 1996. Distribution and paleoecology of central and western North American *Mammuthus*. In *The Proboscidea: Evolution and Palaeoecology of Elephants and Their Relatives*, edited by J. Shoshani and P. Tassy, 280–88. Oxford University Press, Oxford, UK.

Agenbroad, L. D., J. I. Mead, C. V. Haynes Jr., and R. H. Hevly. 2013. Fossil fauna and flora from late Pleistocene Cerros Negros locality, Pinal County, Arizona: With update of *Mammuthus* and all *Mammut* localities from Arizona. *Southwestern Naturalist* 58:50–53.

Agrawal, A. A. 2001. Phenotypic plasticity in the interactions and evolution of species. *Science* 294:321–26.

Akersten, W. A. 1985. *Canine Function in* Smilodon *(Mammalia; Felidae; Machairodontinae)*. Natural History Museum of Los Angeles County Contributions in Science 356.

Akersten, W. A., T. M. Foppe, and G. T. Jefferson. 1988. New source of dietary data for extinct herbivores. *Quaternary Research* 30:92–97.

Albee, B. J., L. M. Shultz, and S. Goodrich. 1988. *Atlas of the Vascular Plants of Utah.* Utah Museum of Natural History Occasional Paper 7.

Alberdi, M. T., and J. L. Prado. 1995. Los équidos de América del sur. In *Evolución biológica y climática de la región pampeana durante los últimos cinco millones de años,* edited by M. A. Alberdi, G. Leone, and E. P. Tonni, 295–308. Museo Nacional de Ciencias Naturales, Madrid.

———. 1998. Comments on: Pleistocene horses from Tarija, Bolivia, and validity of the genus *Onohippidium* (Mammalia: Equidae), by B. J. MacFadden. *Journal of Vertebrate Paleontology* 18:669–72.

Alberdi, M. T., J. L. Prado, E. Ortiz-Jaureguizar, and P. Posadas. 2012. Reply to: A critical appraisal of the phylogenetic proposals for the South American Gomphotheriidae (Proboscidea: Mammalia) by M. A. Cozzuol, D. Mothé, and L. S. Avilla. *Quaternary International* 253:104–6.

Alcaraz, M. A., and A. E. Zurita. 2004. Nuevos registros de cérvidos poco conocidos: *Epieuryceros* cf. *proximus* Castellanos y *Antifer* sp. (Mammalia, Artiodactyla, Cervidae). *Revista del Museo Argentino de Ciencias Naturales* 6 (1): 41–48.

Alcorn, G. 1988. *The Birds of Nevada.* Fairview Press, Fallon, NV.

Allen, B. D., D. W. Love, and R. G. Meyers. 2009. Evidence for late Pleistocene hydrologic and climatic change from Lake Otero, Tularosa Basin, south-central New Mexico. *New Mexico Geology* 31:9–25.

Allen, T. 2012. Researchers find first evidence of Ice Age wolves in Nevada. *Science Daily,* December 16. http://phys.org/news/2012-12-evidence-ice-age-wolves-nevada.html.

Allison, I. S. 1966. *Fossil Lake, Oregon: Its Geology and Fossil Fauna.* Oregon State Monographs, Studies in Geology 9.

Allison, I. S., and C. E. Bond. 1983. Identity and probable age of salmonids from surface deposits at Fossil Lake, Oregon. *Copeia* 1983:563–64.

Ameghino, F. 1898. An existing ground sloth in Patagonia. *Natural Science* 13:324–26.

Anderson, A. 1989. *Prodigious Birds: Moas and Moa-Hunting in Prehistoric New Zealand.* Cambridge University Press, Cambridge.

Anderson, E. 1970. *Quaternary Evolution of the Genus* Martes *(Carnivora, Mustelidae).* Acta Zoologica Fennica 130.

———. 1994. Evolution, prehistoric distribution, and systematics of *Martes.* In *Martens, Sables, and Fishers: Biology and Conservation,* edited by S. W. Buskirk, A. S. Harestad, M. G. Raphael, and R. A. Powell, 13–25. Cornell University Press, Ithaca, NY.

———. 1996. A preliminary report on the Carnivora of Porcupine Cave, Park County, Colorado. In *Palaeoecology and Palaeoenvironments of Late Cenozoic Mammals: Tributes to the Career of C. S. (Rufus) Churcher,* edited by K. M. Stewart and K. L. Seymour, 259–82. University of Toronto Press, Toronto.

———. 2004. The Carnivora from Porcupine Cave. In *Biodiversity Response to Climate Change in the Middle Pleistocene,* edited by A. D. Barnosky, 141–54. University of California Press, Berkeley.

Antón, M. 2013. *Sabertooth.* Indiana University Press, Bloomington.

Antón, M., M. J. Salesa, A. Galobart, and Z. J. Tseng. 2014. The Plio-Pleistocene scimitar-toothed felid genus *Homotherium* Fabrini, 1890 (Machairodontinae, Homotherini): Diversity, palaeogeography and taxonomic implications. *Quaternary Science Reviews* 96:259–68.

Antón, M., M. J. Salesa, J. F. Pastor, I. M. Sánchez, S. Fraile, and J. Morales. 2004. Impli-

cations of the mastoid anatomy of larger extant felids for the evolution and pred-
atory behaviour of sabretoothed cats (Mammalia, Carnivora, Felidae). *Zoological Journal of the Linnean Society of London* 140:207–21.

Anyonge, W. 1993. Body mass in large extant and extinct carnivores. *Journal of Zoology, London* 231:339–50.

———. 1996. Microwear on canines and killing behavior in large carnivores: Saber function in *Smilodon fatalis*. *Journal of Mammalogy* 77:1059–67.

Anyonge, W., and C. Roman. 2006. New body mass estimate for *Canis dirus*, the extinct Pleistocene dire wolf. *Journal of Vertebrate Paleontology* 26:209–12.

Argyll, Duke of. 1883. Metamorphic origin of granite—Prehistoric "giants." *Nature* 27:578–79.

Arnold, J. R., and W. F. Libby. 1951. Radiocarbon dates. *Science* 113:111–20.

Arroyo-Cabrales, J., and O. Carranza Castañeda. 2009. Los cánidos prehistóricos mexi-
canos antes de la llegada del perro. *Arqueobios* 3:34–45.

Arroyo-Cabrales, J., and O. J. Polaco. 2003. Caves and the Pleistocene vertebrate paleon-
tology of Mexico. In *Ice Age Cave Faunas of North America*, edited by B. W. Schubert, J. I. Mead, and R. W. Graham, 273–90. Indiana University Press, Bloomington.

Arroyo-Cabrales, J., O. J. Polaco, E. Johnson, and I. Ferrusquía-Villafranca. 2010. A per-
spective on mammal biodiversity and zoogeography in the Late Pleistocene of México. *Quaternary International* 212:187–97.

Arroyo-Cabrales, J., O. J. Polaco, C. Laurito, E. Johnson, M. T. Alberdi, and A. L. V. Zamora. 2007. The proboscideans (Mammalia) from Mesoamerica. *Quaternary International* 169–70:17–23.

Asevedo, L., G. R. Winck, D. Mothé, and L. S. Avilla. 2012. Ancient diet of the Pleistocene gomphothere *Notiomastodon platensis* (Mammalia, Proboscidea, Gomphoth-
eriidae) from lowland mid-latitudes of South America: Stereomicrowear and tooth calculus analyses combined. *Quaternary International* 255:42–52.

Azzaroli, A. 1998. The genus *Equus* in North America—The Pleistocene species. *Palae-
ontographia Italica* 85:1–60.

Bacon, C. D., D. Silvestro, C. Jaramillo, B. T. Smith, P. Chakrabarty, and A. Antonelli. 2015. Biological evidence supports an early and complex emergence of the Isthmus of Panama. *Proceedings of the National Academy of Sciences* 112:6110–15.

Baldwin, B. G., S. Boyd, B. J. Ertter, R. W. Patterson, T. J. Rosatti, and D. H. Wilken, eds. 2002. *The Jepson Desert Manual: Vascular Plants of Southeastern California*. University of California, Berkeley.

Baldwin, B. G., D. H. Goldman, D. J. Keil, R. Patterson, T. J. Rosatti, and D. H. Wilken, eds. 2012. *The Jepson Manual: Vascular Plants of California*. University of California Press, Berkeley.

Bancroft, H. H. 1890. *History of Nevada: 1540–1888*. History Company, San Francisco.

Banks, W. E., F. d'Errico, A. T. Peterson, M. Kageyama, and G. Colombeau. 2008. Reconstructing ecological niches and geographic distributions of caribou (*Rangifer tarandus*) and red deer (*Cervus elaphus*) during the Last Glacial Maximum. *Quaternary Science Reviews* 27:2568–75.

Bargo, M. S., S. F. Vizcaíno, F. M. Archuby, and R. E. Blanco. 2000. Limb bone propor-
tions, strength and digging in some Lujanian (Late Pleistocene–Early Holocene) mylodontid ground sloths (Mammalia, Xenarthra). *Journal of Vertebrate Paleontology* 20:601–10.

Barlow, C. 2000. *The Ghosts of Evolution*. Basic Books, New York.

Barnes, I., B. Shapiro, A. Lister, T. Kuznetsova, A. Sher, D. Guthrie, and M. G. Thomas. 2007. Genetic structure and extinction of the woolly mammoth, *Mammuthus primigenius. Current Biology* 17:1072–75.

Barnett, R., I. Barnes, M. J. Phillips, L. D. Martin, C. R. Harington, J. A. Leonard, and A. Cooper. 2005. Evolution of the extinct sabretooths and the American cheetah-like cat. *Current Biology* 15:R589–R590.

Barnett, R., B. Shapiro, I. Barnes, et al. (17 authors). 2009. Phylogeography of lions (*Panthera leo* ssp.) reveals three distinct taxa and a late Pleistocene reduction in genetic diversity. *Molecular Ecology* 18:1668–77.

Barnosky, A. D., P. L. Koch, R. S. Feranec, S. L. Wing, and A. B. Shabel. 2004. Assessing the causes of late Pleistocene extinctions on the continents. *Science* 306:70–75.

Barrientos, G., and W. B. Masse. 2014. The archaeology of cosmic impact: Lessons from two mid-Holocene Argentine case studies. *Journal of Archaeological Method and Theory* 21:134–211.

Bartel, J. A. 1994. Cupressaceae Cypress family. *Journal of the Arizona-Nevada Academy of Science* 27:195–200.

Basgall, M. E. 1993. Early Holocene prehistory of the north-central Mojave Desert. PhD diss., University of California, Davis.

Baskaran, N., V. Kannan, K. Thiyagesan, and A. A. Desai. 2011. Behavioural ecology of four-horned antelope (*Tetracerus quadricornis* de Blainville 1816) in the tropical forests of southern India. *Mammalian Biology* 76:741–47.

Beck, C., and G. T. Jones. 2007. Early Paleoarchaic point morphology and chronology. In *Paleoindian or Paleoarchaic? Great Basin Human Ecology at the Pleistocene/Holocene Transition*, edited by K. E. Graf and D. N. Schmitt, 23–41. University of Utah Press, Salt Lake City.

——. 2009. *The Archaeology of the Eastern Nevada Paleoarchaic, Part I: The Sunshine Locality*. University of Utah Anthropological Papers 126.

——. 2010. Clovis and western stemmed: Population migration and the meeting of two technologies in the Intermountain West. *American Antiquity* 75:81–116.

——. 2013. Complexities of the colonization process: A view from the North American West. In *Paleoamerican Odyssey*, edited by K. E. Graf, C. V. Ketron, and M. R. Waters, 273–91. Center for the Study of the First Americans, Texas A&M University, College Station.

Bedetti, C., M. R. Palombo, and R. Sardella. 2001. Last occurrences of large mammals and birds in the late Quaternary of the Italian Peninsula. In *The World of Elephants*, edited by G. Cavarretta, P. Gioia, M. Mussi, and M. R. Palombo, 701–3. Consiglio Nazionale delle Ricerche, Rome.

Beebe, B. F. 1980. Pleistocene peccary, *Platygonus compressus* Le Conte, from Yukon Territory, Canada. *Canadian Journal of Earth Science* 17:1204–9.

Bell, C. J. 2000. Biochronology of North American microtine rodents. In *Quaternary Geochronology: Methods and Applications*, edited by J. S. Noller, J. M. Sowers, and W. R. Lettis, 379–406. AGU Reference Shelf 4. American Geophysical Union, Washington, DC.

Bell, C. J., E. L. Lundelius Jr., A. D. Barnosky, et al. (9 authors). 2004. The Blancan, Irvingtonian, and Rancholabrean Mammal Ages. In *Late Cretaceous and Cenozoic*

Mammals of North America, edited by M. O. Woodburne, 232–314. Columbia University Press, New York.

Bell, C. J., and J. I. Mead. 1998. Late Pleistocene microtine rodents from Snake Creek Burial Cave, White Pine County, Nevada. *Great Basin Naturalist* 58:82–86.

Bengston, G. 2009. Ethnographic context. In *Gypsum Cave Revisited*, edited by A. Gilreath, 9–10. Nevada Cultural Resources Report #CR5-2462-4(P). Far Western Anthropological Research Group, Davis, CA.

Benson, L., and R. A. Darrow. 1981. *Trees and Shrubs of the Southwestern Deserts.* 3rd ed. University of Arizona Press, Tucson.

Ben-Ya'acov, A., A. Solis-Molina, and G. Bufler. 2003. The wild avocado of Monteverde, Costa Rica. *Proceedings of the Fifth World Avocado Conference,* 35–38.

Berger, J., J. K. Young, and K. M. Berger. 2008. Protecting migration corridors: Challenges and optimism for Mongolian saiga. *PLoS Biology* 6 (7): 1365–67.

Berger, R., and W. F. Libby. 1967. UCLA radiocarbon dates VI. *Radiocarbon* 9:477–504.

Berkove, L. I., ed. 2006. *The Sagebrush Anthology: Literature from the Silver Age of the Old West.* University of Missouri Press, Columbia.

Betancourt, J. L., K. A. Rylander, C. Peñalba, and J. L. McVickar. 2001. Late Quaternary vegetation history of Rough Canyon, south-central New Mexico, USA. *Palaeogeography, Palaeoclimatology, Palaeoecology* 165:71–95.

Betancourt, J. L., T. R. Van Devender, and P. S. Martin, eds. 1990. *Packrat Middens: The Last 40,000 Years of Biotic Change.* University of Arizona Press, Tucson.

Bickart, K. J. 1990. The birds of the late Miocene–early Pliocene Big Sandy Formation, Mohave County, Arizona. *Ornithological Monographs* 44:1–72.

Billings, W. D. 1951. Vegetational zonation in the Great Basin of western North America. In *Les bases ecologiques de la régénération de la végétation des zones arides,* 101–22. International Union of Biological Sciences, Series B, No. 9.

Binder, W. J., and B. Van Valkenburgh. 2010. A comparison of tooth wear and breakage in Rancho La Brea sabertooth cats and dire wolves across time. *Journal of Vertebrate Paleontology* 30:255–61.

Blackford, L. C. 1995. Postcranial skeletal analysis of the Pleistocene deer, *Navahoceros fricki* (Cervidae). Master's thesis, Northern Arizona University, Flagstaff.

Blake, W. P. 1884. The Carson-City ichnolites. *Science* 4:273–76.

Bocheński, Z. M., and K. E. Campbell Jr. 2006. The extinct California Turkey, *Meleagris californica*, from Rancho La Brea: Comparative osteology and systematics. *Natural History Museum of Los Angeles County Contributions in Science* 509:1–92.

Boeskorov, G. G. 2006. Arctic Siberia: Refuge of the mammoth fauna in the Holocene. *Quaternary International* 142–43:119–23.

Bond, M. 1999. Quaternary native ungulates of southern South America: A synthesis. *Quaternary of South America and Antarctic Peninsula* 12:177–205.

Bond, M., E. Cerdeño, and G. López. 1995. Los ungulados nativos de América del Sur. In *Evolución biológica y climática de la Región Pampeana durante los últimos cinco millones de años,* edited by M. A. Alberdi, G. Leone, and E. P. Tonni, 259–75. Museo Nacional de Ciencias Naturales, Madrid.

Bond, M., D. Perea, M. Ubilla, and A. Tauber. 2001. *Neolicaphrium recens* Frenguelli, 1921, the only surviving Prototheriidae (Litopterna, Mammalia) into the South American Pleistocene. *Palaeovertebrata* 30:37–50.

Bonnichsen, R., L. Hodges, W. Ream, K. G. Field, D. L. Kirner, K. Selsor, and R. E. Taylor. 2001. Methods for the study of ancient hair: Radiocarbon dates and gene sequences from individual hairs. *Journal of Archaeological Science* 28:775–85.

Booth, D. B. 1987. Timing and processes of deglaciation along the southern margin of the Cordilleran ice sheet. In *North America and Adjacent Oceans during the Last Deglaciation*, edited by W. F. Ruddiman and H. E. Wright Jr., 71–90. *The Geology of North America*, vol. K-3. Geological Society of America, Boulder, CO.

Bradley, B. A., and E. Fleishman. 2008. Relationships between expanding pinyon-juniper cover and topography in the central Great Basin, Nevada. *Journal of Biogeography* 35:951–64.

Brakenridge, G. R. 1981. Terrestrial paleoenvironmental effects of a late Quaternary-Age supernova. *Icarus* 46:81–93.

———. 2011. Core-collapse supernovae and the Younger Dryas/terminal Rancholabrean extinctions. *Icarus* 215:101–6.

Brattstrom, B. H. 1954. Amphibians and reptiles from Gypsum Cave, Nevada. *Bulletin of the Southern California Academy of Sciences* 53 (1): 8–12.

———. 1958. Additions to the Pleistocene herpetofauna of Nevada. *Herpetologica* 14:36.

———. 1961. Some new fossil tortoises from western North America with remarks on the zoogeography and paleoecology of tortoises. *Journal of Paleontology* 35:543–60.

———. 1976. A Pleistocene herpetofauna from Smith Creek Cave, Nevada. *Bulletin of the Southern California Academy of Science* 75:283–84.

Brean, H. 2013. Fossil find is a horse, of course. *Las Vegas Review-Journal*, May 20. http://www.reviewjournal.com/news/las-vegas/fossil-find-horse-course.

Breda, M. 2005. The morphological distinction between the postcranial skeleton of *Cervalces/Alces* and *Megaloceros giganteus* and comparison between the two Alceini genera from the Upper Pliocene–Holocene of western Europe. *Geobios* 38:151–70.

———. 2008. Palaeoecology and palaeoethology of the Plio-Pleistocene genus *Cervalces* (Cervidae, Mammalia) in Eurasia. *Journal of Vertebrate Paleontology* 28:886–99.

Breda, M., and M. Marchetti. 2005. Systematical and biochronological review of Plio-Pleistocene Alceini (Cervidae; Mammalia) from Eurasia. *Quaternary Science Reviews* 24:775–805.

Bregman, R. 1988. Forms of seed dispersal in Cactaceae. *Acta Botanica Neerlandica* 37:395–402.

Bridgham, S. D., H. Cadillo-Quiroz, J. K. Keller, and Q. Zhuang. 2013. Methane emissions from wetlands: Biogeochemical, microbial, and modeling perspectives from local to global scales. *Global Change Biology* 19:1325–46.

Bright, J., D. S. Kaufman, S. L. Forman, W. C. McIntosh, J. I. Mead, and A. Baez. 2010. Comparative dating of a *Bison*-bearing late-Pleistocene deposit, Térapa, Sonora, Mexico. *Quaternary Geochronology* 5:631–43.

Broecker, W. S., G. H. Denton, R. L. Edwards, H. Cheng, R. B. Alley, and A. E. Putnam. 2010. Putting the Younger Dryas cold event into context. *Quaternary Science Reviews* 29:1078–81.

Broecker, W. S., J. L. Kulp, and C. S. Tucek. 1956. Lamont natural radiocarbon measurements III. *Science* 124:154–65.

Broughton, J. M., D. A. Byers, R. A. Bryson, W. Eckerle, and D. B. Madsen. 2008. Did climatic seasonality control late Quaternary artiodactyl densities in western North America? *Quaternary Science Reviews* 27:1916–37.

Brown, B. 1903. A new genus of ground sloth from the Pleistocene of Nebraska. *Bulletin of the American Museum of Natural History* 19:569–83.

Bryan, A. L. 1979. Smith Creek Cave. In *The Archaeology of Smith Creek Canyon, Eastern Nevada*, edited by D. R. Tuohy and D. L. Rendall, 162–253. Nevada State Museum Anthropological Papers 17.

———. 1988. The relationship of the stemmed point and the fluted point traditions in the Great Basin. In *Early Human Occupation in Far Western North America: The Clovis-Archaic Interface*, edited by J. A. Willig, C. M. Aikens, and J. L. Fagan, 53–74. Nevada State Museum Anthropological Papers 21.

Bryan, A. L., and D. R. Tuohy. 1999. Prehistory of the Great Basin/Snake River Plain to about 8,500 years ago. In *Ice Age Peoples of North America*, edited by R. Bonnichsen and K. L. Turnmire, 249–60. Oregon State University Press, Corvallis.

Bunce, M., T. H. Worthy, T. Ford, W. Hoppitt, E. Willerslev, A. Drummond, and A. Cooper. 2003. Extreme reversed sexual size dimorphism in the extinct New Zealand moa *Dinornis*. *Nature* 425:172–75.

Burney, D. A., G. S. Robinson, and L. Pigott Burney. 2003. *Sporormiella* and the late Holocene extinctions in Madagascar. *Proceedings of the National Academy of Sciences* 100:10800–10805.

Burns, J. 2010. Mammalian faunal dynamics in Late Pleistocene Alberta, Canada. *Quaternary International* 217:37–42.

Butler, B. R. 1978. *A Guide to Understanding Idaho Archaeology: The Upper Snake and Salmon River Country*. 3rd ed. Idaho Museum of Natural History, Pocatello.

Byers, J. A. 1997. *American Pronghorn: Social Adaptations and the Ghosts of Predators Past*. University of Chicago Press, Chicago.

Calkins, M. T., E. A. Beever, K. G. Boykin, J. K. Frey, and M. C. Andersen. 2012. Not-so-splendid isolation: Modeling climate-mediated range collapse of a montane mammal *Ochotona princeps* across numerous ecoregions. *Ecography* 35:780–91.

Campbell, K. E., Jr. 1980a. The contributions of Hildegarde Howard. In *Papers in Avian Paleontology Honoring Hildegarde Howard*, edited by K. E. Campbell Jr., xi–xxv. Natural History Museum of Los Angeles County Contributions in Science 330.

———. 1980b. The world's largest flying bird. *Terra* 19 (2): 20–23.

———. 2002. A new species of late Pleistocene lapwing from Rancho La Brea, California. *Condor* 104:170–74.

Campbell, K. E., Jr., and Z. M. Bocheński. 2010. A new genus for the extinct late Pleistocene owl *Strix brea* Howard (Aves: Strigiformes) from Rancho La Brea, California. *Records of the Australian Museum* 62:123–44.

Campbell, K. E., Jr., E. Scott, and K. B. Springer. 1999. A new genus for the Incredible Teratorn (Aves: Teratornithidae). In *Avian Paleontology at the Close of the 20th Century: Proceedings of the 4th International Meeting of the Society of Avian Paleontology and Evolution, Washington, D.C., 4–7 June 1996*, edited by P. Wellnhofer, C. Mourer-Chauviré, D. W. Steadman, and L. D. Martin, 169–75. Smithsonian Contributions to Paleobiology 89.

Campbell, K. E., Jr., and E. P. Tonni. 1981. Preliminary observations on the paleobiology and evolution of teratorns (Aves: Teratornithidae). *Journal of Vertebrate Paleontology* 1:265–72.

———. 1983. Size and locomotion in teratorns (Aves: Teratornithidae). *Auk* 100:390–403.

Campbell, K. L., J. E. E. Roberts, L. N. Watson, et al. (15 authors). 2010. Substitutions in woolly mammoth hemoglobin confer biochemical properties adaptive for cold tolerance. *Nature Genetics* 42:536–40.

Campos, P. F., A. Sher, J. I. Mead, et al. (8 authors). 2010. Clarification of the taxonomic relationship of the extant and extinct ovibovids, *Ovibos*, *Praeovibos*, *Euceratherium* and *Bootherium*. *Quaternary Science Reviews* 29:2123–30.

Campos, P. F., E. Willerslev, J. I. Mead, M. Hofreiter, and M. T. P. Gilbert. 2010. Molecular identification of the extinct mountain goat, *Oreamnos harringtoni* (Bovidae). *Boreas* 39:18–23.

Campos, P. F., E. Willerslev, A. Sher, et al. (20 authors). 2010. Ancient DNA analyses exclude humans as the driving force behind late Pleistocene muskox (*Ovibos moschatus*) population dynamics. *Proceedings of the National Academy of Sciences* 107:5675–80.

Canella, J. A. 2003. *Larrea tridentata* distribution. USGS Global Change Research Program. http://sbsc.wr.usgs.gov/cprs/research/projects/global_change/rangemaps /larreadistributionmap.pdf.

Cannon, M. D., S. D. Livingston, and J. M. Broughton. 2009. Faunal remains from the Sunshine Locality. In *The Archaeology of the Eastern Nevada Paleoarchaic, Part I: The Sunshine Locality*, edited by C. Beck and G. T. Jones, 218–28. University of Utah Anthropological Papers 126.

Cannon, M. D., and D. J. Meltzer. 2004. Early Paleoindian foraging: Examining the faunal evidence for large mammal specialization and regional variability in prey choice. *Quaternary Science Reviews* 23:1955–87.

———. 2008. Explaining variability in early Paleoindian foraging. *Quaternary International* 191:5–17.

Carbone, C., T. Maddox, P. J. Funston, M. G. L. Mills, G. F. Grether, and B. Van Valkenburgh. 2009. Parallels between playbacks and Pleistocene tar seeps suggest sociality in an extinct sabretooth cat, *Smilodon*. *Biology Letters* 5:81–85.

Carbot-Chanona, G. 2010. The first record of *Dasypus* (Xenarthra: Cingulata: Dasipodidae [*sic*]) in the late Pleistocene of México. *Current Research in the Pleistocene* 27:164–66.

Carlson, A. 2010. What caused the Younger Dryas cold event? *Geology* 38:383–84.

Carmona, D., M. J. Lajeunesse, and M. T. J. Johnson. 2011. Plant traits that predict resistance to herbivores. *Functional Ecology* 25:358–67.

The Carson footprints: An amateur professor believes that they are not very ancient. From the *Virginia (Nev.) Enterprise*. 1883. *New York Times*, November 30.

The Carson footprints. From the *Virginia (Nev.) Enterprise*. 1883. *New York Times*, August 12.

The Carson footprints: Second paper of Dr. Harkness of San Francisco. Why he believes them made by man—Evidence of artificial protection of the feet. From the *San Francisco Alta*, August 7. 1883. *New York Times*, August 15.

Cartelle, C. 2000. Preguiças terrícolas, essas desconhecidas. *Ciência Hoje* 27 (161): 18–25.

Cartelle, C., and G. De Iuliis. 2006. *Eremotherium laurillardi* (Lund), the Panamerican giant ground sloth: Taxonomic aspects of the ontogeny of skull and dentition. *Journal of Vertebrate Paleontology* 4:199–209.

Cartelle, C., and W. C. Hartwig. 1996. A new extinct primate among the Pleistocene megafauna of Bahia, Brazil. *Proceedings of the National Academy of Sciences* 93:6405–9.

Chamberlain, C. P., J. R. Waldbauer, K. Fox-Dobbs, et al. (10 authors). 2005. Pleistocene to Recent dietary shifts in California condors. *Proceedings of the National Academy of Sciences* 102:16707–11.

Chatterjee, S., R. J. Templin, and K. E. Campbell Jr. 2007. The aerodynamics of *Argentavis*, the world's largest flying bird from the Miocene of Argentina. *Proceedings of the National Academy of Sciences* 104:12398–403.

Christiansen, P. 1999. What size were *Arctodus simus* and *Ursus spelaeus* (Carnivora: Ursidae)? *Annales Zoologici Fennici* 36:93–102.

Christiansen, P., and J. M. Harris. 2005. Body size of *Smilodon* (Mammalia: Felidae). *Journal of Morphology* 266:369–84.

———. 2009. Craniomandibular morphology and phylogenetic affinities of *Panthera atrox*: Implications for the evolution and paleobiology of the lion lineage. *Journal of Vertebrate Paleontology* 29:934–45.

Churcher, C. S. 1966. The affinities of *Dinobastis serus* Cope 1893. *Quaternaria* 8:263–72.

———. 1991. The status of *Giraffa nebrascensis*, the synonymies of *Cervalces* and *Cervus*, and additional records of *Cervalces scotti*. *Journal of Vertebrate Paleontology* 11:391–97.

Churcher, C. S., and J. D. Pinsof. 1987. Variation in the antlers of North American *Cervalces* (Mammalia: Cervidae): Review of new and previously recorded specimens. *Journal of Vertebrate Paleontology* 7:373–97.

Cione, A. L., E. P. Tonni, M. Bond, et al. (8 authors). 1999. Occurrence charts of Pleistocene mammals in the Pampean area, eastern Argentina. *Quaternary of South America and Antarctic Peninsula* 12:53–59.

Clark, R. D. 1989. *The Odyssey of Thomas Condon*. Oregon Historical Society Press, Portland.

Coates, A. G., and R. F. Stallard. 2013. How old is the Isthmus of Panama? *Bulletin of Marine Science* 89:801–13.

Colbert, E. H., and R. G. Chaffee. 1939. *A Study of* Tetrameryx *and Associated Fossils from Papago Spring Cave, Sonoita, Arizona*. American Museum Novitates 1034.

Cole, K. L. 1990. Reconstruction of past desert vegetation along the Colorado River using packrat middens. *Palaeogeography, Palaeoclimatology, Palaeoecology* 76:349–66.

Cole, K. L., J. F. Fisher, K. Ironside, J. I. Mead, and P. Koehler. 2013. The biogeographic histories of *Pinus edulis* and *Pinus monophylla* over the last 50,000 years. *Quaternary International* 310:96–110.

Cole, K. L., K. Ironside, J. Eischeid, G. Garfin, P. B. Duffy, and C. Toney. 2011. Past and ongoing shifts in Joshua tree distribution support future modeled range contraction. *Ecological Applications* 21:137–49.

Coley, P. D., and T. A. Kursar. 2014. On tropical forests and their pests. *Science* 343:35–36.

Coltrain, J. B., J. M. Harris, T. E. Cerling, J. R. Ehleringer, M.-D. Dearing, J. Ward, and J. Allen. 2004. Rancho La Brea stable isotope biogeochemistry and its implications for the palaeoeconomy of late Pleistocene, coastal southern California. *Palaeogeography, Palaeoclimatology, Palaeoecology* 205:199–219.

Condon, T. 1902. *The Two Islands and What Came of Them*. J. K. Gill, Portland, OR.

———. 1910. *Oregon Geology*. J. K. Gill, Portland, OR.

Condron, A., and P. Winsor. 2012. Meltwater routing and the Younger Dryas. *Proceedings of the National Academy of Sciences* 109:19928–33.

Conklin, E. G. 1944. The early history of the *American Naturalist*. *American Naturalist* 78:29–37.

Connolly, T. J. 1999. *Newberry Crater: A Ten-Thousand-Year Record of Human Occupation and Environmental Change in the Basin-Plateau Borderlands*. University of Utah Anthropological Papers 121.

Cooper, S. M., and N. Owen-Smith. 1986. Effects of plant spinescence on large mammalian herbivores. *Oecologia* 68:446–55.

Cope, E. D. 1878a. Descriptions of new extinct Vertebrata from the upper Tertiary and Dakota Formations. *Bulletin of the U.S. Geological and Geographical Survey of the Territories* 4:379–96.

——. 1878b. Pliocene man. *American Naturalist* 12:125–26.

——. 1878c. Pliocene man. *Proceedings of the American Philosophical Society* 17:292.

——. 1880. Pliocene man. *American Naturalist* 14:60–62.

——. 1882. The recent discoveries of fossil footprints in Carson, Nevada. *American Naturalist* 16:921–23.

——. 1883a. [Contemporaneity of man and Pliocene mammals]. *Proceedings of the Academy of Natural Sciences of Philadelphia* 34:291–92.

——. 1883b. On the fishes of the recent and Pliocene lakes of the western part of the Great Basin, and of the Idaho Pliocene lake. *Proceedings of the Academy of Natural Sciences of Philadelphia* 35:134–67.

——. 1883c. The Carson footprints. *American Naturalist* 17:1153.

——. 1883d. The extinct Rodentia of North America: Pliocene and post-Pliocene Rodentia. *American Naturalist* 17:370–81.

——. 1883e. The Nevada biped tracks. *American Naturalist* 17:69–71.

——. 1884. The extinct Mammalia of the Valley of Mexico. *Proceedings of the American Philosophical Society* 22:1–21.

——. 1885. The occurrence of man in the Upper Miocene of Nebraska. *Proceedings of the American Association for the Advancement of Science* 33:593.

——. 1889a. The Silver Lake of Oregon and its region. *American Naturalist* 23:970–82.

——. 1889b. The vertebrate fauna of the Equus beds. *American Naturalist* 23:160–69.

——. 1893. The genealogy of man. *American Naturalist* 27:321–35.

——. 1895. The antiquity of man in North America. *American Naturalist* 29:593–600.

Corona, A., D. Perea, and H. G. McDonald. 2013. *Catonyx cuvieri* (Xenarthra, Mylodontidae, Scelidotheriinae) from the Late Pleistocene of Uruguay, with comments regarding the systematics of the subfamily. *Journal of Vertebrate Paleontology* 33: 1214–25.

Corona-M., E. 2002. The Pleistocene bird record of México. *Acta Zoologica Cracoviensia* 45:293–306.

Cozzuol, M. A., C. L. Clozato, E. C. Holanda, et al. (8 authors). 2013. A new species of tapir from the Amazon. *Journal of Mammalogy* 94:1331–45.

CPI Inflation Calculator. 2014. http://data.bls.gov/cgi-bin/cpicalc.pl.

Crawley, M. J. 1983. *Herbivory: The Dynamics of Animal-Plant Interactions*. University of California Press, Berkeley.

Crégut-Bonnoure, E. 1984. The Pleistocene Ovibovinae of western Europe: Temporo-spatial expansion and paleoecological implications. In *Proceedings of the First International Muskox Symposium*, edited by D. R. Klein, R. G. White, and S. Keller, 136–44. Biological Papers of the University of Alaska Special Report 4.

Cressman, L. S. 1939. Early man and culture in the northern Great Basin region of south central Oregon. *Carnegie Institution of Washington Year Book* 38:314–17.

———. 1940. Studies on early man in south central Oregon. *Carnegie Institution of Washington Year Book* 39:300–306.

———. 1942. *Archaeological Researches in the Northern Great Basin.* Carnegie Institution of Washington Publication 538.

———. 1946. Early man in Oregon: Stratigraphic evidence. *Scientific Monthly* 62:43–51.

———. 1966. Man in association with extinct fauna in the Great Basin. *American Antiquity* 31:866–67.

Cronquist, A., A. H. Holmgren, N. H. Holmgren, and J. L. Reveal. 1972. *Intermountain Flora: Vascular Plants of the Intermountain West, U.S.A.* Vol. 1. Hafner, New York.

Czaplewski, N. J. 2012. Pleistocene peccaries (Mammalia, Tayassuidae) from western Oklahoma. *Southwestern Naturalist* 57:112–17.

Dalquest, W. W. 1974. A new species of four-horned antilocaprid from Mexico. *Journal of Mammalogy* 55:96–101.

Daniau, A.-L., S. P. Harrison, and P. J. Bartlein. 2010. Fire regimes during the last glacial. *Quaternary Science Reviews* 29:2918–30.

Dansie, A., J. O. Davis, and T. W. Stafford Jr. 1988. The Wizard's Beach recession: Farmdalian (25,500 yr BP) vertebrate fossils co-occur with early Holocene artifacts. In *Early Human Occupation in Far Western North America: The Clovis-Archaic Interface,* edited by J. A. Willig, C. M. Aikens, and J. L. Fagan, 153–200. Nevada State Museum Anthropological Papers 21.

Dansie, A. J., and W. J. Jerrems. 2005. More bits and pieces: A new look at Lahontan chronology and human occupation. In *Paleoamerican Origins: Beyond Clovis,* edited by R. Bonnichsen, B. T. Lepper, D. Stanford, and M. R. Waters, 51–79. Center for the Study of the First Americans, Texas A&M University Press, College Station.

Darwin, C. 1859. *On the Origin of Species by Means of Natural Selection, or the Preservation of Favoured Races in the Struggle for Life.* John Murray, London.

Davidson, G. 1883. *The Carson Fossil Footprints.* Separately printed from the *Mining and Scientific Press.*

Davidson, J. P. 1997. *The Bone Sharp: The Life of Edward Drinker Cope.* Academy of Natural Sciences of Philadelphia Special Publication 17.

Davis, C. A., and G. A. Smith. 1981. *Newberry Cave.* San Bernardino County Museum Association, Redlands, CA.

Davis, E. B. 2007. Family Antilocapridae. In *The Evolution of Artiodactyls,* edited by D. R. Prothero and S. E. Foss, 227–40. Johns Hopkins University Press, Baltimore.

Davis, E. L. 1975. The "exposed archaeology" of China Lake, California. *American Antiquity* 40:39–53.

———. 1978a. Associations of people and a Rancholabrean fauna at China Lake, California. In *Early Man in America from a Circum-Pacific Perspective,* edited by A. L. Bryan, 183–217. Department of Anthropology, University of Alberta Occasional Papers 1.

———, ed. 1978b. *The Ancient Californians: Rancholabrean Hunters of the Mojave Lakes Country.* Natural History Museum of Los Angeles County Science Series 29.

Davis, L. G., S. J. Macfarlan, and C. Henrickson. 2012. A PXRF-based chemostratigraphy and provenience system for the Cooper's Ferry site, Idaho. *Journal of Archaeological Science* 39:663–71.

Davis, O. K. 1987. Spores of the dung fungus *Sporormiella*: Increased abundance in historic sediments and before Pleistocene megafaunal extinction. *Quaternary Research* 28:290–94.

———. 2002. Late Neogene environmental history of the northern Bonneville Basin: A review of palynological studies. In *Great Basin Aquatic Systems History*, edited by R. Hershler, D. B. Madsen, and D. R. Currey, 295–307. Smithsonian Contributions to the Earth Sciences 33.

Davis, O. K., L. Agenbroad, P. S. Martin, and J. I. Mead. 1984. The Pleistocene dung blanket of Bechan Cave, Utah. In *Contributions in Quaternary Vertebrate Paleontology: A Volume in Memorial to John Guilday*, edited by H. H. Genoways and M. R. Dawson, 267–82. Carnegie Museum of Natural History Special Publication 8.

Davis, O. K., and D. S. Shafer. 2006. *Sporormiella* fungal spores, a palynological means of detecting herbivore density. *Palaeogeography, Palaeoclimatology, Palaeoecology* 237:40–50.

Debruyne, R., G. Chu, C. E. King, et al. (21 authors). 2008. Out of America: Ancient DNA evidence for a New World origin of late Quaternary woolly mammoths. *Current Biology* 18:1320–26.

De Esteban-Trivigno, S., M. Mendoza, and M. De Renzi. 2008. Body mass estimation in Xenarthra: A predictive equation suitable for all quadrupedal terrestrial placentals? *Journal of Morphology* 269:1276–93.

De Iuliis, G., F. Pujos, and C. Cartelle. 2009. A new ground sloth (Mammalia: Xenarthra) from the Quaternary of Brazil. *Comptes Rendus Palevol* 8:705–15.

DeMay, I. S. 1941a. Pleistocene bird life of the Carpinteria asphalt, California. *Carnegie Institution of Washington Publication* 530:61–76.

———. 1941b. Quaternary bird life of the McKittrick asphalt, California. *Carnegie Institution of Washington Publication* 530:35–60.

Deméré, T. 2007. Pleistocene Land Mammals of Western San Diego County: New Discoveries of Mastodons and Mammoths. San Diego Association of Geologists Annual Meeting Announcement. http://www.sandiegogeologists.org/newsletters/2007-12_SDAG.pdf.

DeSantis, L. R. G., B. W. Schubert, J. R. Scott, and P. S. Ungar. 2012. Implications of diet for the extinction of saber-toothed cats and American lions. *PLoS One* 7 (12): e52453.

Diamond, J. M. 1992. *The Third Chimpanzee: The Evolution and Future of the Human Animal.* HarperCollins, New York.

Dicke, M. 1999. Evolution of indirect defense of plants. In *The Ecology and Evolution of Inducible Defenses*, edited by R. Tollrian and C. D. Harvell, 62–88. Princeton University Press, Princeton, NJ.

Dillehay, T. D. 1997. *Monte Verde: A Late Pleistocene Settlement in Chile.* Vol. 2, *The Archaeological Context and Interpretation.* Smithsonian Institution Press, Washington, DC.

Dimmit, M. A. 2000. Fabaceae (legume family). In *A Natural History of the Sonoran Desert*, edited by S. J. Phillips and P. W. Comus, 227–39. Arizona-Sonora Desert Museum Press, Tucson.

Dompierre, H., and C. S. Churcher. 1996. Premaxillary shape as an indicator of the diet of seven extinct late Cenozoic New World camels. *Journal of Vertebrate Paleontology* 16:141–48.

Downs, T., H. Howard, T. Clements, and G. A. Smith. 1959. *Quaternary Animals from Schuiling Cave in the Mojave Desert, California.* Los Angeles City Museum Contributions in Science 29.

Dr. Harkness on the Nevada footprints. 1883. *Popular Science Monthly* 23:859–60.

Dundas, R. G. 1999. Quaternary records of the dire wolf, *Canis dirus*, in North and South America. *Boreas* 28:375–85.

Durbin, L. S., S. Hedges, J. W. Duckworth, M. Tyson, A. Lyenga, and A. Venkataraman (IUCN SSC Canid Specialist Group—Dhole Working Group). 2008. *Cuon alpinus*. IUCN Red List of Threatened Species. Version 2013.1. http://www.iucnredlist.org.

Dyke, A. S., J. T. Andrews, P. U. Clark, J. H. England, G. H. Miller, J. Shaw, and J. J. Veillette. 2002. The Laurentide and Innuitian ice sheets during the Last Glacial Maximum. *Quaternary Science Reviews* 21:9–31.

Eames, A. J. 1930. Report on ground sloth coprolite from Doña Ana County, New Mexico. *American Journal of Science* 20:353–56.

Eighty acres af [*sic*] fossil remains near Silver Lake, Oregon. From the *San Francisco Chronicle*. 1877. *New York Times*, June 4.

Elftman, H. O. 1931. *Pleistocene Mammals of Fossil Lake, Oregon*. American Museum Novitates 481.

Elias, S. A. 1990. The timing and intensity of environmental changes during the Paleoindian period in western North America: Evidence from the insect fossil record. In *Megafauna and Man: Discovery of America's Heartland*, edited by L. D. Agenbroad, J. I. Mead, and L. W. Nelson, 11–14. The Mammoth Site of Hot Springs, South Dakota, Scientific Papers 1.

———. 1991. Insects and climate change. *BioScience* 41:552–59.

———. 1995. A paleoenvironmental setting for early Paleoindians in western North America: Evidence from the insect fossil record. In *Ancient Peoples and Landscapes*, edited by E. Johnson, 255–72. Museum of Texas Tech University, Lubbock.

———. 1996. Late Pleistocene and Holocene seasonal temperatures reconstructed from fossil beetle assemblages in the Rocky Mountains. *Quaternary Research* 46:311–18.

Elias, S. A., and B. Crocker. 2008. The Bering Land Bridge: A moisture barrier to the dispersal of steppe-tundra biota? *Quaternary Science Reviews* 27:2473–83.

Elissamburu, A. 2012. Estimación de la masa corporal en géneros del orden Notoungulata. *Estudios Geológicos* 68:91–111.

Emerson, E. 1980. A send-off for Joe Goodman: Mark Twain's "The Carson Fossil-Footprints." *Resources for American Literary Study* 10:71–78.

Emslie, S. D. 1985. The late Pleistocene (Rancholabrean) avifauna of Little Box Elder Cave, Wyoming. *Contributions to Geology, University of Wyoming* 23:63–82.

———. 1987a. Age and diet of fossil California Condors in Grand Canyon, Arizona. *Science* 237:768–70.

———. 1987b. The origin, evolution and extinction of condors in the New World. PhD diss., University of Florida, Gainesville.

———. 1988. The fossil history and phylogenetic relationships of condors (Ciconiiformes: Vulturidae) in the New World. *Journal of Vertebrate Paleontology* 8:212–28.

———. 1990. Additional ^{14}C dates on fossil California Condor. *National Geographic Research* 6:134–35.

———. 1998. *Avian Community, Climate, and Sea-Level Changes in the Plio-Pleistocene of the Florida Peninsula*. Ornithological Monographs 50.

Emslie, S. D., and N. J. Czaplewski. 1985. A new record of giant short-faced bear, *Arctodus simus*, from western North America with a re-evaluation of its paleobiology. *Natural History Museum of Los Angeles County Contributions in Science* 731:1–12.

————. 1999. Two new fossil eagles from the late Pliocene (late Blancan) of Florida and Arizona and their biogeographic implications. In *Avian Paleontology at the Close of the 20th Century: Proceedings of the 4th International Meeting of the Society of Avian Paleontology and Evolution, Washington, D.C., 4–7 June 1996*, edited by S. L. Olson, P. Wellnhofer, C. Mourer-Chauviré, D. W. Steadman, and L. D. Martin, 185–98. Smithsonian Contributions to Paleobiology 89.

Emslie, S. D., and T. H. Heaton. 1987. The late Pleistocene avifauna of Crystal Ball Cave, Utah. *Journal of the Arizona-Nevada Academy of Science* 21:53–60.

Encina, R. L., and M. T. Alberdi. 2011. An updated taxonomic view on the family Gomphotheriidae (Proboscidea) in the final Pleistocene of south-central Chile. *Neues Jahrbuch für Geologie und Paläontologie–Abhandlungen* 262:43–57.

England, J., N. Atkinson, J. Bednarski, A. S. Dyke, D. A. Hodgson, and C. Ó. Cofaigh. 2006. The Innuitian Ice Sheet: Configuration, dynamics and chronology. *Quaternary Science Reviews* 25:689–703.

Enk, J., A. Devault, R. Debruyne, et al. (10 authors). 2011. Complete Columbian mammoth mitogenome suggests interbreeding with woolly mammoths. *Genome Biology* 12:R51.

Enk, J. M., D. R. Yesner, K. J. Crossen, D. W. Veltre, and D. H. O'Rourke. 2009. Phylogeographic analysis of the mid-Holocene mammoth from Qagnax Cave, St. Paul Island, Alaska. *Palaeogeography, Palaeoclimatology, Palaeoecology* 273:184–90.

Ertter, B. 2012. *Rosa woodsii. Jepson eFlora*. Jepson Flora Project. http://ucjeps.berkeley.edu/cgi-bin/get_IJM.pl?tid=41704.

Evans, G. L. 1961. The Friesenhahn Cave. *Bulletin of the Texas Memorial Museum* 2:3–22.

Faith, J. T. 2014. Late Pleistocene and Holocene mammal extinctions on continental Africa. *Earth-Science Reviews* 128:105–21.

Faith, J. T., and T. A. Surovell. 2009. Synchronous extinction of North America's Pleistocene mammals. *Proceedings of the National Academy of Sciences* 106:20641–45.

Faith, J. T., C. A. Tryon, D. J. Peppe, E. J. Beverly, and N. Blegen. 2014. Biogeographic and evolutionary implications of an extinct late Pleistocene impala from the Lake Victoria Basin, Kenya. *Journal of Mammalian Evolution* 21:213–22.

Fariña, R. A. 1996. Trophic relationships among Lujanian mammals. *Evolutionary Theory* 11:125–34.

Fariña, R. A., S. F. Vizcaíno, and A. S. Bargo. 1998. Body mass estimations in Lujanian (late Pleistocene–early Holocene of South America) mammal megafauna. *Mastozoología Neotropical* 5 (2): 87–108.

Fariña, R. A., S. F. Vizcaíno, and G. De Iuliis. 2013. *Megafauna: Giant Beasts of Pleistocene South America*. Indiana University Press, Bloomington.

Farlow, A. O., and J. McClain. 1996. A spectacular specimen of the elk-moose *Cervalces scotti* from Noble County, Indiana, U.S.A. In *Palaeoecology and Palaeoenvironments of Late Cenozoic Mammals: Tributes to the Career of C. S. (Rufus) Churcher*, edited by K. M. Stewart and K. L. Seymour, 322–30. University of Toronto Press, Toronto.

FAUNMAP Working Group. 1994. *FAUNMAP: A Database Documenting Late Quaternary Distributions of Mammal Species in the United States*. Illinois State Museum Scientific Papers 25.

————. 1996. Spatial response of mammals to late Quaternary environmental fluctuations. *Science* 272:1601–6.

Federal judge dismisses most lawsuit claims in fatal mountain goat attack in Olympic National Park. 2012. *Peninsula Daily News* (Port Angeles, WA), August 22.

Feduccia, A. 1974. Another Old World vulture from the New World. *Wilson Bulletin* 86:251–55.

Felger, R. S., M. B. Johnson, and M. F. Wilson. 2001. *The Trees of Sonora, Mexico*. Oxford University Press, New York.

Feranec, R. S. 2003. Stable isotopes, hypsodonty, and the paleodiet of *Hemiauchenia* (Mammalia: Camelidae): A morphological specialization creating ecological generalization. *Paleobiology* 29:230–42.

———. 2004. Isotopic evidence of saber-tooth development, growth rate, and diet from the adult canine of *Smilodon fatalis* from Rancho La Brea. *Palaeogeography, Palaeoclimatology, Palaeoecology* 206:303–10.

———. 2008. Growth differences in the saber-tooth of three felid species. *Palaios* 23:566–69.

Feranec, R. S., and A. L. Kozlowski. 2010. AMS radiocarbon dates from Pleistocene and Holocene mammals housed in the New York State Museum, Albany, New York, USA. *Radiocarbon* 52:205–8.

Feranec, R. S., and B. J. MacFadden. 2000. Evolution of the grazing niche in Pleistocene mammals from Florida: Evidence from stable isotopes. *Palaeogeography, Palaeoclimatology, Palaeoecology* 162:155–69.

Feranec, R. S., N. G. Miller, J. C. Lothrop, and R. W. Graham. 2011. The *Sporormiella* proxy and end-Pleistocene megafaunal extinction: A perspective. *Quaternary International* 245:335–38.

Ferigolo, J. 1999. Late Pleistocene South-American land-mammal extinctions: The infection hypothesis. *Quaternary of South America and Antarctic Peninsula* 12:279–310.

Ferretti, M. F. 2008. A review of South American Proboscideans. In *Neogene Mammals*, edited by S. G. Lucas, G. S. Morgan, J. A. Spielmann, and D. R. Prothero, 381–92. New Mexico Museum of Natural History and Science Bulletin 44.

Ferring, C. R. 2001. *The Archaeology and Paleoecology of the Aubrey Clovis Site (41DN479), Denton County, Texas*. Center for Environmental Archaeology, Department of Geography, University of North Texas, Denton.

Ferrusquía-Villafranca, I., J. Arroyo-Cabrales, E. Martínez-Hernández, J. Gama-Castro, J. Ruiz-González, O. J. Polaco, and E. Johnson. 2010. Pleistocene mammals of Mexico: A critical review of regional chronofaunas, climate change response and biogeographic provinciality. *Quaternary International* 217:53–104.

Figueirido, B., P. Palmqvist, and J. A. Pérez-Claros. 2009. Ecomorphological correlates of craniodental variation in bears and paleobiological implications for extinct taxa: An approach based on geometric morphometrics. *Journal of Zoology* 277:70–80.

Figueirido, B., J. A. Pérez-Claros, V. Torregrosa, A. Martín-Serra, and P. Palmqvist. 2010. Demythologizing *Arctodus simus*, the "short-faced" long-legged and predaceous bear that never was. *Journal of Vertebrate Paleontology* 30:262–75.

Figueirido, B., and L. H. Soibelzon. 2010. Inferring palaeoecology in extinct tremarctine bears (Carnivora, Ursidae) using geometric morphometrics. *Lethaia* 43:209–22.

Finch, W. I., F. C. Whitmore Jr., and J. D. Sims. 1972. *Stratigraphy, Morphology, and Paleoecology of a Fossil Peccary Herd from Western Kentucky*. US Geological Survey Professional Paper 790.

Firby, J. R., J. E. Mawby, and J. O. Davis. 1987. Vertebrate paleontology and geology of the Rye Patch Paleontological locality. In *Studies in Archaeology, Geology and Paleontology at Rye Patch Reservoir, Pershing County, Nevada*, edited by M. K. Rusco and J. O. Davis, 23–40. Nevada State Museum Anthropological Papers 20.

Firestone, R. B. 2009. The case for the Younger Dryas extraterrestrial impact event: Mammoth, megafauna, and Clovis extinction, 12,900 years ago. *Journal of Cosmology* 2:256–85.

Firestone, R. B., and W. Topping. 2001. Terrestrial evidence of a nuclear catastrophe in Paleoindian times. *Mammoth Trumpet* 16 (2): 9–16.

Firestone, R. B., A. West, J. P. Kennett, et al. (26 authors). 2007. Evidence for an extraterrestrial impact 12,900 years ago that contributed to the megafaunal extinctions and the Younger Dryas cooling. *Proceedings of the National Academy of Sciences* 104:16016–21.

Firestone, R. B., A. West, Z. Revay, J. T. Hagstrum, T. Belgya, S. S. Que Hee, and A. R. Smith. 2010. Analysis of the Younger Dryas impact layer. *Journal of Siberian Federal University, Engineering and Technologies* 3:30–62.

Firestone, R. B., A. West, and S. Warwick-Smith. 2006. *The Cycle of Cosmic Catastrophes: How a Stone-Age Comet Changed the Course of World Culture*. Bear and Company, Rochester, VT.

Fisher, D. C. 2001. Season of death, growth rates, and life history of North American mammoths. In *Proceedings of the International Conference on Mammoth Site Studies*, edited by D. West, 121–35. University of Kansas Publications in Anthropology 22.

———. 2008. Taphonomy and paleobiology of the Hyde Park mastodon. In *Mastodon Paleobiology, Taphonomy, and Paleoenviroment in the Late Pleistocene of New York State*, edited by W. D. Allmon and P. L. Nestor, 197–289. Palaeontographica Americana 61.

———. 2009. Paleobiology and extinction of proboscideans in the Great Lakes region of North America. In *American Megafaunal Extinctions at the End of the Pleistocene*, edited by G. Haynes, 55–75. Springer, Dordrecht, Netherlands.

Flueck, W. T., and J. M. Smith-Flueck. 2011. Osteological comparisons of appendicular skeletons: A case study on Patagonian huemul deer and its implications for conservation. *Animal Production Science* 51:327–39.

Footprints of monster men: Curious discoveries that interest scientists of the Pacific coast. From the *San Francisco Chronicle*, August 8. 1882. *New York Times*, August 18.

Forasiepi, A., A. Martinelli, and J. Blanco. 2007. *Bestiario fósil: Mamíferos del Pleistoceno de la Argentina*. Editorial Albatros, Buenos Aires.

Ford, A. T., J. R. Goheen, T. O. Otieno, et al. (9 authors). 2014. Large carnivores make savanna tree communities less thorny. *Science* 346:346–49.

Fortsch, D. E. 1978. The China Lake Rancholabrean faunule. In *The Ancient Californians: Rancholabrean Hunters of the Mojave Lakes Country*, edited by E. L. Davis, 173–76. Natural History Museum of Los Angeles County Science Series 29.

The fossil footprints of Nevada. From the *San Francisco Alta*, August 29. 1882. *New York Times*, September 11.

Fossil human foot-prints in Nevada. 1883. *Popular Science Monthly* 22:137–38.

Fowler, C. S. 1986. Subsistence. In *Great Basin*, edited by W. L. d'Azevedo, 64–97. *Handbook of North American Indians*, vol. 11. Smithsonian Institution Press, Washington, DC.

———. 1995. Some notes on ethnographic subsistence systems in Mojavean environments in the Great Basin. *Journal of Ethnobiology* 15:99–117.

Fox-Dobbs, K., T. S. Stidham, G. J. Bowen, S. D. Emslie, and P. Koch. 2006. Dietary controls on extinction versus survival among avian megafauna in the late Pleistocene. *Geology* 34:685–88.

Franceschi, V. R., and P. A. Nakata. 2005. Calcium oxalate in plants: Formation and function. *Annual Review of Plant Biology* 56:41–71.

Frassinetti, D., and M. T. Alberdi. 2000. Revisión y estudio de los restos fósiles de mastodontes de Chile (Gomphotheriidae): *Cuvieronius hyodon*, Pleistoceno superior. *Estudios Geológicos* 56:197–208.

———. 2005. Presencia del género *Stegomastodon* entre los restos fósiles de mastodontes de Chile (Gomphotheriidae), Pleistoceno superior. *Estudios Geológicos* 61:101–7.

Freedman, A. H., I. Gronau, R. W. Schweizer, et al. (30 authors). 2014. Genome sequencing highlights the dynamic early history of dogs. *PLoS Genetics* 10 (1): e1004016.

Frémont, J. C. 1845. *Report of the Exploring Expedition to the Rocky Mountains in the Year 1842, and to Oregon and North California in the Years 1843–44*. Goles and Seaton, Washington, DC.

Frémont, J. C., and C. Preuss. 1848. *Map of Oregon and Upper California from the Surveys of John Charles Frémont and Other Authorities. Drawn by Charles Preuss under the Order of the Senate of the United States, Washington City 1848*.

Frick, C. 1937. Horned Ruminants of North America. Bulletin of the American Museum of Natural History 69.

Friedel, D. E. 1993. Chronology and climatic controls of Late Quaternary lake-level fluctuations in Chewaucan, Fort Rock, and Alkali Basins, south-central Oregon. PhD diss., University of Oregon, Eugene.

Friscia, A. R., B. Van Valkenburgh, L. Spencer, and J. Harris. 2008. Chronology and spatial distribution of large mammal bones in Pit 91, Rancho La Brea. *Palaios* 23:35–42.

Fulton, T. L., R. W. Norris, R. W. Graham, H. A. Semken Jr., and B. Shapiro. 2013. Ancient DNA supports southern survival of Richardson's collared lemming (*Dicrostonyx richardsoni*) during the last glacial maximum. *Molecular Ecology* 22:2540–48.

Furlong, E. L. 1905. *Preptoceras*, a new ungulate from the Samwel Cave, California. *University of California Publications, Bulletin of the Department of Geology* 4:163–69.

———. 1943. The Pleistocene antelope, *Stockoceros conklingi*, from San Josecito Cave, Mexico. *Carnegie Institution of Washington Publication* 551:1–7.

———. 1946. Generic identification of the Pleistocene antelope from Rancho La Brea. *Carnegie Institution of Washington Publication* 551:136–40.

Fürstenberg-Hägg, J., M. Zagrobelny, and S. Bak. 2013. Plant defense against insect herbivores. *International Journal of Molecular Sciences* 14:10242–97.

Gabrielson, I. N., and S. G. Jewett. 1940. *Birds of Oregon*. Oregon State Monographs, Studies in Zoology 2.

Gama-Campillo, L., and A. Gomez-Pampa. 1992. An ethnoecological approach for the study of *Persea*: A case study in the Maya area. *Proceedings of the Second World Avocado Conference*, 11–17.

García, N., and J. L. Arsuaga. 2003. Last glaciation cold-adapted faunas in the Iberian Peninsula. *Deinsea* 9:159–69.

Gardner, A. L., ed. 2007. *Mammals of South America.* Vol. 1, *Marsupials, Xenarthrans, Shrews, and Bats.* University of Chicago Press, Chicago.

Gardner, D. G. 1994. Injury to the oral mucous membranes caused by the common houseplant, *Dieffenbachia*: A review. *Oral Surgery, Oral Medicine, Oral Pathology* 78:631–33.

Garrott, R. A., and M. D. Oli. 2013. A critical crossroad for BLM's wild horse program. *Science* 341:847–48.

Gazin, C. L. 1933. The status of the extinct American "eland." *Journal of Mammalogy* 14:162–64.

Geiser, S. W. 1953. Ellis William Shuler, Ph.D., LL.D. *Field and Laboratory* 21:5–8.

Geist, V. 1998. *Deer of the World: Their Evolution, Behavior, and Ecology.* Stackpole Books, Mechanicsburg, PA.

Gibbes, C. D. 1882. *Pre-Historic Foot-Prints in the Sandstone Quarry of the Nevada State Prison.* Issued separately from the Proceedings of the California Academy of Sciences, September 4.

Gibson, E., and L. Spencer. 1977. *Fossil Lake Archaeological Survey: Fossil Lake ORV Temporary Closure Area.* Report submitted to the Bureau of Land Management, Lakeview, OR.

Gilbert, C., A. Ropiquet, and A. Hassanin. 2006. Mitochondrial and nuclear phylogenies of Cervidae (Mammalia, Ruminantia): Systematics, morphology, and biogeography. *Molecular Phylogenetics and Evolution* 40:101–17.

Gilbert, G. K. 1890. *Lake Bonneville.* US Geological Survey Monograph 1.

Gilbert, M. T. P., D. I. Drautz, A. M. Lesk, et al. (33 authors). 2008. Intraspecific phylogenetic analysis of Siberian woolly mammoths using complete mitochrondrial genomes. *Proceedings of the National Academy of Sciences* 105:8327–32.

Gilbert, M. T. P., D. L. Jenkins, A. Götherstrom, et al. (13 authors). 2008. DNA from pre-Clovis human coprolites in Oregon. North America. *Science* 320:786–89.

Gilbow, D. W. 1981. Inference of human activity from faunal remains. Master's thesis, Washington State University, Pullman.

Gill, F., and D. Donsker, eds. 2014. IOC World Bird List (version 4.2). http://www.worldbirdnames.org/. doi:10.14344/IOC.ML.4.2.

Gill, F. L., M. P. Crump, R. Schouten, and I. D. Bull. 2009. Lipid analysis of a ground sloth coprolite. *Quaternary Research* 72:284–88.

Gill, J. L., J. W. Williams, S. T. Jackson, K. B. Lininger, and G. S. Robinson. 2009. Pleistocene megafaunal collapse, novel plant communities, and enhanced fire regimes in North America. *Science* 326:1100–1103.

Gill, T. 1897. Edward Drinker Cope, naturalist—A chapter in the history of science. *American Naturalist* 31:831–63.

Gillette, D. D., and D. B. Madsen. 1992. The short-faced bear *Arctodus simus* from the late Quaternary in the Wasatch Mountains of central Utah. *Journal of Vertebrate Paleontology* 12:107–12.

———. 1993. The Columbian mammoth, *Mammuthus columbi*, from the Wasatch Mountains of central Utah. *Journal of Paleontology* 67:669–80.

Gillette, D. D., H. G. McDonald, and M. C. Hayden. 1999. The first record of Jefferson's ground sloth, *Megalonyx jeffersonii*, in Utah (Pleistocene, Rancholabrean Land

Mammal Age). In *Vertebrate Paleontology in Utah*, edited by D. D. Gillette, 509–21. Utah Geological Survey Miscellaneous Publication 99-1.

Gillette, D. D., and W. E. Miller. 1999. Catalogue of new Pleistocene mammalian sites and recovered fossils from Utah. In *Vertebrate Paleontology in Utah*, edited by D. D. Gillette, 523–30. Utah Geological Survey Miscellaneous Publication 99-1.

Gillette, D. D., and C. E. Ray. 1981. Glyptodonts of North America. *Smithsonian Contributions to Paleobiology* 40.

Gillette, D. D., and P. M. Whisler. 1986. Notes and comments on the late Pleistocene glyptodont, *Glyptotherium floridanum*, from Florida. *Florida Scientist* 49:55–62.

Gilmour, D. M., V. L. Butler, J. E. O'Connor, E. B. Davis, B. J. Culleton, D. J. Kennett, and G. Hodgins. 2015. Chronology and ecology of late Pleistocene megafauna in the northern Willamette Valley, Oregon. *Quaternary Research* 83:127–36.

Gilreath, A. 2009. *Gypsum Cave Revisited*. Nevada Cultural Resources Report #CR5-2462-4(P). Far Western Anthropological Research Group, Davis, CA.

Gleason, H. A. 1926. The individualistic concept of the plant association. *Bulletin of the Torrey Botanical Club* 53:7–26.

Global Raptor Information Network. 2013. http://globalraptors.org/grin/indexAlt.asp.

Glowiak, E. M., and S. M. Rowland. 2008. Gypsum Cave revisited: A faunal and taphonomic analysis of a Rancholabrean-to-Holocene fauna in southern Nevada. *Geological Society of America Abstracts with Programs* 40 (1): 50.

Gobetz, K. E., and S. R. Bozarth. 2001. Implications for late Pleistocene mastodon diet from opal phytoliths in tooth calculus. *Quaternary Research* 55:115–22.

Godsoe, W., E. Strand, C. I. Smith, J. B. Yoder, T. C. Esque, and O. Pellmyr. 2009. Divergence in an obligate mutualism is not explained by divergent climatic requirements. *New Phytologist* 183:589–99.

Godsoe, W., J. B. Yoder, C. I. Smith, C. S. Drummond, and O. Pellmyr. 2010. Absence of population-level phenotype matching in an obligate pollination mutualism. *Journal of Evolutionary Biology* 23:2739–46.

Godsoe, W., J. B. Yoder, C. I. Smith, and O. Pellmyr. 2008. Coevolution and divergence in the Joshua tree/Yucca moth mutualism. *American Naturalist* 171:816–23.

Goebel, T., B. Hockett, K. D. Adams, D. Rhode, and K. Graf. 2011. Climate, environment, and humans in North America's Great Basin during the Younger Dryas, 12,900–11,600 calendar years ago. *Quaternary International* 242:479–501.

Goebel, T., H. L. Smith, L. DiPietro, et al. (11 authors). 2013. Serpentine Hot Springs, Alaska: Results of excavations and implications for the age and significance of northern fluted points. *Journal of Archaeological Science* 40:4222–33.

Goldstein, I., X. Velez-Liendo, S. Paisley, and D. L. Garshelis (IUCN SSC Bear Specialist Group). 2008. *Tremarctos ornatus*. IUCN Red List of Threatened Species. Version 2014.1. http://www.iucnredlist.org/details/22066/0.

Goodwin, H. T. 1995. Pliocene-Pleistocene biogeographic history of prairie dogs, genus *Cynomys* (Sciuridae). *Journal of Mammalogy* 76:100–122.

Goss, R. J. 1983. *Deer Antlers: Regeneration, Function, and Evolution*. Academic Press, New York.

Graham, R. W. 1976a. Late Wisconsin mammalian faunas and environmental gradients of the eastern United States. *Paleobiology* 2:343–50.

————. 1976b. Pleistocene and Holocene mammals, taphonomy, and paleoecology of the Friesenhahn Cave local fauna, Bexar County, Texas. PhD diss., University of Texas, Austin.

————. 1985a. Diversity and community structure of the late Pleistocene mammal fauna of North America. *Acta Zoologica Fennica* 170:181–92.

————. 1985b. Response of mammalian communities to environmental changes during the late Quaternary. In *Community Ecology*, edited by J. Diamond and T. J. Case, 300–313. Harper and Row, New York.

————. 1986. Plant-animal interactions and Pleistocene extinctions. In *The Dynamics of Extinction*, edited by D. K. Elliot, 131–54. John Wiley and Sons, New York.

————. 1990. Evolution of new ecosystems at the end of the Pleistocene. In *Megafauna and Man: Discovery of America's Heartland*, edited by L. D. Agenbroad, J. I. Mead, and L. W. Nelson, 54–60. The Mammoth Site of Hot Springs, South Dakota, Scientific Papers 1.

————. 1992. *Palaeolama mirifica* from the central Mississippi River Valley: Paleoecological and evolutionary implications. *Journal of Vertebrate Paleontology* 12 (3): 31A.

————. 2003. Pleistocene tapir from Hill Top Cave, Trigg County, Kentucky, and a review of Plio-Pleistocene tapirs of North America and their paleobiology. In *Ice Age Cave Faunas of North America*, edited by B. W. Schubert, J. I. Mead, and R. W. Graham, 87–118. Indiana University Press, Bloomington.

Graham, R. W., J. O. Farlow, and J. E. Vandike. 1996. Tracking ice age felids: Identification of tracks of *Panthera atrox* from a cave in southern Missouri, U.S.A. In *Palaeoecology and Palaeoenvironments of Late Cenozoic Mammals: Tributes to the Career of C. S. (Rufus) Churcher*, edited by K. M. Stewart and K. L. Seymour, 331–45. University of Toronto Press, Toronto.

Graham, R. W., C. V. Haynes, D. L. Johnson, and M. Kay. 1981. Kimmswick: A Clovis-mastodon association in eastern Missouri. *Science* 213:1115–17.

Graham, R. W., and M. Kay. 1988. Taphonomic comparisons of cultural and noncultural faunal deposits at the Kimmswick and Barnhart sites, Jefferson County, Missouri. In *Late Pleistocene and Early Holocene Paleoecology and Archaeology of the Eastern Great Lakes Region*, edited by R. S. Laub, N. G. Miller, and D. W. Steadman, 227–40. Buffalo Society of Natural Sciences Bulletin 33.

Graham, R. W., and E. L. Lundelius Jr. 1984. Coevolutionary disequilibrium and Pleistocene extinctions. In *Quaternary Extinctions: A Prehistoric Revolution*, edited by P. S. Martin and R. G. Klein, 223–49. University of Arizona Press, Tucson.

————. 2010. FAUNMAP II: New data for North America with a temporal extension for the Blancan, Irvingtonian and Early Rancholabrean. FAUNMAP II Database. http://www.ucmp.berkeley.edu/faunmap/about/data.html.

Graham, R. W., E. L. Lundelius Jr., and L. Meissner. 2013. Friesenhahn Cave: Late Pleistocene paleoecology and predator-prey relationships of mammoths with an extinct scimitar cat. In *Late Cretaceous to Quaternary Strata and Fossils of Texas: Field Excursions Celebrating 125 Years of GSA and Texas Geology*, GSA south-central section meeting, Austin, TX, April 2013, edited by B. B. Hunt and E. J. Catlos, 15–31. Geological Society of America Field Guide 30.

Graham, R. W. and J. I. Mead. 1987. Environmental fluctuations and evolution of mammalian faunas during the last deglaciation in North America. In *North America and Adjacent Oceans during the Last Deglaciation*, edited by W. F. Ruddiman and

H. E. Wright Jr., 371–402. *Geology of North America*, vol. K-3. Geological Society of America, Boulder, CO.

Graham, R. W., and H. A. Semken Jr. 1976. Paleoecological significance of the short-tailed shrew (*Blarina*) with a systematic discussion of *Blarina ozarkensis*. *Journal of Mammalogy* 57:433–49.

Grayson, D. K. 1977. Pleistocene avifaunas and the overkill hypothesis. *Science* 195: 691–93.

———. 1980. Vicissitudes and overkill: The development of explanations of Pleistocene extinctions. *Advances in Archaeological Method and Theory* 3:357–403.

———. 1983. *The Establishment of Human Antiquity*. Academic Press, New York.

———. 1984a. Explaining Pleistocene extinctions: Thoughts on the structure of a debate. In *Quaternary Extinctions: A Prehistoric Revolution*, edited by P. S. Martin and R. G. Klein, 807–23. University of Arizona Press, Tucson.

———. 1984b. Nineteenth century explanations of Pleistocene extinctions: A review and analysis. In *Quaternary Extinctions: A Prehistoric Revolution*, edited by P. S. Martin and R. G. Klein, 5–39. University of Arizona Press, Tucson.

———. 1984c. Time of extinction and nature of adaptation of the noble marten, *Martes nobilis*. In *Contributions in Quaternary Vertebrate Paleontology: A Volume in Memorial to John E. Guilday*, edited by H. H. Genoways and M. R. Dawson, 233–40. Carnegie Museum of Natural History Special Publication 8.

———. 1985. The paleontology of Hidden Cave: Birds and mammals. In *The Archaeology of Hidden Cave, Nevada*, edited by D. H. Thomas, 125–61. American Museum of Natural History Anthropological Papers 61 (1).

———. 1986. Eoliths, archaeological ambiguity, and the generation of "middle-range" research. In *American Archaeology: Past and Future*, edited by D. J. Meltzer, D. D. Fowler, and J. A. Sabloff, 77–133. Smithsonian Institution Press, Washington, DC.

———. 1987. An analysis of the chronology of late Pleistocene mammalian extinctions in North America. *Quaternary Research* 28:281–89.

———. 1991. Late Pleistocene extinctions in North America: Taxonomy, chronology, and explanations. *Journal of World Prehistory* 5:193–231.

———. 1993. *The Desert's Past: A Natural Prehistory of the Great Basin*. Smithsonian Institution Press, Washington, DC.

———. 2000a. Mammalian responses to middle Holocene climatic change in the Great Basin of the western United States. *Journal of Biogeography* 27:181–92.

———. 2000b. The Homestead Cave mammals. In *Late Quaternary Paleoecology in the Bonneville Basin*, edited by D. B. Madsen, 67–89. Utah Geological Survey Bulletin 130.

———. 2001. The archaeological record of human impacts on animal populations. *Journal of World Prehistory* 15:1–68.

———. 2005. A brief history of Great Basin pikas. *Journal of Biogeography* 32:2101–11.

———. 2006a. Holocene bison in the Great Basin, western USA. *Holocene* 16:913–25.

———. 2006b. Ice Age extinctions [Essay review of P. S. Martin's *Twilight of the Mammoths*]. *Quarterly Review of Biology* 81:259–64.

———. 2007. Deciphering North American Pleistocene extinctions. *Journal of Anthropological Research* 63:185–213.

———. 2011. *The Great Basin: A Natural Prehistory*. University of California Press, Berkeley.

Grayson, D. K., and F. Delpech. 2003. Ungulates and the Middle-to-Upper Paleolithic transition at Grotte XVI (Dordogne, France). *Journal of Archaeological Science* 30:1633–48.

———. 2005. Pleistocene reindeer and global warming. *Conservation Biology* 19:557–62.

———. 2008. The large mammals of Roc de Combe (Lot, France): The Châtelperronian and Aurignacian assemblages. *Journal of Anthropological Archaeology* 27:338–62.

Grayson, D. K. and J. L. Fisher. 2009. Holocene elk (*Cervus elaphus*) in the Great Basin. In *Past, Present and Future Issues in Great Basin Archaeology: Papers in Honor of Don D. Fowler*, edited by B. Hockett, 67–83. Bureau of Land Management Cultural Resource Series 20. Reno, NV.

Grayson, D. K., and C. Maser. 1978. First record for the Long-Tailed Jaeger in eastern Oregon. *Murrelet* 59:75–77.

Grayson, D. K., and D. J. Meltzer. 2002. Clovis hunting and large mammal extinction: A critical review of the evidence. *Journal of World Prehistory* 16:313–59.

———. 2003. A requiem for North American overkill. *Journal of Archaeological Science* 30:585–93.

———. 2015. Revisiting Paleoindian exploitation of extinct North American mammals. *Journal of Archaeological Science* 56:177–93.

Green, J. L. 2009. Dental microwear in the orthodentine of the Xenarthra (Mammalia) and its use in reconstructing the paleodiet of extinct taxa: The case study of *Nothrotheriops shastensis* (Xenarthra, Tardigrada, Nothrotheriidae). *Zoological Journal of the Linnean Society* 156:201–22.

Green, J. L., G. M. Semprebon, and N. Solounias. 2005. Reconstructing the paleodiet of Florida *Mammut americanum* via low-magnification stereomicroscopy. *Palaeogeography, Palaeoclimatology, Palaeoecology* 223:34–48.

Griffin, J. B., D. J. Meltzer, B. D. Smith, and W. C. Sturtevant. 1988. A mammoth fraud in science. *American Antiquity* 53:578–82.

Grubb, P. J. 1992. A positive distrust in simplicity: Lessons from plant defences and from competition among plants and among animals. *Journal of Ecology* 80:585–610.

Gruhn, R., and A. L. Bryan. 1981. A response to McGuire's cautionary tale about the association of man and extinct fauna in Great Basin cave sites. *Quaternary Research* 16:117–21.

Guedes, P. G., and L. O. Salles. 2005. New insights on the phylogenetic relationships of the two giant extinct New World monkeys (Primates, Platyrrhini). *Arquivos do Museu Nacional, Rio de Janeiro* 63:147–59.

Guérin, C., and M. Faure. 1999. *Palaeolama (Hemiauchenia) niedae* nov. sp., nouveau Camelidae du nordeste brésilien et sa place parmi les Lamini d'Amerique du Sud. *Geobios* 32:629–59.

———. 2004. *Macrauchenia patachonica* Owen (Mammalia, Litopterna) de la région de São Raimundo Nonato (Puauí, Nordeste brésilien) et la diversité des Macraucheniidae pléistocènes. *Geobios* 37:516–35.

Guérin, C. and M. Patou-Mathis, eds. 1996. *Les grands mammifères Plio-Pléistocènes d'Europe*. Masson, Paris.

Guilday, J. E. 1967. Differential extinction during Late-Pleistocene and recent times. In *Pleistocene Extinctions: The Search for a Cause*, edited by P. S. Martin and H. E. Wright Jr., 121–40. Yale University Press, New Haven, CT.

Guilday, J. E., H. W. Hamilton, and A. D. McCrady. 1971. The Welsh Cave peccaries (*Platygonus*) and associated fauna, Kentucky Pleistocene. *Annals of the Carnegie Museum* 43:249–320.

Guimarães, P. R., Jr., M. Galetti, and P. Jordano. 2008. Seed dispersal anachronisms: Rethinking the fruits extinct megafauna ate. *PLoS One* 3 (3): e1745.

Gustafson, C. E. 1985. The Manis Mastodon Site. *National Geographic Society Research Reports* 1979:283–95.

Gustafson, C. E., D. Gilbow, and R. D. Daugherty. 1979. The Manis Mastodon Site: Early man on the Olympic Peninsula. *Canadian Journal of Archaeology* 3:157–64.

Gustafson, C. E., and C. Manis. 1984. *The Manis Mastodon Site: An Adventure in Prehistory*. Manis Enterprises, Sequim, WA.

Guthrie, D. A. 2009. An updated catalogue of the birds from the Carpinteria asphalt, Pleistocene of California. *Bulletin of the Southern California Academy of Sciences* 108:52–62.

Guthrie, R. D. 1990. *Frozen Fauna of the Mammoth Steppe*. University of Chicago Press, Chicago.

———. 1992. New paleoecological and paleoethological information on the extinct helmeted muskoxen from Alaska. *Annales Zoologici Fennici* 28:175–86.

———. 2001. Origin and causes of the mammoth steppe: A story of cloud cover, woolly mammal tooth pits, buckles, and inside-out Beringia. *Quaternary Science Reviews* 20:549–74.

———. 2003. Rapid body size decline in Alaskan Pleistocene horses before extinction. *Nature* 426:169–71.

———. 2006. New carbon dates link climatic change with human colonization and Pleistocene extinctions. *Nature* 441:207–9.

Guthrie, R. D., A. V. Sher, and C. R. Harington. 2001. New radiocarbon dates on saiga antelopes (*Saiga tatarica*) from Alaska, Canada, and Siberia: Their paleoecological significance. In *People and Wildlife in Northern North America: Essays in Honor of R. Dale Guthrie*, edited by S. G. Gerlach and M. S. Murray, 50–57. BAR International Series 944.

Gutiérrez, M., M. T. Alberdi, J. L. Prado, and D. Perea. 2005. Late Pleistocene *Stegomastodon* (Mammalia, Proboscidea) from Uruguay. *Neues Jahrbuch für Geologie und Paläontologie. Monatshefte* 2005 (11): 641–62.

Haas, S. K. 2005. *Panthera leo. Mammalian Species* 762:1–11.

Hafner, D. J. 1994. Pikas and permafrost: Post-Wisconsin zoogeography of *Ochotona* in the southern Rocky Mountains, U.S.A. *Arctic and Alpine Research* 26:375–82.

Hafner, D. J., and R. M. Sullivan. 1995. Historical and ecological biogeography of Nearctic pikas (Lagomorpha: Ochotonidae). *Journal of Mammalogy* 76:302–21.

Hager, M. W. 1972. A late Wisconsin-Recent vertebrate fauna from the Chimney Rock animal trap, Larimer County, Colorado. *University of Wyoming Contributions in Geology* 11 (2): 63–71.

Halenar, L. B. 2011. Reconstructing the locomotor repertoire of *Protopithecus braziliensis*. I. Body size. *Anatomical Record* 294:2024–47.

Halenar, L. B., and A. L. Rosenberger. 2013. A closer look at the "*Protopithecus*" fossil assemblage: New genus and species from Bahia, Brazil. *Journal of Human Evolution* 65:374–90.

Hall, E. R. 1981. *The Mammals of North America*. 2nd ed. John Wiley and Sons, NY.

Hallett, D. J., L. V. Hills, and J. J. Clague. 1997. New accelerator mass spectrometry radiocarbon ages for the Mazama tephra layer from Kootenay National Park, British Columbia, Canada. *Canadian Journal of Earth Sciences* 34:1202–9.

Hanley, M. E., B. B. Lamont, M. M. Fairbanks, and C. M. Rafferty. 2007. Plant structural traits and their role in anti-herbivore defence. *Perspectives in Plant Ecology, Evolution, and Systematics* 8:157–78.

Hansen, R. M. 1978. Shasta ground sloth food habits, Rampart Cave, Arizona. *Paleobiology* 4:302–19.

———. 1980. Late Pleistocene plant fragments in the dungs of herbivores at Cowboy Cave. In *Cowboy Cave*, edited by J. D. Jennings, 179–89. University of Utah Anthropological Papers 104.

Hargrave, J. E. 2009. Lithostratigrapy and fossil avifaunas of the Pleistocene Fossil Lake Formation, Fossil Lake, Oregon, and the Oligocene Etadunna Formation, Lake Palankarinna, South Australia. PhD diss., University of Oklahoma, Norman.

Harington, C. R. 1968. A Pleistocene muskox (*Symbos*) from Dease Lake, British Columbia. *Canadian Journal of Earth Sciences* 5:1161–65.

———. 1981. Pleistocene saiga antelopes in North America and their paleoenvironmental implications. In *Quaternary Paleoclimate*, edited by W. C. Mahaney, 193–225. Geo Abstracts, Norwich, UK.

———. 2011. Quaternary cave faunas of Canada: A review of the vertebrate remains. *Journal of Cave and Karst Studies* 73:162–80.

Harington, C. R., and J. Cinq-Mars. 1995. Radiocarbon dates on saiga antelope (*Saiga tatarica*) fossils from Yukon and the Northwest Territories. *Arctic* 48:1–7.

Harkness, H. W. 1882. *Footprints Found at the Carson State Prison*. Issued separately from the Proceedings of the California Academy of Sciences, August 7.

Harrington, J. 1943. Adventure in a Nevada cave. *Desert Magazine* 6 (7): 18–19.

Harrington, Marie. 1985. *On the Trail of Forgotten People: A Personal Account of the Life and Career of Mark Raymond Harrington*. Great Basin Press, Reno, NV.

Harrington, Mark R. 1928. The oldest camp-fires. *Masterkey* 28:233–34.

———. 1930. The Gypsum Cave murder case: A detective story in the making. *Masterkey* 4:37–43.

———. 1931. Lights and shadows in Gypsum Cave: The final phases. *Masterkey* 4:233–35.

———. 1932a. Report of the curator in charge. *Masterkey* 6:9–21.

———. 1932b. When was America discovered? *Scientific American* (July): 7–10.

———. 1933. *Gypsum Cave, Nevada*. Southwest Museum Papers 8.

———. 1934a. American horses and ancient men in Nevada. *Masterkey* 8:165–69.

———. 1934b. The meaning of Gypsum Cave. *Texas Archaeological and Paleontological Society Bulletin* 6:58–69.

———. 1934c. Researches by M. R. Harrington. *Carnegie Institution of Washington Year Book* 33:306.

———. 1935. Report of the curator. *Masterkey* 9:57–59.

———. 1936. Smith Creek Cave. *Masterkey* 10:193.

———. 1940. Man and beast in Gypsum Cave. *Desert Magazine* 3 (6): 3–5, 34.

———. 1951. The magic of carbon 14. *Masterkey* 25:25–26.

Harrington, Mark R., and R. D. Simpson. 1961. *Tule Springs, Nevada, with Other Evidences of Pleistocene Man in North America*. Southwest Museum Papers 8.

Harris, A. H. 1985. *Late Pleistocene Vertebrate Paleoecology of the West.* University of Texas Press, Austin.

———. 1993. Quaternary vertebrates of New Mexico. In *Pleistocene Paleontology in New Mexico,* edited by S. G. Lucas and J. Zidek, 179–97. New Mexico Museum of Natural History and Science Bulletin 2.

———. 2003. The Pleistocene vertebrate fauna from Pendejo Cave. In *Pendejo Cave,* edited by R. S. MacNeish and J. G. Libby, 37–65. University of New Mexico Press, Albuquerque.

———. 2014. Pleistocene Vertebrates of Southwestern USA and Northwestern Mexico. http://www.utep.edu/leb/pleistNM/default.htm.

Harris, A. H., and J. Hearst. 2012. Late Wisconsin mammalian fauna from Dust Cave, Guadalupe Mountains National Park, Culberson County, Texas. *Southwestern Naturalist* 57 (2): 202–6.

Harrison, J. A. 1979. *Revision of the Camelinae (Artiodactyla, Tylopoda) and Description of the New Genus Alforjas.* University of Kansas Paleontological Contributions 95.

Hartwig, W. C. 1995. A giant New World monkey from the Pleistocene of Brazil. *Journal of Human Evolution* 28:189–95.

Haskin and Sells. 1931. Auditor's report. *Masterkey* 4:251–55.

Hattori, E. M. 1982. *The Archaeology of Falcon Hill, Winnemucca Lake, Washoe County, Nevada.* Nevada State Museum Anthropological Papers 18.

———. 2008. Mysterious crescents. In *The Great Basin: People and Place in Ancient Times,* edited by C. S. Fowler and D. D. Fowler, 39. School for Advanced Research Press, Santa Fe, NM.

Haury, E. W. 1953. Artifacts with mammoth remains, Naco, Arizona. 1. Discovery of the Naco mammoth and the associated projectile points. *American Antiquity* 19:1–14.

Haury, E. W., E. B. Sayles, and W. W. Wasley. 1959. The Lehner Mammoth Site, southeastern Arizona. *American Antiquity* 25:2–30.

Hay, O. P. 1927. *The Pleistocene of the Western Region of North America and Its Vertebrated Animals.* Carnegie Institution of Washington Publication 322B.

Haynes, C. V., Jr. 1964. Fluted projectile points: Their age and dispersion. *Science* 145:1408–13.

———. 2008. Younger Dryas "black mats" and the Rancholabrean termination in North America. *Proceedings of the National Academy of Sciences* 105:6520–25.

Haynes, C. V., Jr., and B. B. Huckell, eds. 2007. *Murray Springs: A Clovis Site with Multiple Activity Areas in the San Pedro Valley, Arizona.* University of Arizona Anthropological Papers 71.

Haynes, G. 1991. *Mammoths, Mastodonts, and Elephants: Biology, Behavior, and the Fossil Record.* Cambridge University Press, Cambridge.

Hayward, C. L., C. Cottam, A. M. Woodbury, and H. H. Frost. 1976. *Birds of Utah.* Great Basin Naturalist Memoirs 1.

Heaton, T. H. 1985. Quaternary paleontology and paleoecology of Crystal Ball Cave, Millard County, Utah: With emphasis on mammals and description of a new species of fossil skunk. *Great Basin Naturalist* 45:337–90.

———. 1990. Quaternary mammals of the Great Basin: Extinct giants, Pleistocene relicts, and recent immigrants. In *Causes of Evolution: a Paleontological Perspective,* edited by R. M. Ross and W. D. Allmon, 422–65. University of Chicago Press, Chicago.

Heintzman, P. D., G. D. Zazula, J. A. Cahill, A. V. Reyes, R. D. E. MacPhee, and B. Shapiro. 2015. Genomic data from extinct North American *Camelops* revise camel evolutionary history. *Molecular Biology and Evolution.* doi:10.1093/molbev/msv128.

Heizer, R. F., and R. Berger. 1970. Radiocarbon age of the Gypsum Cave culture. *Contributions of the University of California Archaeological Research Facility* 7:13–18.

Henrickson, J. 1972. A taxonomic revision of the Fouquieriaceae. *Aliso* 7:439–537.

Heptner, V. G., A. A. Nasimovich, and A. G. Bannikov. 1988. *Mammals of the Soviet Union.* Vol. 1, *Artiodactyla and Perissodactyla.* Smithsonian Institution Libraries and National Science Foundation, Washington, DC.

Hernández Fernández, M., and E. S. Vrba. 2005. A complete estimate of the phylogenetic relationships in Ruminantia: A dated species-level supertree of the extant ruminants. *Biological Reviews* 80:269–302.

Herridge, V. L., and A. M. Lister. 2012. Extreme insular dwarfism evolved in a mammoth. *Proceedings of the Royal Society B* 279:3193–3200.

Hertel, F. 1994. Diversity in body size and feeding morphology within past and present vulture assemblages. *Ecology* 75:1074–84.

———. 1995. Ecomorphological indicators of feeding behavior in recent and fossil raptors. *Auk* 112:890–903.

Hesketh Prichard, H. 1902. *Through the Heart of Patagonia.* D. Appleton, New York.

Hillenius, W. J. 1992. The evolution of nasal turbinates and mammalian endothermy. *Paleobiology* 18:17–29.

Hockett, B., and E. Dillingham. 2004. *Paleontological Investigations at Mineral Hill Cave.* Contributions to the Study of Cultural Resources Technical Report 18. US Department of the Interior Bureau of Land Management, Reno, NV.

Hockett, B., and D. L. Jenkins. 2013. Identifying stone tool cut marks and the pre-Clovis occupation of the Paisley Caves. *American Antiquity* 78:762–78.

Hodnett, J-P. M., J. I. Mead, and A. Baez. 2009. Dire wolf, *Canis dirus* (Mammalia; Carnivora; Canidae), from the late Pleistocene (Rancholabrean) of east-central Sonora, Mexico. *Southwestern Naturalist* 54:74–81.

Hoffmann, R., C. Stock, L. Miller, R. W. Chaney, and H. L. Mason. 1927. The finding of Pleistocene material in an asphalt pit at Carpinteria, California. *Science* 66:155–57.

Hofreiter, M., H. N. Poinar, W. G. Spaulding, K. Bauer, P. S. Martin, G. Possnert, and S. Pääbo. 2000. A molecular analysis of ground sloth diet through the last glaciation. *Molecular Ecology* 9:1975–84.

Holliday, V. T., and D. J. Meltzer. 2010. The 12.9-ka ET impact hypothesis and North American Paleoindians. *Current Anthropology* 51:575–607.

Holliday, V. T., T. Surovell, D. J. Meltzer, D. K. Grayson, and M. Boslough. 2014. The Younger Dryas impact hypothesis: A cosmic catastrophe. *Journal of Quaternary Science* 29:515–30.

Holman, J. A. 1961. Osteology of living and fossil New World quails (Aves, Galliformes). *Bulletin of the Florida State Museum, Biological Sciences* 6 (2): 131–234.

Holmgren, C. A., J. L. Betancourt, M. C. Peñalba, et al. (8 authors). 2014. Evidence against a Pleistocene desert refugium in the Lower Colorado River Basin. *Journal of Biogeography* 41:1769–80.

Holmgren, C. A., J. L. Betancourt, and K. A. Rylander. 2006. A 36,000-yr vegetation history from the Peloncillo Mountains, southeastern Arizona, USA. *Palaeogeography, Palaeoclimatology, Palaeoecology* 240:405–22.

Holmgren, C. A., J. Norris, and J. L. Betancourt. 2007. Inferences about winter temperatures and summer rains from the late Quaternary record of C$_4$ perennial grasses and
C$_3$ desert shrubs in the northern Chihuahuan Desert. *Journal of Quaternary Science*
22:141–61.

Honey, J. G., J. A. Harrison, D. R. Prothero, and M. S. Stevens. 1998. Camelidae. In *Evolution of Tertiary Mammals of North America*, edited by C. M. Janis, K. M. Scott, and
L. L. Jacobs, 439–62. Cambridge University Press, Cambridge.

Hoorn, C., and S. Flantua. 2015. An early start for the Panama land bridge. *Science*
348:186–87.

Hopkins, S. S. B. 2008. Reassessing the mass of exceptionally large rodents using tooth-
row length and area as proxies for body mass. *Journal of Mammalogy* 89:232–43.

Hoppe, K. A. 2004. Late Pleistocene mammoth herd structure, migration patterns, and
Clovis hunting strategies inferred from isotopic analyses of multiple death assemblages. *Paleobiology* 30:129–45.

Hoppe, K. A., and P. L. Koch. 2006. The biogeochemistry of the Aucilla River fauna. In
First Floridians and Last Mastodons: The Page-Ladson Site in the Aucilla River, edited
by S. D. Webb, 379–401. Springer, Dordrecht, Netherlands.

———. 2007. Reconstructing the migration pattern of late Pleistocene mammals from
northern Florida, USA. *Quaternary Research* 68:347–52.

Hoppe, K. A., P. L. Koch, R. W. Carlson, and S. D. Webb. 1999. Tracking mammoths
and mastodons: Reconstruction of migratory behavior using strontium isotopes.
Geology 27:439–42.

Howard, H. 1932. *Eagles and Eagle-Like Vultures of the Pleistocene of Rancho La Brea*.
Carnegie Institution of Washington Publication 429.

———. 1935. A new species of eagle from a Quaternary cave deposit in eastern Nevada.
Condor 37:206–9.

———. 1936. Further studies upon the birds of the Pleistocene of Rancho La Brea.
Condor 38:32–36.

———. 1946. A review of the Pleistocene birds of Fossil Lake, Oregon. *Carnegie Institution of Washington Publication* 551:143–95.

———. 1947. A preliminary survey of trends in avian evolution from Pleistocene to
Recent time. *Condor* 49:10–13.

———. 1952. The prehistoric avifauna of Smith Creek Cave, Nevada, with a description of
a new gigantic raptor. *Bulletin of the Southern California Academy of Sciences* 51:50–54.

———. 1955. *Fossil Birds from Manix Lake, California*. US Geological Survey Professional
Paper 264-J.

———. 1962. *Fossil Birds with Especial Reference to the Birds of Rancho La Brea, Los Angeles
County, Revised*. Museum Science Series 17, Paleontology 10.

———. 1964a. *A New Species of the "Pigmy Goose," Anabernicula, from the Oregon Pleistocene, with a Discussion of the Genus*. American Museum Novitates 2200.

———. 1964b. Fossil Anseriformes. In *The Waterfowl of the World*, vol. 4, edited by
J. Delacour, 233–326. Country Life, London.

———. 1968. Limb measurements of the extinct vulture, *Coragyps occidentalis* with a description of a new subspecies. In *Collected Papers in Honor of Lyndon Lane Hargrave*,
edited by A. H. Schroeder, 115–27. Papers of the Archaeological Society of New
Mexico 1.

———. 1971a. In Memoriam: Loye Holmes Miller. *Auk* 88:276–85.

———. 1971b. Quaternary avian remains from Dark Canyon Cave, New Mexico. *Condor* 73:237–40.

———. 1972. The Incredible Teratorn again. *Condor* 74:341–44.

———. 1974. *Postcranial Elements of the Extinct Condor* Breagyps clarki *(Miller)*. Natural History Museum of Los Angeles County Contributions in Science 256.

Howard, H., and A. H. Miller. 1933. Bird remains from cave deposits in New Mexico. *Condor* 35:15–18.

Howden, H. F., and O. L. Cartwright. 1963. Scarab beetles of the genus *Onthophagus* Latreille north of Mexico (Coleoptera: Scarabaeidae). *Proceedings of the United States National Museum* 114:1–135.

Howe, H. F. 1985. Gomphothere fruits: A critique. *American Naturalist* 125:853–65.

Hoyle, B. G., D. C. Fisher, H. W. Borns Jr., L. L. Churchill-Dickson, C. C. Dorion, and T. K. Weddle. 2004. Late Pleistocene mammoth remains from coastal Maine, USA. *Quaternary Research* 61:277–88.

Hu, A., G. A. Meehl, B. L. Otto-Bliesner, et al. (9 authors). 2010. Influence of Bering Strait flow and North Atlantic circulation on glacial sea-level changes. *Nature Geoscience* 3:118–21.

Hubbard, W. J., D. C. Fisher, and P. N. Kordulias. 2000. Mastodon body weight estimates from footprints and a scale model. *Current Research in the Pleistocene* 17:122–24.

Hulbert, R. C., Jr. 1995. The giant tapir, *Tapirus haysii*, from the Leisey Shell Pit 1A and other Florida Irvingtonian localities. In *Paleontology and Geology of the Leisey Shell Pits, Early Pleistocene of Florida*, edited by R. C. Hulbert, G. S. Morgan, and S. D. Webb, 515–51. Bulletin of the Florida Museum of Natural History 37.

Hulbert, R. C., and J. J. Becker. 2001. Reptilia 3: Birds. In *The Fossil Vertebrates of Florida*, edited by R. C. Hulbert, 152–65. University Press of Florida, Gainesville.

Hulbert, R. C., G. S. Morgan, and A. Kerner. 2009. Collared peccary (Mammalia, Artiodactyla, Tayassuidae, *Pecari*) from the late Pleistocene of Florida. In *Papers on Geology, Vertebrate Paleontology, and Biostratigraphy in Honor of Michael O. Woodburne*, edited by L. B. Albright III, 543–55. Museum of Northern Arizona Bulletin 65.

Human material from Gypsum Cave carbon-dated to about 3000 years. 1967. *Masterkey* 41:66.

Huynen, L., C. D. Millar, R. P. Scofield, and D. M. Lambert. 2003. Nuclear DNA sequences detect species limits in ancient moa. *Nature* 425:175–78.

Important notice to subscribers. 1877. *American Naturalist* 11:758.

International Commission on Stratigraphy. 2013. International Chronostratigraphic Chart, version 2013/01. http://www.stratigraphy.org/index.php/ics-chart-timescale.

Jackman, E. R., and R. A. Long. 1964. *The Oregon Desert*. Caxton Printers, Caldwell, ID.

Jackson, D. 1970. Maps of the John Charles Frémont Expeditions. In *The Expeditions of John Charles Frémont*, edited by D. Jackson and M. L. Spence, 5–16. University of Illinois Press, Urbana.

James, S. R. 2004. Mineral Hill Cave avifauna. In *Paleontological Investigations at Mineral Hill Cave*, edited by B. Hockett and E. Dillingham, 136–42. Contributions to the Study of Cultural Resources Technical Report 18. US Department of the Interior Bureau of Land Management, Reno, NV.

Janis, C. M., and E. Manning. 1998. Antilocapridae. In *Evolution of Tertiary Mammals of*

North America, edited by C. M. Janis, K. M. Scott, and L. L. Jacobs, 491–507. Cambridge University Press, Cambridge.

Jansen, P. A., B. T. Hirsch, W-J. Emsens, V. Zamora-Gutierrez, M. Wikelski, and R. Kays. 2012. Thieving rodents as substitute dispersers of megafaunal seeds. *Proceedings of the National Academy of Sciences* 109:12610–15.

Janzen, D. H. 1986. Chihuahuan Desert nopaleras: Defaunated big mammal vegetation. *Annual Review of Ecology and Systematics* 17:595–636.

Janzen, D. H., and P. S. Martin. 1982. Neotropical anachronisms: The fruits the gomphotheres ate. *Science* 215:19–27.

Jass, C. N. 2007. New perspectives on Pleistocene biochronology and biotic change in the east-central Great Basin: An examination of the vertebrate fauna from Cathedral Cave, Nevada. PhD diss., University of Texas, Austin.

———. 2011. Caves, arvicoline rodents, and chronologic resolution. *Paleontologia Electronica* 14 (3): 40A.

Jass, C. N., J. I. Mead, and L. E. Logan. 2000. Harrington's extinct mountain goat (*Oreamnos harringtoni* Stock 1936) from Muskox Cave, New Mexico. *Texas Journal of Science* 52:121–32.

Jefferson, G. T. 1985. *Review of the Late Pleistocene Avifauna from Lake Manix, Central Mojave Desert, California*. Natural History Museum of Los Angeles County Contributions in Science 362.

———. 1987. The Camp Cady Local Fauna: Paleoenvironment of the Lake Manix Basin. *San Bernardino County Museum Association Quarterly* 34 (3–4): 3–35.

———. 1989a. *Late Cenozoic Tapirs (Mammalia: Perissodactyla) of Western North America*. Natural History Museum of Los Angeles County Contributions in Science 406.

———. 1989b. Late Pleistocene and earliest Holocene fossil localities and vertebrate taxa from the western Mojave Desert. In *The West-Central Mojave Desert: Quaternary Studies between Kramer and Afton Canyon*, edited by R. E. Reynolds, 27–40. San Bernardino County Museum Association, Redlands, CA.

———. 1990. Rancholabrean age vertebrates from the eastern Mojave Desert, California. In *At the End of the Mojave: Quaternary Studies in the Eastern Mojave Desert*, edited by R. E. Reynolds, S. G. Wells, and R. H. Brady III, 109–15. San Bernardino County Museum Association, Redlands, CA.

———. 1991a. *A Catalogue of Late Quaternary Vertebrates from California: Part Two, Mammals*. Natural History Museum of Los Angeles County Technical Reports 7.

———. 1991b. Rancholabrean age vertebrates from the southeastern Mojave Desert, California. In *Crossing the Borders: Quaternary Studies in Eastern California and Southwestern Nevada*, edited by R. E. Reynolds, 163–76. San Bernardino County Museum Association, Redlands, CA.

———. 2003. Stratigraphy and paleontology of the middle to late Pleistocene Manix Formation and paleoenvironments of the central Mojave River, southern California. In *Paleoenvironments and Paleohydrology of the Mojave and Southern Great Basin Deserts*, edited by Y. Enzel, S. G. Wells, and N. Lancaster, 43–60. Geological Society of America Special Papers 368.

Jefferson, G. T., H. G. McDonald, W. A. Akersten, and S. J. Miller. 2002. Catalogue of late Pleistocene and Holocene fossil vertebrates from Idaho. In *And Whereas...*

Papers on the Vertebrate Paleontology of Idaho Honoring John A. White, vol. 2, edited by W. A. Akersten, M. E. Thompson, D. J. Meldrum, R. A. Rapp, and H. G. McDonald, 157–92. Idaho Museum of Natural History Occasional Papers 37.

Jefferson, G. T., H. G. McDonald, and S. D. Livingston. 2004. *Catalogue of Late Quaternary and Holocene Fossil Vertebrates from Nevada*. Nevada State Museum Occasional Papers 6.

Jefferson, G. T., and A. E. Tejada-Flores. 1993. The late Pleistocene record of *Homotherium* (Felidae: Machairodontinae) in the southwestern United States. *PaleoBios* 15 (3): 37–46.

Jefferson, T. 1799. A memoir on the discovery of certain bones of a quadruped of the clawed kind in the western parts of Virginia. *Transactions of the American Philosophical Society* 4:246–60.

Jehl, J. R., Jr. 1967. Pleistocene birds from Fossil Lake, Oregon. *Condor* 69:24–27.

Jenkins, D. L. 2007. Distribution and dating of cultural and paleontological remains at the Paisley Five Mile Point Caves in the northern Great Basin. In *Paleoindian or Paleoarchaic? Great Basin Human Ecology at the Pleistocene/Holocene Transition*, edited by K. E. Graf and D. N. Schmitt, 57–81. University of Utah Press, Salt Lake City.

Jenkins, D. L., L. G. Davis, T. W. Stafford Jr., et al. (20 authors). 2012. Clovis age western stemmed projectile points and human coprolites at the Paisley Caves. *Science* 337:223–28.

Jenkins, D. L., L. G. Davis, T. W. Stafford Jr., et al. (18 authors). 2014. Geochronology, archaeological context, and DNA at the Paisley Caves. In *Paleoamerican Odyssey*, edited by K. E. Graf, C. V. Ketron, and M. R. Waters, 485–510. Texas A&M University Press, College Station.

Jenkins, S. H., and E. Busher. 1979. *Castor canadensis*. *Mammalian Species* 120:1–8.

Jennings, S. A., and D. L. Elliott-Fisk. 1993. Packrat midden evidence of late Quaternary vegetation change in the White Mountains, California-Nevada. *Quaternary Research* 39:214–21.

Jiménez-Hidalgo, E., O. Carranza-Castañeda, and M. Montellano-Ballesteros. 2004. A Pliocene record of *Capromeryx* (Mammalia: Antilocapridae) in Mexico. *Journal of Paleontology* 78:1179–86.

Johnson, C. N. 2002. Determinants of loss of mammal species during the Late Quaternary "megafauna" extinctions: Life history and ecology, but not body size. *Proceedings of the Royal Society of London B* 269:2221–27.

Johnson, L., and J. Johnson. 1987. *Escape from Death Valley*. University of Nevada Press, Reno.

Jones, G. T., and C. Beck. 1999. Paleoarchaic archaeology in the Great Basin. In *Models for the Millennium: Great Basin Anthropology Today*, edited by C. Beck, 83–95. University of Utah Press, Salt Lake City.

Jordan, D. S. 1907. The fossil fishes of California with supplementary notes on other species of extinct fishes. *University of California Publications, Bulletin of the Department of Geology* 5 (7): 95–144.

Joughin, I., B. E. Smith, and B. Medley. 2014. Marine ice sheet collapse potentially underway for the Thwaites Glacier basin, West Antarctica. *Science* 344:735–38.

Kahlke, H. D. 1990. On the evolution, distribution and taxonomy of fossil elk/moose. *Quartärpaläontologie* 8:83–106.

Kahlke, R.-D. 1991. Pleistocene distributional and evolutionary history of the genus *Saiga* Gray, 1843 (Mammalia, Artiodactyla, Bovidae) in the Palaearctic. *Vertebrata PalAsiatica* 24:315–22.

Karban, R. 2010. The ecology and evolution of induced resistance against herbivores. *Functional Ecology* 25:339–47.

Kartesz, J. T. 1988. A flora of Nevada. PhD diss., University of Nevada, Reno.

Kelly, T. S. 1995. A Pleistocene mammalian fauna from Adrian Valley, Lyon County, west central Nevada. *Current Research in the Pleistocene* 12:99–102.

Kennett, D. J., J. P. Kennett, G. J. West, et al. (10 authors). 2008. Wildfire and abrupt ecosystem disruption on California's Northern Channel Islands at the Ållerød–Younger Dryas boundary (13.0–12.9 ka). *Quaternary Science Reviews* 27:2530–45.

Kennett, J. P., D. J. Kennett, B. J. Culleton, et al. (26 authors). 2015. Bayesian chronological analyses consistent with synchronous age of 12,835–12,735 Cal BP for Younger Dryas boundary on four continents. *Proceedings of the National Academy of Sciences.* doi:10.1073/pnas.1507146112.

Kharlamova, A., S. Saveliev, A. Kurtova, et al. (8 authors). 2014. Preserved brain of the woolly mammoth *Mammuthus primigenius* (Blum. 1799) from the Yakutian Permafrost. *Sixth International Conference on Mammoths and Their Relatives, Grevena—Siatista Special Volume* 102:80–81.

King, J. E., and J. J. Saunders. 1984. Environmental insularity and the extinction of the American mastodont. In *Quaternary Extinctions: A Prehistoric Revolution*, edited by P. S. Martin and R. G. Klein, 315–39. University of Arizona Press, Tucson.

Kinzie, C. R., S. S. Que Hee, A. Stich, et al. (26 authors). 2014. Nanodiamond-rich layer across three continents consistent with major cosmic impact at 12,800 cal BP. *Journal of Geology* 122 (5). doi:10.1086/677046.

Kitchen, D. W. 1974. *Social Behavior and Ecology of the Pronghorn*. Wildlife Monographs 38.

Kleman, J., K. Jansson, H. De Angelis, A. Stroeven, C. Hättestrand, G. Alm, and N. Glasser. 2010. North American ice sheet build-up during the last glacial cycle, 115–21kyr. *Quaternary Science Reviews* 29:2036–51.

Klippel, W. E., and P. W. Parmalee. 1984. Armadillos in North American late Pleistocene contexts. In *Contributions in Quaternary Vertebrate Paleontology: A Volume in Memorial to John E. Guilday*, edited by H. H. Genoways and M. R. Dawson, 149–60. Carnegie Museum of Natural History Special Publications 8.

Koch, P. L., and A. D. Barnosky. 2006. Late Quaternary extinctions: State of the debate. *Annual Review of Ecology and Systematics* 37:215–50.

Koch, P. L., K. A. Hoppe, and S. D. Webb. 1998. The isotopic ecology of late Pleistocene mammals in North America. Part I. Florida. *Chemical Geology* 152:119–38.

Koehler, P. A., and R. S. Anderson. 1995. Thirty thousand years of vegetation changes in the Alabama Hills, Owens Valley, California. *Quaternary Research* 43:238–48.

Kohl, K. D., R. B. Weiss, J. Cox, C. Dale, and M. D. Dearing. 2014. Gut microbes of mammalian herbivores facilitate intake of plant toxins. *Ecology Letters* 17:1238–46.

Kohn, M. J., M. P. McKay, and J. L. Knight. 2005. Dining in the Pleistocene: Who's on the menu? *Geology* 33:649–52.

Kooyman, B., L. V. Hills, S. Tolman, and P. McNeil. 2012. Late Pleistocene western camel (*Camelops hesternus*) hunting in southwestern Canada. *American Antiquity* 77:115–24.

Koricheva, J., H. Nykänen, and E. Gianoli. 2004. Meta-analysis of trade-offs among plant antiherbivore defenses: Are plants jacks-of-all-trades, masters of all? *American Naturalist* 163:E64–E75.

Kozlowski, T. T. 1971. *Growth and Development of Trees*. Vol. 1, *Seed Germination, Ontogeny, and Shoot Growth*. Academic Press, New York.

Krause, J., T. Unger, A. Noçon, et al. (18 authors). 2008. Mitochrondrial genomes reveal an explosive radiation of extinct and extant bears near the Miocene-Pliocene boundary. *BMC Evolutionary Biology* 8:220.

Krausman, P. R., and S. M. Morales. 2005. *Acinonyx jubatus*. *Mammalian Species* 771:1–6.

Kropf, M., J. I. Mead, and R. S. Anderson. 2007. Dung, diet, and the paleoenvironment of the extinct shrub-ox (*Euceratherium collinum*) on the Colorado Plateau, USA. *Quaternary Research* 67:143–51.

Kubiak, H. 1982. Morphological characters of the mammoth: An adaptation to the Arctic-Steppe environment. In *Paleoecology of Beringia*, edited by D. M. Hopkins, J. V. Matthews Jr., C. E. Schweger, and S. B. Young, 281–89. Academic Press, New York.

Kurtén, B. 1967. *Pleistocene Bears of North America. II. Genus* Arctodus, *Short-Faced Bears*. Acta Zoologica Fennica 117.

———. 1975. A new Pleistocene genus of American mountain deer. *Journal of Mammalogy* 56:507–8.

———. 1976a. Fossil puma (Mammalia: Felidae) in North America. *Netherlands Journal of Zoology* 26:505–34.

———. 1976b. *The Cave Bear Story*. Columbia University Press, New York.

———. 1979. The stilt-legged deer *Sangoma* of the North-American Pleistocene. *Boreas* 8:313–21.

———. 1984. Geographic differentiation in the Rancholabrean dire wolf (*Canis dirus* Leidy) in North America. In *Contributions in Quaternary Vertebrate Paleontology: A Volume in Memorial to John E. Guilday*, edited by H. H. Genoways and M. R. Dawson, 218–27. Carnegie Museum of Natural History Special Publications 8.

Kurtén, B., and E. Anderson. 1980. *Pleistocene Mammals of North America*. Columbia University Press, New York.

Kuzmin, Y. V. 2010. Extinction of the woolly mammoth (*Mammuthus primigenius*) and woolly rhinoceros (*Coelodonta antiquitatis*) in Eurasia: Review of chronological and environmental issues. *Boreas* 39:247–61.

Lachniet, M. S., R. F. Denniston, Y. Asmerom, and V. J. Polyak. 2014. Orbital control of western North America atmospheric circulation and climate over two glacial cycles. *Nature Communications* 5:3805. doi:10.1038/ncomms4895.

Land, B. 1992. Secrets of the Black Rock Desert. *Silver and Blue* (December): 14–17, 48.

Landon, M. N., ed. 1999. *The Journals of George Q. Cannon*. Vol. 1, *To California in '49*. Deseret Book, Salt Lake City.

Lane, C. S., A. Brauer, S. P. E. Blockley, and P. Dulski. 2013. Volcanic ash reveals time-transgressive abrupt climate change during the Younger Dryas. *Geology* 41:1251–54.

Lanner, R. M. 1984. *Trees of the Great Basin*. University of Nevada Press, Reno.

Larivière, S. 2001. *Ursus americanus*. *Mammalian Species* 647:1–11.

Larson, F. 1940. The role of the bison in maintaining the short grass plains. *Ecology* 21:113–21.

Larson, P. R. 1999. The Columbian mammoth (*Mammuthus columbi*) from Escalante Valley, Iron County, Utah—Discovery and implications. In *Vertebrate Paleontology in Utah*, edited by D. D. Gillette, 531–36. Utah Geological Survey Miscellaneous Publications 99-1.

Laudermilk, J. D. 1931. On the origin of desert varnish. *American Journal of Science* 21:51–66.

———. 1933. The identification of cloth ash. *Journal of Criminal Law and Criminology* 24:503–16.

Laudermilk, J. D., and P. A. Munz. 1934. Plants in the dung of *Nothrotherium* from Gypsum Cave, Nevada. *Carnegie Institution of Washington Publication* 453:31–37.

———. 1938. Plants in the dung of *Nothrotherium* from Rampart and Muav Caves, Arizona. *Carnegie Institution of Washington Publication* 487:271–81.

Le Conte, J. 1882. *On Certain Remarkable Tracks, Found in the Rocks of Carson Quarry.* Issued separately from the Proceedings of the California Academy of Sciences, August 27.

———. 1883. Carson footprints. *Nature* 28:101–2.

Lenz, L. W. 2001. Seed dispersal in *Yucca brevifolia* (Agavaceae): Present and past, with consideration of the future of the species. *Aliso* 20:61–74.

———. 2007. Reassessment of *Yucca brevifolia* and recognition of *Y. jaegeriana* as a distinct species. *Aliso* 24:97–104.

Lepper, B. T., T. A. Frolking, D. C. Fisher, et al. (8 authors). 1991. Intestinal contents of a late Pleistocene mastodont from midcontinental North America. *Quaternary Research* 36:120–25.

Leslie, D. M., Jr., and K. Sharma. 2009. *Tetracerus quadricornis* (Artiodactyla: Bovidae). *Mammalian Species* 843:1–11.

Letts, B. C. 2011. Using ancient DNA to investigate the influence of environmental change on animal populations. PhD diss., Pennsylvania State University, University Park.

Leviton, A. E., and M. L. Aldrich. 1997. *Theodore Henry Hittell's The California Academy of Sciences: A Narrative History, 1853–1906.* California Academy of Sciences, San Francisco.

Lev-Yadun, S. 2001. Aposematic (warning) coloration associated with thorns in higher plants. *Journal of Theoretical Biology* 210:385–88.

———. 2003. Weapon (thorn) automimicry and mimicry of aposematic colorful thorns in plants. *Journal of Theoretical Biology* 224:183–88.

Lewis, G. E. 1970. New discoveries of Pleistocene bisons and peccaries in Colorado. *US Geological Survey Professional Paper* 700-B:B137–B140.

———. 1990. New information on high-elevation mammoths, and on a prongbuck with supernumerary horns. *Journal of Vertebrate Paleontology* 10 (3). Supplement. Abstracts of Papers:32A.

Lister, A. M. 1993. Evolution of mammoth and moose: The Holarctic perspective. In *Morphological Change in Quaternary Mammals of North America*, edited by R. A. Martin and A. D. Barnosky, 178–204. Cambridge University Press, New York.

Lister, A. M., and P. Bahn. 2007. *Mammoths: Giants of the Ice Age.* Rev. ed. University of California Press, Berkeley.

Lister, A. M., and A. V. Sher. 2001. The origin and evolution of the woolly mammoth. *Science* 294:1094–97.

Lister, A. M., A. V. Sher, H. van Essen, and G. Wei. 2005. The pattern and process of mammoth evolution in Eurasia. *Quaternary International* 126–28:49–64.

Little, D. P. 2006. Evolution and circumscription of the true cypresses (Cupressaceae: *Cupressus*). *Systematic Botany* 31:461–80.

Livingston, S. D. 1991. The DeLong Mammoth Locality, Black Rock Desert, Nevada. *Current Research in the Pleistocene* 8:94–97.

Lomolino, M. V., B. R. Riddle, and J. H. Brown. 2006. *Biogeography*. 3rd ed. Sinauer Associates, Sunderland, MA.

Long, A., and P. S. Martin. 1974. Death of American ground sloths. *Science* 186:638–40.

Long, A., and A. B. Muller. 1981. Arizona radiocarbon dates X. *Radiocarbon* 23:191–217.

Long, C. A., and C. J. Yahnke. 2011. End of the Pleistocene: Elk-moose (*Cervalces*) and caribou (*Rangifer*) in Wisconsin. *Journal of Mammalogy* 92:1127–35.

Lord, R. D. 2007. *Mammals of South America*. Johns Hopkins University Press, Baltimore.

Lorenzen, E. D., D. Nogués-Bravo, L. Orlando, et al. (55 authors). 2011. Species-specific responses of Late Quaternary megafauna to climate and humans. *Nature* 479:359–64.

Louderback, L. A., D. K. Grayson, and M. Llobera. 2011. Middle- Holocene climates and human population densities in the Great Basin, western USA. *Holocene* 21:366–73.

Louderback, L. A., B. M. Pavlik, and A. M. Spurling. 2013. Ethnographic and archaeo-logical evidence corroborating *Yucca* as a food source, Mojave Desert, USA. *Journal of Ethnobiology* 33:281–97.

Louderback, L. A., and D. Rhode. 2009. 15,000 years of vegetation change in the Bonne-ville Basin: The Blue Lake record. *Quaternary Science Reviews* 28:308–26.

Lovejoy, A. O. 1964. *The Great Chain of Being*. Harvard University Press, Cambridge, MA.

Lucas, S. G. 2008. Taxonomic nomenclature of *Cuvieronius* and *Haplomastodon*, proboscideans from the Plio-Pleistocene of the New World. In *Neogene Mammals*, edited by S. G. Lucas, G. S. Morgan, J. A. Spielmann, and D. R. Prothero, 409–16. New Mexico Museum of Natural History and Science Bulletin 44.

Lucas, S. G., B. D. Allen, G. S. Morgan, R. G. Myers, D. W. Love, and D. Bustos. 2007. Mammoth footprints from the upper Pleistocene of the Tularosa Basin, Doña Ana County, New Mexico. In *Cenozoic Vertebrate Tracks and Traces*, edited by S. G. Lucas, J. A. Spielmann, and M. G. Lockley, 149–54. New Mexico Museum of Natural History and Science Bulletin 42.

Lucas, S. G., and G. E. Alvarado. 2010. Fossil proboscidea from the upper Cenozoic of Central America: Taxonomy, evolutionary and paleobiogeographic significance. *Revista Geológica de América Central* 42:9–42.

Lucas, S. G., G. E. Alvarado, R. García, E. Espinoza, J. C. Cisneros, and U. Martens. 2007. Vertebrate paleontology. In *Central America: Geology, Resources, and Hazards*, vol. 1, edited by J. Bundschuh and G. E. Alvarado, 443–51. Taylor and Francis, London.

Lucas, S. G., and G. S. Morgan. 2005. Ice Age proboscideans of New Mexico. In *New Mexico's Ice Ages*, edited by S. G. Lucas, G. S. Morgan, and K. E. Zeigler, 255–61. New Mexico Museum of Natural History and Science Bulletin 28.

Lull, R. S. 1921. Fauna of the Dallas Sand Pits. *American Journal of Science* series 5, vol. 2:159–76.

———. 1929. *A Remarkable Ground Sloth*. Memoirs of the Peabody Museum 3 (2).

———. 1930. The ground sloth, *Nothrotheriops*. *American Journal of Science* series 5, vol. 20:344–52.

Lundelius, E. L., Jr. 1960. Mylohyus nasutus: *Long-nosed Peccary of the Texas Pleistocene*. Bulletin of the Texas Memorial Museum 1.

———. 1972. *Fossil Vertebrates from the Late Pleistocene Ingleside Fauna, San Patricio*

County, Texas. University of Texas at Austin Bureau of Economic Geology Report of Investigations 77.

Lundelius, E. L., Jr., V. M. Bryant, R. Mandel, K. J. Thies, and A. Thoms. 2013. The first occurrence of a toxodont (Mammalia, Notoungulata) in the United States. *Journal of Vertebrate Paleontology* 33:229–32.

Lundelius, E. L., Jr., R. W. Graham, E. Anderson, J. Guilday, J. A. Holman, D. W. Steadman, and S. D. Webb. 1983. Terrestrial vertebrate faunas. In *Late-Quaternary Environments of the United States*. Vol. 1, *The Late Pleistocene*, edited by S. C. Porter, 311–53. University of Minnesota Press, Minneapolis.

Lyell, C. 1832. *Principles of Geology, Being an Attempt to Explain the Former Changes of the Earth's Surface by Reference to Causes Now in Operation*, vol. 2. John Murray, London.

Lyman, R. L. 1987. Zooarchaeology and taphonomy: A general consideration. *Journal of Ethnobiology* 7:93–117.

———. 1998. *White Goats, White Lies: The Abuse of Science in Olympic National Park*. University of Utah Press, Salt Lake City.

Lynch, V. J., O. C. Bedoya-Reina, A. Ratan, et al. (8 authors). 2015. Elephantid genomes reveal the molecular bases of woolly mammoth adaptations to the Arctic. *Cell Reports* 12:217–28.

MacFadden, B. J., B. A. Purdy, K. Church, and T. W. Stafford Jr. 2012. Humans were contemporaneous with late Pleistocene mammals in Florida: Evidence from rare earth elemental analyses. *Journal of Vertebrate Paleontology* 32:708–16.

Mack, R. N., and J. N. Thompson. 1982. Evolution in steppe with few large, hoofed mammals. *American Naturalist* 119:757–73.

MacPhee, R. D. E., and A. D. Greenwood. 2007. Continuity and change in the extinction dynamics of late Quaternary muskox (*Ovibos*): Genetic and radiometric evidence. *Bulletin of the Carnegie Museum* 39:203–12.

———. 2013. Infectious disease, endangerment, and extinction. *International Journal of Evolutionary Biology*, Article ID 571939.

MacPhee, R. D. E., and P. A. Marx. 1997. The 40,000 year plague: Humans, hyperdisease, and first-contact extinctions. In *Natural Change and Human Impact in Madagascar*, edited by S. M. Goodman and B. D. Patterson, 169–217. Smithsonian Institution Press, Washington, DC.

MacPhee, R. D. E., A. N. Tikhonov, D. Mol, et al. (8 authors). 2002. Radiocarbon chronologies and extinction dynamics of the late Quaternary mammalian megafauna from the Taimyr Peninsula, Russian Federation. *Journal of Archaeological Science* 29:1017–42.

Madsen, D. B. 2000a. A high-elevation Allerød—Younger Dryas megafauna from the west-central Rocky Mountains. In *Intermountain Archaeology*, edited by D. B. Madsen and M. D. Metcalf, 100–115. University of Utah Anthropological Papers 122.

———. 2000b. *Late Quaternary Paleoecology in the Bonneville Basin*. Utah Geological Survey Bulletin 130.

Madsen, D. B., D. R. Currey, and J. H. Madsen. 1976. *Man, Mammoth, and Lake Fluctuations in Utah*. Antiquities Section Selected Papers 5. Division of State History, Salt Lake City.

Magnanou, E., J. R. Malenke, and M. D. Dearing. 2009. Expression of biotransformation genes in woodrat (*Neotoma*) herbivores on novel and ancestral diets: Identification of candidate genes responsible for dietary shifts. *Molecular Ecology* 18:2401–14.

Maher, L. J., Jr. 1964. *Ephedra* pollen in sediments of the Great Lakes Region. *Ecology*
45:391–95.

Manly, W. L. 1894. *Death Valley in '49*. Pacific Tree and Vine, San Jose, CA. Reprint,
Chalfant Press, Bishop, CA, 1977.

Mann, D. H., P. Groves, M. L. Kunz, R. E. Reanier, and B. V. Gaglioti. 2013. Ice-age
megafauna in Arctic Alaska: Extinction, invasion, survival. *Quaternary Science
Reviews* 70:91–108.

Marché, J. D., II. 1986. "Extraordinary petrifactions": The fossil footprints at Nevada
State Prison. *Terra* 24 (5): 12–18.

Marean, C. W., and C. L. Ehrhardt. 1995. Paleoanthropological and paleoecological
implications of the taphonomy of a sabertooth's den. *Journal of Human Evolution*
29:515–47.

Marlon, J., P. J. Bartlein, M. K. Walsh, et al. (23 authors). 2009. Wildfire responses to
abrupt climate change in North America. *Proceedings of the National Academy of
Sciences* 106:2519–24.

Marsh, O. C. 1883. On the supposed human foot-prints recently found in Nevada. *American Journal of Science* 26:139–40.

Marshall, D. B., M. G. Hunter, and A. L. Contreras, eds. 2006. *Birds of Oregon*. Oregon
State University Press, Corvallis.

Martin, J. E. 1996. First occurrence of *Cynomys* from west of the Rocky Mountains.
Journal of Vertebrate Paleontology 16 (3) Supplement:51A.

Martin, J. E., D. Patrick, A. J. Kihm, F. F. Foit Jr., and D. E. Grandstaff. 2005. Lithostratigraphy, tephrochronology, and rare earth element geochemistry of fossils at the
classical Pleistocene Fossil Lake area, south central Oregon. *Journal of Geology*
113:139–55.

Martin, L. D., J. O. Babiarz, and V. L. Naples. 2011. Introduction. In *The Other Saber-
Tooths: Scimitar-Tooth Cats of the Western Hemisphere*, edited by V. L. Naples, L. D.
Martin, and J. P. Babiarz, 3–17. Johns Hopkins University Press, Baltimore.

Martin, L. D., V. L. Naples, and J. P. Babiarz. 2011. Revision of the New World Homotheriini. In *The Other Saber-Tooths: Scimitar-Tooth Cats of the Western Hemisphere*,
edited by V. L. Naples, L. D. Martin, and J. P. Babiarz, 185–93. Johns Hopkins University Press, Baltimore.

Martin, P. S. 1958. Pleistocene ecology and biogeography of North America. In *Zoogeography*, edited by C. L. Hubbs, 375–420. American Association for the Advancement of Science, Washington, DC.

———. 1963. *The Last 10,000 Years: A Fossil Pollen Record of the American Southwest*.
University of Arizona Press, Tucson.

———. 1967. Prehistoric overkill. In *Pleistocene Extinctions: The Search for a Cause*,
edited by P. S. Martin and H. E. Wright Jr., 75–120. Yale University Press, New
Haven, CT.

———. 1973. The discovery of America. *Science* 179:969–974.

———. 1984. Prehistoric overkill: The global model. In *Quaternary Extinctions: A Prehistoric Revolution*, edited by P. S. Martin and R. G. Klein, 354–403. University of
Arizona Press, Tucson.

———. 1990. Who or what destroyed our mammoths? In *Megafauna and Man: Discovery of America's Heartland*, edited by L. D. Agenbroad, J. I. Mead, and L. W. Nelson,
109–17. The Mammoth Site of Hot Springs, South Dakota, Scientific Papers 1.

———. 1999. Deep history and a wilder west. In *Ecology of Sonoran Desert Plants and Plant Communities*, edited by R. H. Robichaux, 256–90. University of Arizona Press, Tucson.

———. 2000. Forward. In *The Ghosts of Evolution*, edited by C. Barlow, ix–xi. Basic Books, New York.

———. 2005. *Twilight of the Mammoths*. University of Arizona Press, Tucson.

Martin, P. S., and D. A. Burney. 1999. Bring back the elephants! *Wild Earth* 9 (1): 57–64.

Martin, P. S., B. E. Sabels, and D. Shutler Jr. 1961. Rampart Cave coprolite and ecology of the Shasta ground sloth. *American Journal of Science* 259:102–27.

Martin, P. S., and D. W. Steadman. 1984. Extinction of birds in the late Pleistocene of North America. In *Quaternary Extinctions: A Prehistoric Revolution*, edited by P. S. Martin and R. G. Klein, 466–77. University of Arizona Press, Tucson.

———. 1999. Prehistoric extinctions on islands and continents. In *Extinctions in Near Time*, edited by R. D. E. MacPhee, 17–55. Kluwer Academic/Plenum Publishers, New York.

Martínez-Vilalta, J., and W. T. Pockman. 2002. The vulnerability to freezing-induced xylem cavitation of *Larrea tridentata* (Zygophyllaceae) in the Chihuahuan Desert. *American Journal of Botany* 89:1916–24.

Matheus, P. E. 1995. Diet and co-ecology of Pleistocene short-faced bears and brown bears in eastern Beringia. *Quaternary Research* 44:447–53.

———. 2001. Pleistocene predators and people in eastern Beringia: Did short-faced bears really keep humans out of North America? In *People and Wildlife in Northern North America: Essays in Honor of R. Dale Guthrie*, edited by S. G. Gerlach and M. S. Murray, 79–101. BAR International Series 944.

Matthew, W. D. 1902. List of the Pleistocene fauna from Hay Springs, Nebraska. *Bulletin of the American Museum of Natural History* 16:317–22.

Mawby, J. E. 1967. Fossil vertebrates of the Tule Springs Site, Nevada. In *Pleistocene Studies in Southern Nevada*, edited by H. M. Wormington and D. Ellis, 105–28. Nevada State Museum Anthropological Papers 13.

Mayer, J. J., and R. M. Wetzel. 1986. *Catagonus wagneri*. *Mammalian Species* 259:1–5.

McAuliffe, J. R., and T. R. Van Devender. 1998. A 22,000-year record of vegetation change in the north-central Sonoran Desert. *Palaeogeography, Palaeoclimatology, Palaeoecology* 141:253–75.

McCain, C. M., and S. R. B. King. 2014. Body size and activity times mediate mammalian responses to climate change. *Global Change Biology* 20:1760–69.

McCain, E. B., and J. L. Childs. 2008. Evidence of resident jaguars (*Panthera onca*) in the southwestern United States and the implications for conservation. *Journal of Mammalogy* 89:1–10.

McCall, S., V. Naples, and L. Martin. 2003. Assessing behavior in extinct animals: Was *Smilodon* social? *Brain, Behavior, and Evolution* 61:159–64.

McCormick, A. C., S. B. Unsicker, and J. Gershenzon. 2012. The specificity of herbivore-induced plant volatiles in attracting herbivore enemies. *Trends in Plant Science* 17:303–10.

McCornack, E. C. 1914. A study of Oregon Pleistocene. *University of Oregon Bulletin* 12 (2): 1–16.

———. 1920. *Contributions to the Pleistocene History of Oregon*. University of Oregon Leaflet Series 6 (3), part 2.

———. 1928. *Thomas Condon: Pioneer Geologist of Oregon*. University of Oregon Press, Eugene.

McDonald, H. G. 1998. The Massacre Rocks local fauna from the Pleistocene of south-eastern Idaho. In *And Whereas…Papers on the Vertebrate Paleontology of Idaho Honoring John A. White*, vol. 1, edited by W. A. Akersten, H. G. McDonald, D. J. Meldrum, and M. E. T. Flint, 156–72. Idaho Museum of Natural History Occasional Papers 36.

———. 2002. *Platygonus compressus* from Franklin County, Idaho, and a review of the genus in Idaho. In *And Whereas…Papers on the Vertebrate Paleontology of Idaho Honoring John A. White*, vol. 2, edited by W. A. Akersten, M. E. Thompson, D. J. Meldrum, R. A. Rapp, and H. G. McDonald, 141–49. Idaho Museum of Natural History Occasional Papers 37.

———. 2003. Sloth remains from North American caves and associated karst features. In *Ice Age Cave Faunas of North America*, edited by B. W. Schubert, J. I. Mead, and R. W. Graham, 1–16. Indiana University Press, Bloomington.

———. 2007. Biomechanical inferences of locomotion in ground sloths: Integrating morphological and track data. In *Cenozoic Vertebrate Tracks and Traces*, edited by S. G. Lucas, J. A. Spielmann, and M. G. Lockley, 201–8. New Mexico Museum of Natural History and Science Bulletin 42.

McDonald, H. G., and R. A. Bryson. 2010. Modeling Pleistocene local climatic parameters using macrophysical climate modeling and the paleoecology of Pleistocene megafauna. *Quaternary International* 217:131–37.

McDonald, H. G., and G. De Iuliis. 2008. Fossil history of sloths. In *The Biology of the Xenarthra*, edited by S. F. Vizcaíno and W. J. Loughry, 39–55. University Press of Florida, Gainesville.

McDonald, H. G., and G. T. Jefferson. 2008. Distribution of Pleistocene *Nothrotheriops* (Xenarthra, Nothrotheriidae) in North America. *Natural History Museum of Los Angeles County Science Series* 41:313–31.

McDonald, H. G., and E. L. Lundelius Jr. 2009. The giant ground sloth *Eremotherium laurillardi* (Xenarthra, Megatheriidae) in Texas. In *Papers on Geology, Vertebrate Paleontology, and Biostratigraphy in Honor of Michael O. Woodburne*, edited by L. B. Albright III, 407–21. Museum of Northern Arizona Bulletin 65.

McDonald, H. G., and G. S. Morgan. 2011. Ground sloths of New Mexico. In *Fossil Record 3*, edited by R. M. Sullivan, S. G. Lucas, and J. A. Spielmann, 652–63. New Mexico Museum of Natural History and Science Bulletin 53.

McDonald, H. G., and S. Pelikan. 2006. Mammoths and mylodonts: Exotic species from two different continents in North American Pleistocene faunas. *Quaternary International* 142–43:229–41.

McDonald, H. G., and D. Perea. 2002. The large scelidothere *Catonyx tarijensis* (Xenarthra, Mylodontidae) from the Pleistocene of Uruguay. *Journal of Vertebrate Paleontology* 22:677–83.

McDonald, H. G., A. D. Rincón, and T. J. Gaudin. 2013. A new genus of megalonychid sloth (Mammalia, Xenarthra) from the late Pleistocene (Lujanian) of Sierra de Perija, Zulia State, Venezuela. *Journal of Vertebrate Paleontology* 33:1226–38.

McDonald, H. G., T. W. Stafford Jr., and D. M. Gnidovec. 2015. Youngest radiocarbon age for Jefferson's ground sloth, *Megalonyx jeffersonii* (Xenarthra, Megalonychidae). *Quaternary Research* 83:355–59.

McDonald, J. N. 1984. An extinct muskox mummy from near Fairbanks, Alaska: A progress report. In *Proceedings of the First International Muskox Symposium*, edited by D. R. Klein, R. G. White, and S. Keller, 148–52. Biological Papers of the University of Alaska Special Report 4.

McDonald, J. N., and C. S. Bartlett Jr. 1983. An associated musk ox skeleton from Saltville, Virginia. *Journal of Vertebrate Paleontology* 2:453–70.

McDonald, J. N., S. W. Neusius, and V. L. Clay. 1987. An associated partial skeleton of *Symbos cavifrons* (Artiodactyla: Bovidae) from Montezuma County, Colorado. *Journal of Paleontology* 61:831–43.

McDonald, J. N., and C. E. Ray. 1989. *The Autochthonous North American Musk Oxen* Bootherium, Symbos, *and* Gidleya *(Mammalia: Artiodactyla: Bovidae)*. Smithsonian Contributions to Paleobiology 66.

McDonald, J. N., C. E. Ray, and F. Grady. 1996. Pleistocene caribou (*Rangifer tarandus*) in the eastern United States: New records and range extensions. In *Palaeoecology and Palaeoenvironments of Late Cenozoic Mammals: Tributes to the Career of C. S. (Rufus) Churcher*, edited by K. M. Stewart and K. L. Seymour, 406–30. University of Toronto Press, Toronto.

McElroy, M. W., R. C. Carr, D. S. Ensor, and G. R. Markowski. 1982. Size distribution of fine particles from coal combustion. *Science* 215:13–19.

McFadden, B. J. 2005. Diet and habitat of toxodont megaherbivores (Mammalia, Notoungulata) from the late Quaternary of South and Central America. *Quaternary Research* 64:113–24.

McGuire, K. R. 1980. Cave sites, faunal analysis, and big-game hunters of the Great Basin: A caution. *Quaternary Research* 14:263–68.

———. 1982. Reply to Gruhn and Bryan's comments on "Cave sites, faunal analysis, and big-game hunters of the Great Basin: A caution." *Quaternary Research* 18:240–42.

McHenry, C. R., S. Wroe, P. D. Clausen, K. Moreno, and E. Cunningham. 2007. Supermodeled sabercat, predatory behavior in *Smilodon fatalis* revealed by high-resolution 3D computer simulation. *Proceedings of the National Academy of Sciences* 104:16010–15.

McHorse, B. K., J. D. Orcutt, and E. B. Davis. 2012. The carnivoran fauna of Rancho La Brea: Average or aberrant. *Palaeogeography, Palaeoclimatology, Palaeoecology* 329–30:118–23.

McKenna, M. C., and S. K. Bell. 1997. *Classification of Mammals above the Species Level.* Columbia University Press, New York.

McNeil, P., L. V. Hills, B. Kooyman, and S. M. Tolman. 2005. Mammoth tracks indicate a declining Late Pleistocene population in southwestern Alberta, Canada. *Quaternary Science Reviews* 24:1253–59.

McNeil, P., L. V. Hills, M. S. Tolman, and B. Kooyman. 2007. Significance of latest Pleistocene tracks, trackways, and trample grounds from southern Alberta, Canada. In *Cenozoic Vertebrate Tracks and Traces*, edited by S. G. Lucas, J. A. Spielmann, and M. G. Lockley, 209–24. New Mexico Museum of Natural History and Science Bulletin 42.

McWethy, D. B., C. Whitlock, J. M. Wilmshurst, M. S. McGlone, and X. Li. 2009. Rapid deforestation of South Island, New Zealand, by early Polynesian fires. *Holocene* 19:883–97.

Meachen, J. A. 2003. A new species of *Hemiauchenia* (Camelidae: Lamini) from the Plio-Pleistocene of Florida. Master's thesis, University of Florida, Gainesville.

———. 2005. A new species of *Hemiauchenia* (Artiodactyla, Camelidae) from the late Blancan of Florida. *Bulletin of the Florida Museum of Natural History* 45:435–47.

Meachen, J. A., and D. P. Hallman. 2002. Resource partitioning in two sympatric Pleistocene llamas from Florida using stable isotopes and post-cranial morphometrics. *Journal of Vertebrate Paleontology* 22 (3) Abstracts:87A.

Meachen-Samuels, J. A. 2012. Morphological convergence of the prey-killing arsenal of sabertooth predators. *Paleobiology* 38:1–14.

Meachen-Samuels, J. A., and W. J. Binder. 2010. Sexual dimorphism and ontogenetic growth in the American lion and sabertoothed cat from Rancho La Brea. *Journal of Zoology* 280:271–79.

Meachen-Samuels, J. A., and B. Van Valkenburgh. 2010. Radiographs reveal exceptional forelimb strength in the sabertooth cat, *Smilodon fatalis*. *PLoS One* 5 (7): e11412.

Mead, J. I. 1983. Harrington's extinct mountain goat (*Oreamnos harringtoni*) and its environment in the Grand Canyon, Arizona. PhD diss., University of Arizona, Tucson.

———. 1987. Quaternary records of pika, *Ochotona*, in North America. *Boreas* 16:165–71.

———. 2003. Late Pleistocene faunas from caves in the eastern Grand Canyon, Arizona. In *Ice Age Cave Faunas of North America*, edited by B. W. Schubert, J. I. Mead, and R. W. Graham, 64–86. Indiana University Press, Bloomington.

Mead, J. I., L. D. Agenbroad, O. K. Davis, and P. S. Martin. 1986. Dung of *Mammuthus* in the arid southwest, North America. *Quaternary Research* 25:121–27.

Mead, J. I., L. D. Agenbroad, A. M. Phillips III, and L. T. Middleton. 1987. Extinct mountain goat (*Oreamnos harringtoni*) in southeastern Utah. *Quaternary Research* 27:323–331.

Mead, J. I., A. Baez, S. L. Swift, et al. (9 authors). 2006. Tropical marsh and savanna of the late Pleistocene in northeastern Sonora, Mexico. *Southwestern Naturalist* 51:226–39.

Mead, J. I., C. J. Bell, and L. K. Murray. 1992. *Mictomys borealis* (northern bog lemming) and the Wisconsin paleoecology of the east-central Great Basin. *Quaternary Research* 37:229–38.

Mead, J. I., N. J. Czaplewski, and L. D. Agenbroad. 2005. Rancholabrean (late Pleistocene) mammals and localities of Arizona. In *Vertebrate Paleontology of Arizona*, edited by R. D. McCord, 139–80. Mesa Southwest Museum Bulletin 11.

Mead, J. I., and F. Grady. 1996. *Ochotona* (Lagomorpha) from late Quaternary cave deposits in eastern North America. *Quaternary Research* 45:93–101.

Mead, J. I., and M. C. Lawler. 1994. Skull, mandible, and metapodials of the extinct Harrington's mountain goat (*Oreamnos harringtoni*). *Journal of Vertebrate Paleontology* 14:562–76.

Mead, J. I., P. S. Martin, R. C. Euler, et al. (8 authors). 1986. Extinction of Harrington's mountain goat. *Proceedings of the National Academy of Sciences* 83:836–39.

Mead, J. I., M. K. O'Rourke, and T. M. Foppe. 1986. Dung and diet of the extinct Harrington's mountain goat (*Oreamnos harringtoni*). *Journal of Mammalogy* 67:284–93.

Mead, J. I., and A. M. Phillips III. 1981. The late Pleistocene and Holocene fauna and flora of Vulture Cave, Grand Canyon, Arizona. *Southwestern Naturalist* 26:257–88.

Mead, J. I., and S. L. Swift. 2012. Late Pleistocene (Rancholabrean) dung deposits of the Colorado Plateau, western North America. In *Vertebrate Coprolites*, edited by A. P.

Hunt, J. Milàn, S. G. Lucas, and J. A. Spielmann, 337–42. New Mexico Museum of Natural History and Science Bulletin 57.

Mead, J. I., R. S. Thompson, and T. R. Van Devender. 1982. Late Wisconsinan and Holocene fauna from Smith Creek Canyon, Snake Range, Nevada. *Transactions of the San Diego Society of Natural History* 20:1–26.

Mead, J. I., R. S. White, A. Baez, M. G. Hollenshead, S. L. Swift, and M. C. Carpenter. 2010. Late Pleistocene (Rancholabrean) *Cynomys* (Rodentia, Sciuridae: prairie dog) from northwestern Sonora, Mexico. *Quaternary International* 217:138–42.

Meade, G. E. 1961. The saber-toothed cat *Dinobastis serus*. *Bulletin of the Texas Memorial Museum* 2:23–60.

Mech, L. D. 1974. *Canis lupus. Mammalian Species* 37:1–6.

Medeiros, J. S., and W. T. Pockman. 2011. Drought increases freezing tolerance of both leaves and xylem of *Larrea tridentata. Plant, Cell and Environment* 34:43–51.

Meiri, M., A. M. Lister, M. J. Collins, et al. (13 authors). 2014. Faunal record identifies Bering isthmus conditions as constraint to end-Pleistocene migration to the New World. *Proceedings of the Royal Society B.* doi:10.1098/rsbp.2013.2167.

Melott, A. L., and B. C. Thomas. 2009. Late Ordovician geographic patterns of extinction compared with simulations of astrophysical ionizing radiation damage. *Paleobiology* 35:311–20.

Melott, A. L., B. C. Thomas, G. Dreschoff, and C. K. Johnson. 2010. Cometary airbursts and atmospheric chemistry: Tunguska and a candidate Younger Dryas event. *Geology* 38:355–58.

Meltzer, D. J. 1997. Monte Verde and the Pleistocene peopling of the Americas. *Science* 276:754–55.

———. 2006. *Folsom: New Archaeological Investigations of a Classic Paleoindian Bison Kill.* University of California Press, Berkeley.

———. 2009. *First Peoples in a New World: Colonizing Ice Age America.* University of California Press, Berkeley.

———. 2014. Clovis at the end of the world. *Proceedings of the National Academy of Sciences* 111:12276–77.

———. 2015. *The Great Paleolithic War: How Science Forged an Understanding of America's Ice Age Past.* University of Chicago Press, Chicago.

Meltzer, D. J., D. K. Grayson, G. Ardila, et al. (9 authors). 1997. On the Pleistocene antiquity of Monte Verde, southern Chile. *American Antiquity* 62:659–63.

Meltzer, D. J., V. T. Holliday, M. D. Cannon, and D. S. Miller. 2014. Chronological evidence fails to support claim for an isochronous widespread layer of cosmic impact indicators dated to 12,800 years ago. *Proceedings of the National Academy of Sciences.* doi:10.1073/pnas.1401150111.

Meltzer, D. J., and W. C. Sturtevant. 1983. The Holly Oak shell game: An historic archaeological fraud. In *Lulu Linear Punctated: Essays in Honor of George Irving Quimby,* edited by R. C. Dunnell and D. K. Grayson, 325–52. University of Michigan Anthropological Papers 72.

Menégaz, A. N., and E. Ortiz Jaureguizar. 1995. Los Artíodáctilos. In *Evolución biológica y climática de la región pampeana durante los últimos cinco millones de años,* edited by M. T. Alberdi, G. Leone, and E. P. Tonni, 311–37. Museo Nacional de Ciencias Naturales, Madrid.

Mensing, S. A. 2001. Late-glacial and early Holocene vegetation and climate change near Owens Lake, eastern California. *Quaternary Research* 55:57–65.

Mensing, S. A., L. V. Benson, M. Kashgarian, and S. Lund. 2004. A Holocene pollen record of persistent droughts from Pyramid Lake, Nevada, USA. *Quaternary Research* 62:29–38.

Merino, M. L., and R. V. Rossi. 2010. Origin, systematics, and morphological radiation. In *Neotropical Cervidology: Biology and Medicine of Latin American Deer*, edited by J. M. B. Duarte and S. González, 2–11. IUCN/FUNEP, Gland, Switzerland.

Merriam, J. C. 1906. Recent discoveries of Quaternary mammals in southern California. *Science* 24:248–50.

———. 1910. New mammalia from Rancho La Brea. *University of California Publications, Bulletin of the Department of Geology* 5 (25): 391–95.

———. 1911. Note on a gigantic bear from the Pleistocene of Rancho La Brea. *University of California Publications, Bulletin of the Department of Geology* 6 (6): 163–66.

———. 1915. An occurrence of mammalian remains in a Pleistocene lake deposit at Astor Pass, near Pyramid Lake, Nevada. *University of California Publications, Bulletin of the Department of Geology* 8:377–84.

———. 1930. An interesting letter. *Masterkey* 4:61.

Merriam, J. C., and C. Stock. 1921. Occurrence of Pleistocene vertebrates in an asphalt deposit near McKittrick, California. *Science* 54:566–67.

———. 1932. *The Felidae of Rancho La Brea*. Carnegie Institution of Washington, Publication 422.

Metcalfe, J. Z., F. J. Longstaffe, J. A. M. Ballenger, and C. V. Haynes Jr. 2011. Isotopic paleoecology of Clovis mammoths from Arizona. *Proceedings of the National Academy of Sciences* 108:17916–20.

Meyer de Schauensee, R. 1970. *A Guide to the Birds of South America*. Livingston Publishing, Wynnewood, PA.

Meyers, J. I. 2007. Basicranial analysis of *Martes* and the extinct *Martes nobilis* (Carnivora: Mustelidae) using geometric morphometrics. Master's thesis, Northern Arizona University, Flagstaff.

Millar, C. I., and R. D. Westfall. 2010. Distribution and climatic relationships of the American pika (*Ochotona princeps*) in the Sierra Nevada and western Great Basin, U.S.A.; Periglacial landforms as refugia in warming climates. *Arctic, Antarctic, and Alpine Research* 42:76–88.

Miller, A. H. 1932. An extinct icterid from Shelter Cave, New Mexico. *Auk* 49:38–41.

———. 1947. A new genus of icterid from Rancho La Brea. *Condor* 49:22–24.

Miller, G. J. 1968. *On the Age Distribution of* Smilodon californicus *Bovard from Rancho La Brea*. Los Angeles County Museum Contributions in Science 131.

Miller, L. H. 1909. *Teratornis*, a new avian genus from Rancho La Brea. *University of California Publications, Bulletin of the Department of Geology* 5 (21): 305–17.

———. 1910a. The condor-like vultures of Rancho La Brea. *University of California Publications, Bulletin of the Department of Geology* 6 (1): 1–19.

———. 1910b. Wading birds from the Quaternary asphalt beds of Rancho La Brea. *University of California Publications, Bulletin of the Department of Geology* 5 (30): 439–48.

———. 1911a. A series of eagle tarsi from the Pleistocene of Rancho La Brea. *University of California Publications, Bulletin of the Department of Geology* 6 (12): 305–16.

———. 1911b. Additions to the avifauna of the Pleistocene deposits at Fossil Lake, Oregon. *University of California Publications, Bulletin of the Department of Geology* 6 (4): 79–87.

——. 1912. Contributions to avian paleontology from the Pacific coast of North America. *University of California Publications, Bulletin of the Department of Geology* 7 (5): 61–115.

——. 1916. Two vulturid raptors from the Pleistocene of Rancho La Brea. *University of California Publications, Bulletin of the Department of Geology* 9 (6): 105–9.

——. 1931. The California Condor in Nevada. *Condor* 33:32.

Miller, L. H., and I. S. DeMay. 1942. The fossil birds of California. *University of California Publications in Zoology* 47 (4): 47–142.

Miller, L. H., and H. Howard. 1937. The status of the extinct condor-like birds of the Rancho La Brea Pleistocene. *Publications of the University of California at Los Angeles in Biological Science* 1 (9): 169–76.

Miller, S. J., 1979. The archaeological fauna of four sites in Smith Creek Canyon. In *The Archaeology of Smith Creek Canyon, Eastern Nevada*, edited by D. R. Tuohy and D. L. Rendall, 272–329. Nevada State Museum Anthropological Papers 17.

Miller, W. E. 1976. Late Pleistocene vertebrates of the Silver Creek Local Fauna from north central Utah. *Great Basin Naturalist* 36:387–424.

——. 1982. Pleistocene vertebrates from deposits of Lake Bonneville, Utah. *National Geographic Society Research Reports* 14:473–78.

——. 1987. *Mammut americanum*, Utah's first record of the American mastodon. *Journal of Paleontology* 61:168–83.

——. 2002. Quaternary vertebrates of the northeastern Bonneville Basin and vicinity of Utah. In *Great Salt Lake: An Overview of Change*, edited by J. W. Gwynn, 54–69. Utah Department of Natural Resources Special Publication, Salt Lake City.

Milner-Gulland, E. J., M. V. Kholodova, A. Bekenov, O. M. Bukreeva, Iu. A. Grachev, L. Amgalan, and A. A. Lushchekina. 2001. Dramatic declines in saiga antelope populations. *Oryx* 35:340–45.

Miño-Boilini, A. R., and A. A. Carlini. 2009. The Scelidotheriinae Ameghino, 1904 (Phyllophaga, Xenarthra) from the Ensenadan–Lujanian Stage/Ages (Early Pleistocene to Early-Middle Pleistocene–Early Holocene) of Argentina. *Quaternary International* 210:93–101.

Minor, R., and L. Spencer. 1977. *Site of a Probable Camelid Kill Site at Fossil Lake, Oregon*. Report submitted to the Bureau of Land Management, Lakeview, OR.

Mithöfer, A., and W. Boland. 2012. Plant defense against herbivores: Chemical aspects. *Annual Review of Plant Biology* 63:431–50.

Mol, D., Y. Coppens, A. N. Tikhonov, et al. (14 authors). 2001. The Jarkov Mammoth: 20,000-year-old carcass of a Siberian woolly mammoth *Mammuthus primigenius* (Blumenbach, 1799). In *The World of Elephants*, edited by G. Cavarretta, P. Gioia, M. Mussi, and M. R. Palombo, 305–9. Consiglio Nazionale delle Ricerche, Rome.

Montellano-Ballesteros, M., and G. Carbot-Chanona. 2009. *Panthera leo atrox* (Mammalia: Carnivora: Felidae) in Chiapas, Mexico. *Southwestern Naturalist* 54:217–22.

Montes, C., G. Bayona, A. Cardona, et al. (10 authors). 2012. Arc-continent collision and orocline formation: Closing of the Central American seaway. *Journal of Geophysical Research* 117. doi:10.1029/2011JB008959.

Montes, C., A. Cardona, C. Jaramillo, et al. (11 authors). 2015. Middle Miocene closure of the Central American Seaway. *Science* 348:226–29.

Morejohn, G. V., and D. C. Dailey. 2004. *The Identity and Postcranial Osteology of* Odo-coileus lucasi *(Hay) 1927: A Plio-Pleistocene Deer from California and Idaho.* Sierra College Natural History Museum Bulletin 1.

Moreno, A. 2014. Lags within the Younger Dryas. *Nature Geoscience* 7:87–88.

Moreno, F. P. 1899a. Note on the discovery of *Miolania* and *Glossotherium* (*Neomylodon*) in Patagonia. *Geological Magazine New Series* 6:385–88.

——. 1899b. On a portion of mammalian skin, named *Neomylodon listai,* from a cavern near Consuelo Cove, Last Hope Inlet, Patagonia. 1. Account of the discovery. *Proceedings of the Zoological Society of London* 1899:144–48.

——. 1902. Account of the discovery. In *Through the Heart of Patagonia,* edited by H. Hesketh Prichard, 301–4. D. Appleton, New York.

Morey, D. F. 2014. In search of Paleolithic dogs: A quest with mixed results. *Journal of Archaeological Science* 52:300–307.

Morgan, G. S. 2002. Late Rancholabrean mammals from southernmost Florida, and the Neotropical influence in Florida Pleistocene faunas. In *Cenozoic Mammals of Land and Sea: Tributes to the Career of Clayton E. Ray,* edited by R. J. Emry, 15–38. Smithsonian Contributions to Paleobiology 93.

——. 2008. Vertebrate fauna and geochronology of the Great American Biotic Interchange in North America. In *Neogene Mammals,* edited by S. G. Lucas, G. S. Morgan, J. A. Spielmann, and D. R. Prothero, 93–140. New Mexico Museum of Natural History and Science Bulletin 44.

Morgan, G. S., and S. D. Emslie. 2010. Tropical and western influences in vertebrate faunas from the Pliocene and Pleistocene of Florida. *Quaternary International* 217:143–58.

Morgan, G. S., and S. G. Lucas. 2005. Pleistocene vertebrate faunas in New Mexico from alluvial, fluvial, and lacustrine deposits. In *New Mexico's Ice Ages,* edited by S. G. Lucas, G. S. Morgan, and K. E. Zeigler, 185–248. New Mexico Museum of Natural History and Science Bulletin 28.

——. 2006. Pleistocene vertebrates from southeastern New Mexico. In *Caves and Karst of Southeastern New Mexico,* edited by L. Lund, V. W. Lueth, W. Raatz, P. Boston, and D. W. Love, 317–36. New Mexico Geological Society Guidebook, 57th Field Conference.

Morgan, J. K., and N. H. Morgan. 1995. A new species of *Capromeryx* (Mammalia: Artiodactyla) from the Taunton Local Fauna of Washington, and the correlation with other Blancan faunas of Washington and Idaho. *Journal of Vertebrate Paleontology* 15:160–70.

Morlan, R. E. 1987. The Pleistocene archaeology of Beringia. In *The Evolution of Human Hunting,* edited by M. H. Nitecki and D. V. Nitecki, 267–307. Plenum, New York.

Morrison, J. L., and J. F. Dwyer. 2012. Crested Caracara (*Caracara cheriway*). In *The Birds of North America Online,* edited by A. Poole. Cornell Laboratory of Ornithology. http://bna.birds.cornell.edu/bna/species/249.

Moss, M. L., and J. M. Erlandson. 2013. Waterfowl and lunate crescents in western North America: The archaeology of the Pacific Flyway. *Journal of World Prehistory* 26:173–211.

Mothé, D., and L. Avilla. 2015. Mythbusting evolutionary issues on South American Gomphotheriidae (Mammalia: Proboscidea). *Quaternary Science Reviews* 110:23–35.

Mothé, D., L. S. Avilla, and M. A. Cozzuol. 2013. The South American gomphotheres

(Mammalia, Proboscidea, Gomphotheriidae): Taxonomy, phylogeny, and bio-geography. *Journal of Mammalian Evolution* 20:23–32.

Mothé, D., L. S. Avilla, M. Cozzuol, and G. R. Winck. 2012. Taxonomic revision of the Quaternary gomphotheres (Mammalia: Proboscidea: Gomphotheriidae) from the South American lowlands. *Quaternary International* 276–77: 2–7.

Moura, A. E., C. J. van Rensberg, M. Pilot, et al. (12 authors). 2014. Killer whale nuclear genome and mtDNA reveal widespread population bottleneck during the last glacial maximum. *Molecular Biology and Evolution* 31:1121–31.msu058

Mozingo, H. N. 1987. *Shrubs of the Great Basin*. University of Nevada Press, Reno.

Muhs, D. R., K. R. Simmons, L. T. Groves, J. P. McGeehin, R. R. Schumann, and L. D. Agenbroad. 2015. Late Quaternary sea-level history and the antiquity of mammoths (*Mammuthus exilis* and *Mammuthus columbi*), Channel Islands National Park, California, USA. *Quaternary Research* 83:502–21.

Munson, P. J. 1991. A herd of fossil peccaries (*Platygonus compressus*) from late Wisconsinan deposits in central Indiana. *Proceedings of the Indiana Academy of Science* 99:203–10.

Munz, P. A. 1959. *A California Flora*. University of California Press, Berkeley.

Murray, L. K., C. J. Bell, M. T. Dolan, and J. I. Mead. 2005. Late Pleistocene fauna from the southern Colorado Plateau, Navajo County, Arizona. *Southwestern Naturalist* 50:363–74.

Mysterious tracks in stone: The foot-prints of a mastodon and of a human being believed to be imprinted in the same rock. From the *San Francisco Call*, August 4. 1882. *New York Times*, August 13.

Nalawade-Chavan, S., G. Zazula, F. Brock, J. Southon, R. MacPhee, and P. Druckenmiller. 2014. New single amino acid hydroxyproline radiocarbon dates for two problematic American mastodon fossils from Alaska. *Quaternary Geochronology* 20:23–28.

Neilson, E. H., J. Q. D. Goodger, I. E. Woodrow, and B. L. Møller. 2013. Plant chemical defense: At what cost? *Trends in Plant Science* 18:250–58.

Nelson, M. E., and J. H. Madsen Jr. 1978. Late Pleistocene musk oxen from Utah. *Transactions of the Kansas Academy of Science* 81:277–95.

———. 1979. The Hay-Romer camel debate: Fifty years later. *University of Wyoming Contributions to Geology* 18:47–50.

———. 1980. A summary of Pleistocene, fossil vertebrate localities in the northern Bonneville Basin of Utah. In *Great Salt Lake: A Scientific, Historical, and Economic Overview*, edited by J. W. Gwynn, 97–113. Utah Geological and Mineral Survey Bulletin 116.

———. 1983. A giant short-faced bear (*Arctodus simus*) from the Pleistocene of northern Utah. *Transactions of the Kansas Academy of Science* 86:1–9.

———. 1987. A review of Lake Bonneville shoreline faunas (Late Pleistocene) of northern Utah. In *Cenozoic Geology of Western Utah*, edited by R. S. Kopp and R. E. Cohenour, 319–33. Utah Geological Association Publication 16.

The Nevada footprints. 1884. *Hartford (CT) Daily Courant*, January 5.

The Nevada footprints: Impressions taken of a series of tracks in the state prison quarry. From the *San Francisco Bulletin*, July 23. 1883. *New York Times*, July 31.

Nevada Legislature. 2014. http://www.leg.state.nv.us/Session/77th2013/Reports/history.cfm?ID=799.

Newsom, L. A., and M. C. Mihlbachler. 2006. Mastodons (*Mammut americanum*) diet foraging patterns based on analysis of dung deposits. In *First Floridians and Last Mastodons: The Page-Ladson Site in the Aucilla River*, edited by S. D. Webb, 263–331. Springer, Dordrecht, Netherlands.

Nikolskiy, P. A., L. D. Sulerzhitsky, and V. V. Pitulko. 2011. Last straw versus blitzkrieg overkill: Climate-driven changes in the Arctic Siberian mammoth population and the late Pleistocene extinction problem. *Quaternary Science Reviews* 30:2309–28.

Nobel, P. S. 1978. Surface temperatures of cacti: Influences of environmental and morphological factors. *Ecology* 59:986–96.

———. 1980. Morphology, surface temperatures, and northern limits of columnar cacti in the Sonoran Desert. *Ecology* 61:1–7.

———. 1983. Spine influences on PAR interception, stem temperature, and nocturnal acid accumulation by cacti. *Plant, Cell and Environment* 6:153–59.

Nobel, P. S., and M. E. Loik. 1999. Form and function of cacti. In *Ecology of Sonoran Desert Plants and Plant Communities*, edited by R. H. Robichaux, 143–63. University of Arizona Press, Tucson.

Nordenskjold, E. 1899. La grotte de *Glossotherium* (*Neomylodon*) en Patagonie. *Comptes Rendus des Séances de l'Académie des Sciences* 129:1216–17.

Nowak, C. L., R. S. Nowak, R. J. Tausch, and P. E. Wigand. 1994a. A 30000 year record of vegetation dynamics at a semi-arid locale in the Great Basin. *Journal of Vegetation Science* 5:579–90.

———. 1994b. Tree and shrub dynamics in northwestern Great Basin woodland and shrub steppe during the Late-Pleistocene and Holocene. *American Journal of Botany* 81:265–77.

Nowak, R. M. 1979. *North American Quaternary* Canis. Museum of Natural History, University of Kansas, Monograph 6.

———. 1999. *Walker's Mammals of the World*. 6th ed. Johns Hopkins University Press, Baltimore.

Nunez, E. E., B. J. MacFadden, J. I. Mead, and A. Baez. 2010. Ancient forests and grasslands in the desert: Diet and habitat of Late Pleistocene mammals from north-central Sonora, Mexico. *Palaeogeography, Palaeoclimatology, Palaeoecology* 297: 391–400.

Nussbaum, J. L. 1930. *Annual Report of the Department Archaeologist and Superintendent of Mesa Verde National Park to the Secretary of the Interior: 1930*. US Government Printing Office, Washington, DC.

———. 1931. *Annual Report of the Department Consulting Archaeologist and Superintendent of Mesa Verde National Park to the Secretary of the Interior: 1931*. US Government Printing Office, Washington, DC.

Nye, A. S. 2007. Pleistocene peccaries from Guy Wilson Cave, Sullivan County, Tennessee. Master's thesis, East Tennessee State University, Johnson City.

Nyström, V., L. Dalén, S. Vartanyan, K. Lidén, N. Ryman, and A. Angerbjörn. 2010. Temporal genetic change in the last remaining population of woolly mammoth. *Proceedings of the Royal Society B* 277:2331–37.

O'Brien, M. J., M. T. Boulanger, M. Collard, B. Buchanan, L. Tarle, L. G. Straus, and M. I. Eren. 2014. On thin ice: Problems with Stanford and Bradley's proposed Solutrean colonization of North America. *Antiquity* 88:606–13.

O'Gara, B. W. 1990. The pronghorn (*Antilocapra americana*). In *Horns, Pronghorns, and Antlers*, edited by G. A. Bubenik and A. B. Bubenik, 231–64. Springer-Verlag, New York.

———. 2004. Physical characteristics. In *Pronghorn Ecology and Management*, edited by B. W. O'Gara and J. D. Yoakum, 109–43. University Press of Colorado, Boulder.

O'Gara, B. W., and C. Janis. 2004. The fossil record. In *Pronghorn Ecology and Management*, edited by B. W. O'Gara and J. D. Yoakum, 27–39. University Press of Colorado, Boulder.

O'Keefe, F. R., E. V. Fet, and J. M. Harris. 2009. *Compilation, Calibration, and Synthesis of Faunal and Floral Radiocarbon Dates, Rancho La Brea, California*. Natural History Museum of Los Angeles County Contributions in Science 518.

Oliveira, É. V., F. J. Prevosti, and J. C. Pereira. 2005. *Protocyon troglodytes* (Lund) (Mammalia, Carnivora) in the late Pleistocene of Rio Grande do Sul and their paleoecological significance. *Revista Brasileira de Paleontologia* 8:215–20.

Olson, S. L. 2007. The "Walking Eagle" *Wetmoregyps daggetti* Miller: A scaled-up version of the Savanna Hawk (*Buteogallus meridionalis*). *Ornithological Monographs* 63:110–14.

Olson, S. L., and P. C. Rasmussen. 2001. Miocene and Pliocene birds from the Lee Creek Mine, North Carolina. In *Geology and Paleontology of the Lee Creek Mine, North Carolina, III*, edited by C. E. Ray and D. J. Bohaska, 233–365. Smithsonian Contributions to Paleobiology 90.

Olson, V. A., and S. T. Turvey. 2013. The evolution of sexual dimorphism in New Zealand giant moa (*Dinornis*) and other ratites. *Proceedings of the Royal Society B* 280. doi:10.1098/rspb.2013.0401

Oregon Blue Book. 2014. Governors of Oregon. http://bluebook.state.or.us/state /elections/elections24.htm.

Oregon State Archives. 2014. Oregon legislators and staff guide. http://arcweb.sos.state .or.us/pages/records/legislative/recordsguides/histleg/.

Oregon State University. 2015. Consumer Price Index (CPI) conversion factors for years 1774 to estimated 2024 to convert to estimated dollars of 2015. http://liberal arts.oregonstate.edu/sites/liberalarts.oregonstate.edu/files/polisci/faculty-research /sahr/inflation-conversion/pdf/cv2015.pdf.

Oren, D. C. 2001. Does the endangered xenarthran fauna of Amazonia include remnant ground sloths? *Edentata* 2001 (4): 2–5.

Orlando, L., D. Male, M. T. Alberdi, J. L. Prado, A. Prieto, A. Cooper, and C. Hänni. 2008. Ancient DNA clarifies the evolutionary history of American late Pleistocene equids. *Journal of Molecular Evolution* 66:533–38.

Orlando, L., J. L. Metcalf, M. T. Alberdi, et al. (22 authors). 2009. Revising the recent evolutionary history of equids using ancient DNA. *Proceedings of the National Academy of Sciences* 106:21754–59.

Orr, P. C. 1956. *Pleistocene Man in Fishbone Cave, Pershing County, Nevada*. Nevada State Museum Department of Archeology Bulletin 2.

———. 1969. *Felis trumani*: A new radiocarbon dated cat skull from Crypt Cave, Nevada. *Bulletin of the Santa Barbara Museum of Natural History Department of Geology* 2:1–8.

———. 1974. Notes on the archaeology of the Winnemucca Caves. *Nevada State Museum Anthropological Papers* 16:47–59.

Orsi, R. J. 2005. *Sunset Limited: The Southern Pacific Railroad and the Development of the American West, 1850–1930*. University of California Press, Berkeley.

Ortiz, P. E., U. F. J. Pardiñas, and S. J. Steppan. 2000. A new fossil phyllotine (Rodentia: Muridae) from northwestern Argentina and relationships of the *Reithrodon* group. *Journal of Mammalogy* 81:37–51.

Osborn, H. F. 1929. Biographical memoir of Edward Drinker Cope 1840–1897. *National Academy of Sciences Biographical Memoirs* 13 (3): 125–317.

———. 1931. *Cope: Master Naturalist*. Princeton University Press, Princeton, NJ.

Owen-Smith, N., and S. M. Cooper. 1987. Palatability of woody plants to browsing ruminants in a South African savanna. *Ecology* 68:319–31.

Packard, E. L. 1952. *Fossil Edentates of Oregon*. Oregon State Monographs, Studies in Geology 8.

Pardiñas, U. F. J., G. D'Elía, and P. E. Ortiz. 2002. Sigmodontinos fósiles (Rodentia, Muroidea, Sigmodontinae) de América del Sur: Estado actual de su conocimiento y prospectiva. *Mastozoología Neotropical* 9:209–52.

Parker, N. E., and J. W. Williams. 2011. Influences of climate, cattle density, and lake morphology on *Sporormiella* abundances in modern lake sediments in the US Great Plains. *Holocene* 22:475–83.

Parkinson, J. 1804. *Organic Remains of a Former World*. Vol. 1. J. Robson and others, London.

Parmalee, P. W., and R. W. Graham. 2002. Additional records of the giant beaver, *Castoroides*, from the mid-south: Alabama, Tennessee, and South Carolina. In *Cenozoic Mammals of Land and Sea: Tributes to the Career of Clayton E. Ray*, edited by R. J. Emry, 65–71. Smithsonian Contributions to Paleobiology 93.

Pasenko, M. 2011. A specimen of *Mammut americanum* (Proboscidea, Mammalia) from Yavapai County, west-central Arizona. *Journal of the Arizona-Nevada Academy of Sciences* 42 (2): 61–64.

Pasenko, M. R., and L. D. Agenbroad. 2012. Late Pleistocene mammalian fauna from Prescott Valley, west-central Arizona. *Southwestern Naturalist* 57:74–86.

Pasenko, M. R., and B. W. Schubert. 2004. *Mammuthus jeffersonii* (Proboscidea, Mammalia) from northern Illinois. *PaleoBios* 24 (3): 19–24.

Pavlik, B. M. 2008. *The California Deserts: An Ecological Rediscovery*. University of California Press, Berkeley.

Pellmyr, O. 2003. Yuccas, yucca moths, and coevolution: A review. *Annals of the Missouri Botanical Garden* 90:30–55.

Pellmyr, O., and K. A. Segraves. 2003. Pollinator divergence within an obligate mutualism: Two yucca moth species (Lepidoptera; Prodoxidae: *Tegeticula*) on the Joshua tree (*Yucca brevifolia*; Agavaceae). *Annals of the Entomological Society of America* 96:716–22.

Pellmyr, O., K. A. Segraves, D. M. Althoff, M. Balcázar-Lara, and J. Leebens-Mack. 2007. The phylogeny of yuccas. *Molecular Phylogenetics and Evolution* 43:493–501.

Pickford, M., J. Morales, and D. Soria. 1995. Fossil camels from the Upper Miocene of Europe: Implications for biogeography and faunal change. *Geobios* 28:641–50.

Pigati, J. S., J. A. Rech, J. Quade, and J. Bright. 2014. Desert wetlands in the geological record. *Earth-Science Reviews* 132:67–81.

Pinsof, J. D. 1995. Additional occurrences of *Cervalces scotti* from the Pleistocene of New York. *Current Research in the Pleistocene* 12:104–5.

Pinson, A. O. 1997. *Report on the Archaeological Survey of a Portion of Fossil Lake, Lake County, Oregon.* Report submitted to the Bureau of Land Management, Lakeview, OR.

Pitulko, V. V., P. A. Nikolsky, E. Y. Girya, et al. (9 authors). 2004. The Yana RHS Site: Humans in the Arctic before the Last Glacial Maximum. *Science* 303:52–56.

Pockman, W. T., and J. S. Sperry. 1997. Freezing-induced xylem cavitation and the northern limit of *Larrea tridentata. Oecologia* 109:19–27.

Poinar, H. N., M. Hofrieter, W. G. Spaulding, et al. (9 authors). 1998. Molecular coproscopy: Dung and diet of the extinct ground sloth *Nothrotheriops shastensis. Science* 281:402–6.

Polaco, O. J., J. Arroyo-Cabrales, E. Corona-M., and J. G. López-Oliva. 2001. The American mastodon in Mexico. In *The World of Elephants,* edited by G. Cavarretta, P. Gioia, M. Mussi, and M. R. Palombo, 237–42. Consiglio Nazionale delle Ricerche, Rome.

Polyak, V. J., Y. Asmerom, S. J. Burns, and M. S. Lachniet. 2012. Climatic backdrop to the terminal Pleistocene extinction of North American mammals. *Geology* 40:1023–26.

Poole, J. H. 1994. Sex differences in the behavior of African elephants. In *The Differences between the Sexes,* edited by R. V. Short and E. Balaban, 331–46. Cambridge University Press, Cambridge.

Pope, A. (1733–1734) 1963. An essay on man. In *The Poems of Alexander Pope,* edited by J. Butt, 501–47. Yale University Press, New Haven, CT.

Porpino, K. de O., J. C. Fernicola, and L. P. Bergqvist. 2010. Revisiting the intertropical Brazilian species *Hoplophorus euphractus* (Cingulata: Glyptodontoidea) and the phylogenetic affinities of *Hoplophorus. Journal of Vertebrate Paleontology* 30:911–27.

Prado, J. L., M. T. Alberdi, B. Azanza, B. Sánchez, and D. Frassinetti. 2001. The Pleistocene gomphotheres (Proboscidea) from South America: Diversity, habitats, and feeding ecology. In *The World of Elephants,* edited by G. Cavarretta, P. Gioia, M. Mussi, and M. R. Palombo, 337–40. Consiglio Nazionale delle Ricerche, Rome.

Prado, J. L., J. Arroyo-Cabrales, E. Johnson, M. T. Alberdi, and O. J. Polaco. 2012. New World proboscidean extinctions: Comparisons between North and South America. *Archaeological and Anthropological Sciences.* doi:10.1007/s12520-012-0094-3.

Prescott, G. W., D. R. Williams, A. Balmford, R. E. Green, and A. Manica. 2012. Quantitative global analysis of the role of climate and people in explaining late Quaternary megafaunal extinctions. *Proceedings of the National Academy of Sciences* 109:4527–31.

Preuss, C. 1958. *Exploring with Frémont: The Private Diaries of Charles Preuss, Cartographer for John C. Frémont on His First, Second, and Fourth Expeditions to the Far West.* Translated and edited by E. G. Gudde and E. K. Gudde. University of Oklahoma Press, Norman.

Prevosti, F. J., and A. D. Rincón. 2007. A new fossil canid assemblage from the late Pleistocene of northern South America: The canids of the Inciarte Asphalt Pit (Zulia, Venezuela), fossil record and biogeography. *Journal of Paleontology* 81:1053–65.

Prevosti, F. J., E. P. Tonni, and J. C. Bidegain. 2009. Stratigraphic range of the large canids (Carnivora, Canidae) in South America, and its relevance to Quaternary biostratigraphy. *Quaternary International* 210:76–81.

Prevosti, F. J., and S. F. Vizcaíno. 2006. Paleoecology of the large carnivore guild from the late Pleistocene of Argentina. *Acta Palaeontologica Polonica* 51:407–22.

Prevosti, F. J., A. E. Zurita, and A. A. Carlini. 2005. Biostratigraphy, systematics, and paleoecology of *Protocyon* Giebel, 1855 (Carnivora, Canidae) in South America. *Journal of South American Earth Sciences* 20:5–12.

Prychid, C. J., and P. J. Rudall. 1999. Calcium oxalate crystals in monocotyledons: A review of their structure and systematics. *Annals of Botany* 84:725–39.

Pujos, F. 2000. *Scelidodon chiliensis* (Xenarthra, Mammalia) du Pléistocène terminal de "Pampa de los Fósiles" (Nord-Pérou). *Quaternaire* 11:197–206.

Pujos, F., G. De Iuliis, C. Argot, and L. Werdelin. 2007. A peculiar climbing Megalonychidae from the Pleistocene of Peru and its implications for sloth history. *Zoological Journal of the Linnean Society* 149:179–235.

Purdy, B. A. 2012. The mammoth engraving from Vero Beach, Florida: Ancient or Recent? *Florida Anthropologist* 65:205–17.

Purdy, B. A., K. S. Jones, J. J. Mecholsky, et al. (12 authors). 2011. Earliest art in the Americas: Incised image of a proboscidean on a mineralized extinct animal bone from Vero Beach, Florida. *Journal of Archaeological Science* 38:2908–13.

Quade, J. 1986. Late Quaternary environmental changes in the Upper Las Vegas Valley, Nevada. *Quaternary Research* 26:340–57.

Quade, J., R. M. Forester, W. L. Pratt, and C. Carter. 1998. Black mats, spring-fed streams, and late-glacial-age recharge in the southern Great Basin. *Quaternary Research* 49:129–48.

Quade, J., R. M. Forester, and J. E. Whelan. 2003. Late Quaternary paleohydrologic and paleotemperature change in southern Nevada. In *Paleoenvironments and Paleohydrology of the Mojave and Southern Great Basin Deserts*, edited by Y. Enzel, S. G. Wells, and N. Lancaster, 165–88. Geological Society of America Special Paper 368.

Quade, J., and W. L. Pratt. 1989. Late Wisconsin groundwater discharge environments of the southwestern Indian Springs Valley, southern Nevada. *Quaternary Research* 31:351–70.

Quammen, D. 2014. People of the horse. *National Geographic* 225 (3): 104–27.

Raper, D., and M. Bush. 2009. A test of *Sporormiella* representation as a predictor of megaherbivore presence and abundance. *Quaternary Research* 71:490–96.

Rappaport, A. B., and H. Hochman. 1988. Cystic calculi as a cause of recurrent rectal prolapse in a sloth (*Cholepus* sp.) [*sic*]. *Journal of Zoo Animal Medicine* 19:235–36.

Rasmussen, M., S. L. Anzick, M. R. Waters, et al. (42 authors). 2014. The genome of a Late Pleistocene human from a Clovis burial site in western Montana. *Nature* 506: Pages:225–29.

Rasmussen, S. O., K. K. Andersen, A. N. Svensson, et al. (16 authors). 2006. A new Greenland ice core chronology for the last glacial termination. *Journal of Geophysical Research* 111:D06102.

Rawn-Schatzinger, V. 1992. *The Scimitar Cat* Homotherium serum *Cope*. Illinois State Museum Reports of Investigations 47.

Ray, C. E. 1967. Pleistocene mammals from Ladds, Bartow County, Georgia. *Bulletin of the Georgia Academy of Science* 25:120–50.

Ray, C. E., C. S. Denny, and M. Rubin. 1970. A peccary, *Platygonus compressus* LeConte, from drift of Wisconsinan age in northern Pennsylvania. *American Journal of Science* 268:78–94.

Ray, C. E., and A. E. Sanders. 1984. Pleistocene tapirs in the eastern United States. In *Contributions in Quaternary Vertebrate Paleontology: A Volume in Memorial to John E.*

Guilday, edited by H. H. Genoways and M. R. Dawson, 283–315. Carnegie Museum of Natural History Special Publication 8.

The recent discoveries of fossil footprints in Carson, Nevada. 1882. *American Naturalist* 16:921–23.

Retrum, J. B. 2010. A paleoclimatic and paleohydrologic reconstruction of Pleistocene Fossil Lake, Oregon. PhD diss., University of Oklahoma, Norman.

Reynolds, P. S. 2002. How big is a giant? The importance of methods in estimating body size of extinct mammals. *Journal of Mammalogy* 83:321–32.

Reynolds, R. E., R. L. Reynolds, and C. J. Bell. 1991. The Devil Peak sloth. In *Crossing the Borders: Quaternary Studies in Eastern California and Southwestern Nevada*, edited by R. E. Reynolds, 115–16. San Bernardino County Museum Association, Redlands, CA.

Rhode, D. 1987. The mountains and the lake: Prehistoric lacustrine-upland settlement relationships in the Walker watershed, western Nevada. PhD diss., University of Washington, Seattle.

———. 2000. Middle and late Wisconsin vegetation in the Bonneville Basin. In *Late Quaternary Paleoecology in the Bonneville Basin*, edited by D. B. Madsen, 137–47. Utah Geological Survey Bulletin 130.

———. 2001. Packrat middens as a tool for reconstructing historic ecosystems. In *The Historical Ecology Handbook: A Restorationist's Guide to Reference Ecosystems*, edited by D. Egan and E. A. Howell, 257–93. Island Press, Washington, DC.

———. 2002. *Native Plants of Southern Nevada: An Ethnobotany*. University of Utah Press, Salt Lake City.

———. 2003. Coprolites from Hidden Cave, revisited: Evidence for site occupation history, diet and sex of occupants. *Journal of Archaeological Science* 30:909–22.

Rhode, D., and L. A. Louderback. 2007. Dietary plant use in the Bonneville Basin during the terminal Pleistocene/early Holocene transition. In *Paleoindian or Paleoarchaic? Great Basin Human Ecology at the Pleistocene/Holocene Transition*, edited by K. E. Graf and D. N. Schmitt, 231–47. University of Utah Press, Salt Lake City.

Rhode, D., and D. B. Madsen. 1995. Late Wisconsin/early Holocene vegetation in the Bonneville Basin. *Quaternary Research* 44:246–56.

Rhodes, A. N., J. W. Urbance, H. Youga, et al. (8 authors). 1998. Identification of bacterial isolates obtained from intestinal contents associated with 12,000-year-old mastodon remains. *Applied and Environmental Microbiology* 674:651–58.

Richards, R. L., C. S. Churcher, and W. D. Turnbull. 1996. Distribution and size variation in North American short-faced bears, *Arctodus simus*. In *Palaeoecology and Palaeoenvironments of Late Cenozoic Mammals: Tributes to the Career of C. S. (Rufus) Churcher*, edited by K. M. Stewart and K. L. Seymour, 191–246. University of Toronto Press, Toronto.

Rick, T. C., C. A. Hofman, T. J. Braje, J. E. Maldonado, T. S. Sillett, K. Danchisko, and J. M. Erlandson. 2012. Flightless ducks, giant mice, and pygmy mammoths: Late Quaternary extinctions on California's Channel Islands. *World Archaeology* 44:3–20.

Rignot, E., J. Mouginot, M. Morlighem, H. Seroussi, and B. Scheuchl. 2014. Widespread, rapid grounding line retreat of Pine Island, Thwaites, Smith and Kohler glaciers, West Antarctica from 1992 to 2011. *Geophysical Research Letters* 41:3502–9.

Rincón, A. D., F. J. Prevosti, and G. E. Parra. 2011. New saber-toothed cat records (Felidae: Machairodontinae) for the Pleistocene of Venezuela, and the Great American Biotic Interchange. *Journal of Vertebrate Paleontology* 31:468–78.

Ripoll, M. P., J. V. M. Pérez, A. S. Serra, J. E. A. Tortosa, and I. C. Montañana. 2010. Presence of the genus *Cuon* in upper Pleistocene and initial Holocene sites of the Iberian Peninsula: New remains identified in archaeological contexts of the Mediterranean region. *Journal of Archaeological Science* 37:437–50.

Robinson, G. S., L. P. Burney, and D. A. Burney. 2005. Landscape paleoecology and megafaunal extinction in southeastern New York. *Ecological Monographs* 75:295–315.

Rodríguez-de la Rosa, R. A., J. R. Guzmán-Gutiérrez, and C. Ortega-Hurtado de Mendoza. 2011. A new occurrence of toxodonts in the Pleistocene of México. *Current Research in the Pleistocene* 28:29–30.

Römpler, H., N. Rohland, C. Lalueza-Fox, et al. (9 authors). 2006. Nuclear gene indicates coat-color polymorphism in mammoths. *Science* 313:62.

Roosevelt, Q., and J. W. Burden. 1934. *A New Species of Antilocaprine,* Tetrameryx onusrosagris, *from a Pleistocene Cave Deposit in Southern Arizona.* American Museum Novitates 754.

Ross, R. C. 1935. A new genus and species of pigmy goose from the McKittrick Pleistocene. *Transactions of the San Diego Society of Natural History* 8:107–14.

Roth, V. L. 1996. Pleistocene dwarf elephants of the California islands. In *The Proboscidea: Evolution and Palaeoecology of Elephants and Their Relatives,* edited by J. Shoshani and P. Tassy, 249–53. Oxford University Press, Oxford, UK.

Rothschild, B. M. 2003. Pathology in Hiscock Site vertebrates, and its bearing on hyperdisease among North American mastodons. In *The Hiscock Site: Late Pleistocene and Holocene Paleoecology and Archaeology of Western New York State,* edited by R. S. Laub, 171–75. Bulletin of the Buffalo Society of Natural Sciences 37.

Rothschild, B. M., and R. Laub. 2006. Hyperdisease in the late Pleistocene: Validation of an early 20th century hypothesis. *Naturwissenschaften* 93:557–64.

Rothschild, B. M., and L. D. Martin. 2003. Frequency of pathology in a large natural sample from Natural Trap Cave with special remarks on erosive disease in the late Pleistocene. *Reumatismo* 55:58–65.

———. 2006. Did ice-age bovids spread tuberculosis? *Naturwissenschaften* 93:565–69.

Rowland, S. M., and D. Tanke. 2007. The career of James E. Thurston and the extinction of the professional field collector in North American vertebrate paleontology. *Geological Society of America Abstracts with Programs* 39 (6): 382.

Russell, A. P., and H. N. Bryant. 2001. Claw retraction and protraction in the Carnivora: The cheetah (*Acinonyx jubatus*) as an atypical felid. *Journal of Zoology* 254:67–76.

Russell, B. D., and A. H. Harris. 1986. A new leporine (Lagomorpha: Leporidae) from Wisconsinan deposits of the Chihuahuan Desert. *Journal of Mammalogy* 67:632–39.

Ryser, F. A., Jr. 1985. *Birds of the Great Basin.* University of Nevada Press, Reno.

Saint-André, P.-A., F. Pujos, C. Cartelle, G. De Iuliis, T. J. Gaudin, H. G. McDonald, and B. M. Quispe. 2010. Nouveaux paresseux terrestres (Mammalia, Xenarthra, Mylodontidae) du Néogène de l'Altiplano bolivien. *Geodiversitas* 32 (2): 255–306.

Salinas, M. L., T. Ogura, and L. Soffchi. 2001. Irritant contact dermatitis caused by needle-like calcium oxalate crystals, raphides, in *Agave tequilana* among workers in tequila distilleries and agave plantations. *Contact Dermatitis* 44:94–96.

Salminen, J-P., and M. Karonen. 2011. Chemical ecology of tannins and other phenolics: We need a change in approach. *Functional Ecology* 25:325–38.

San Bernardino County Museum. 2012. Sabre-toothed cat fossils discovered in Las Vegas. http://www.blm.gov/wo/st/en/prog/more/CRM/paleontology/paleo _news/NV_Sabre-toothed.html.

Sanchez, G., V. T. Holliday, E. T. Gaines, et al. (10 authors). 2014. Human (Clovis)/ gomphothere (*Cuvieronius* sp.) association ~13,390 cal years BP in Sonora, Mexico. *Proceedings of the National Academy of Sciences* 111:10972–77.

Sanders, A. E. 2002. *Additions to the Pleistocene Mammal Faunas of South Carolina, North Carolina, and Georgia.* American Philosophical Society Transactions 92 (5).

Sanders, A. E., R. E. Weems, and L. B. Albright III. 2009. Formalization of the middle Pleistocene "Ten Mile Hill Beds" in South Carolina with evidence for placement of the Irvingtonian-Rancholabrean boundary. In *Papers on Geology, Vertebrate Paleontology, and Biostratigraphy in Honor of Michael O. Woodburne,* edited by L. B. Albright III, 363–69. Museum of Northern Arizona Bulletin 65.

Saunders, J. J. 1977. *Late Pleistocene Vertebrates of the Western Ozark Highland, Missouri.* Illinois State Museum Reports of Investigations 33.

———. 1996. North American Mammutidae. In *The Proboscidea: Evolution and Paleoecology of Elephants and Their Relatives,* edited by J. Shosani and P. Tassy, 271–79. Oxford University Press, Oxford, UK.

Saunders, J. J., E. C. Grimm, C. C. Widga, et al. (10 authors). 2010. Paradigms and proboscideans in the southern Great Lakes region, USA. *Quaternary International* 217:175–87.

Saysette, J. E. 1999. Postcranial estimators of body mass in pecorans with emphasis on *Capromeryx* (Mammalia: Artiodactyla). PhD diss., Colorado State University, Fort Collins.

Schaedler, J. M., L. Krook, J. A. Wootton, et al. (10 authors). 1992. Studies of collagen in bone and dentin matrix of a Columbian mammoth (Late Pleistocene) of central Utah. *Matrix* 12:297–307.

Scherer, C. S., J. Ferigolo, A. M. Ribeiro, and C. Cartelle Guerra. 2007. Contribution to the knowledge of *Hemiauchenia paradoxa* (Artiodactyla, Camelidae) from the Pleistocene of southern Brazil. *Revista Brasileira de Paleontologia* 10:35–52.

Scherer, C. S., V. G. Pitana, and A. M. Ribeiro. 2009. Protacheriidae and Macraucheniidae (Litopterna, Mammalia) from the Pleistocene of Rio Grande do Sul State, Brazil. *Revista Brasileira de Paleontologia* 12:231–46.

Scherer, J. A. B. 1930. A ground-sloth lair. *Masterkey* 3 (8): 5–11.

Schmidt, G. D., D. W. Duszynski, and P. S. Martin. 1992. Parasites of the extinct Shasta ground sloth, *Nothrotheriops shastensis*, in Rampart Cave, Arizona. *Journal of Parasitology* 78:811–16.

Schmidt, J. O. 1989. Spines and venoms in flora and fauna in the Arizona Upland Sonoran Desert and how they act as defenses against vertebrates. In *Special Biotic Relationships in the Arid Southwest,* edited by J. S. Schmidt, 107–29. University of New Mexico Press, Albuquerque.

Schubert, B. W. 2010. Late Quaternary chronology and extinction of North American giant short-faced bears (*Arctodus simus*). *Quaternary International* 217:188–94.

Schubert, B. W., and R. W. Graham. 2000. Terminal Pleistocene armadillo (*Dasypus*) remains from the Ozark Plateau, Missouri, USA. *Paleobios* 20:1–6.

Schubert, B. W., R. W. Graham, H. G. McDonald, E. C. Grimm, and T. W. Stafford Jr. 2004. Latest Pleistocene paleoecology of Jefferson's ground sloth (*Megalonyx jeffersonii*) and elk-moose (*Cervalces scotti*) in northern Illinois. *Quaternary Research* 61:231–40.

Schubert, B. W., R. C. Hulbert Jr., B. J. MacFadden, M. Searle, and S. Searle. 2010. Giant short-faced bears (*Arctodus simus*) in Pleistocene Florida, USA, a substantial range extension. *Journal of Paleontology* 84:79–87.

Schubert, B. W., and J. E. Kaufmann. 2003. A partial short-faced bear skeleton from an Ozark Cave with comments on the paleobiology of the species. *Journal of Cave and Karst Studies* 65:101–10.

Schubert, B. W., and S. C. Wallace. 2009. Late Pleistocene giant short-faced bears, mammoths, and large carcass scavenging in the Saltville Valley of Virginia, USA. *Boreas* 38:482–92.

Schulte, P., L. Alegret, I. Arenillas, et al. (41 authors). 2010. The Chicxulub asteroid impact and mass extinction at the Cretaceous-Paleogene boundary. *Science* 327:1214–18.

Schultz, C. B., and E. B. Howard. 1936. The fauna of Burnet Cave, Guadalupe Mountains, New Mexico. *Proceedings of the Academy of Natural Sciences of Philadelphia* 87:273–98.

Scott, E. 2010. Extinctions, scenarios, and assumptions: Changes in latest Pleistocene large herbivore abundance and distribution in western North America. *Quaternary International* 217:225–39.

Scott, E., T. W. Stafford Jr., R. W. Graham, and L. D. Martin. 2010. Morphology and metrics, isotopes and dates: Determining the validity of *Equus laurentius* Hay, 1913. *Journal of Vertebrate Paleontology* 30:1840–47.

Scott, W. B. 1913. *A History of Land Mammals in the Western Hemisphere*. MacMillan, New York.

Sedor, F. A., P. A. Born, and F. M. S. dos Santos. 2004. Fósseis pleistocênicos de *Scelidodon* (Mylodontidae) e *Tapirus* (Tapiridae) em cavernas paranaenses (PR, Sul do Brasil). *Acta Biológica Paranaense* 33:121–28.

Semken, H. A., Jr., R. W. Graham, and T. W. Stafford Jr. 2010. AMS ^{14}C analysis of late Pleistocene non-analog faunal components from 21 cave deposits in southeastern North America. *Quaternary International* 217:240–55.

Semprebon, G. M., and F. Rivals. 2010. Trends in the paleodietary habits of fossil camels from the Tertiary and Quaternary of North America. *Palaeogeography, Palaeoclimatology, Palaeoecology* 295:131–45.

Sen, S. 2010. Camelids do not occur in the late Miocene mammal locality of Çobanpinar, Turkey. *Russian Journal of Theriology* 9 (2): 87–91.

Shafer, S. L., P. J. Bartlein, and R. S. Thompson. 2001. Potential changes in the distributions of western North America tree and shrub taxa under future climate scenarios. *Ecosystems* 4:200–215.

Shapiro, B., R. W. Graham, and B. Letts. 2014. A revised evolutionary history of armadillos (*Dasypus*) in North America based on ancient mitochondrial DNA. *Boreas* 44:14–23.

Shipman, P., D. C. Fisher, and J. J. Rose. 1984. Mastodon butchery: Microscopic evidence of carcass processing and bone tool use. *Paleobiology* 10:358–65.

Shoshani, J. 1996. Para- or monophyly of the gomphotheres and their position within Proboscidea. In *The Proboscidea: Evolution and Palaeoecology of Elephants and Their*

Relatives, edited by J. Shoshani and P. Tassy, 149–77. Oxford University Press, Oxford, UK.

———. 2003. *Mammut* hyoid elements from the Hiscock Site: Description and implications. In *The Hiscock Site: Late Pleistocene and Holocene Paleoecology and Archaeology of Western New York State*, edited by R. S. Laub, 114–20. Bulletin of the Buffalo Society of Natural Sciences 37.

Shoshani, J., and G. H. Marchant. 2001. Hyoid apparatus: A little known complex of bones and its "contribution" to proboscidean evolution. In *The World of Elephants*, edited by G. Cavarretta, P. Gioia, M. Mussi, and M. R. Palombo, 668–75. Consiglio Nazionale delle Ricerche, Rome.

Shoshani, J., and P. Tassy. 2005. Advances in proboscidean taxonomy and classification, anatomy and physiology, and ecology and behavior. *Quaternary International* 126–28:5–20.

Shreve, F., and I. L. Wiggins. 1964. *Vegetation and Flora of the Sonoran Desert*. Stanford University Press, Stanford, CA.

Shufeldt, R. W. 1891a. Fossil birds from the Equus Beds of Oregon. *American Naturalist* 25:818–21.

———. 1891b. On a collection of fossil birds from the Equus Beds of Oregon. *American Naturalist* 25:359–62.

———. 1891c. Tertiary fossils of North American birds. *Auk* 8:365–68.

———. 1892. A study of the fossil avifauna of the Equus Beds of the Oregon desert. *Journal of the Academy of Natural Sciences of Philadelphia* 9:389–425.

———. 1907. *The Negro: A Menace to American Civilization*. Gorham Press, Boston.

———. 1912. Prehistoric birds of Oregon. *Overland Monthly* 60 (second series): 536–42.

———. 1913a. New and extinct birds and other species from the Pleistocene of Oregon. *Science* 37:306–7.

———. 1913b. Review of the fossil fauna of the desert region of Oregon, with a description of additional material collected there. *Bulletin of the American Museum of Natural History* 32 (6): 123–78.

———. 1913c. Studies of the fossil birds of the Oregon desert. *Auk* 30:36–39.

Simmons, A. 1999. *Faunal Extinctions in an Island Society*. Kluwer Academic/Plenum Publishers, New York.

Simpson, D. A. 1965. Mark Harrington: Father of Nevada archaeology. *Nevada Historical Quarterly* 8 (3–4): 3–23.

Simpson, G. G. 1941. *Large Pleistocene Felines of North America*. American Museum Novitates 1136.

———. 1980. *Splendid Isolation*. Yale University Press, New Haven, CT.

Sinclair, W. J. 1904. The exploration of the Potter Creek Cave. *University of California Publications in American Archaeology and Ethnology* 2 (1): 1–27.

———. 1905. New mammalia from the Quaternary caves of California. *University of California Publications, Bulletin of the Department of Geology* 4:145–61.

Sinclair, W. J., and E. L. Furlong. 1904. *Euceratherium*, a new ungulate from the Quaternary caves of California. *University of California Publications, Bulletin of the Department of Geology* 3:411–18.

Sistiaga, A., F. Berna, R. Laursen, and P. Goldberg. 2014. Steroidal biomarker analysis of a 14,000 years old putative human coprolite from Paisley Cave, Oregon. *Journal of Archaeological Science* 41:813–17.

Skinner, M. F. 1942. The fauna of Papago Springs Cave, Arizona, and a study of *Stockoceros*, with three new antilocaprines from Nebraska and Arizona. *Bulletin of the American Museum of Natural History* 80:143–220.

Skrupa, S. 2014. Footprints in stone: Pleistocene age footprints preserved at old Nevada State Prison. *Reno (NV) Gazette-Journal*, February 26. http://www.rgj.com /article/20140226/LIV/302260001/Footprints-stone-Pleistocene-Age-footprints -preserved-old-Nevada-State-Prison.

Smith, C. I., C. S. Drummond, W. Godsoe, J. B. Yoder, and O. Pellmyr. 2009. Host specificity and reproductive success of yucca moths (*Tegeticula* spp. Lepidoptera: Prodoxidae) mirror patterns of gene flow between host plant varieties of the Joshua tree (*Yucca brevifolia*: Agavaceae). *Molecular Ecology* 18:5218–29.

Smith, C. I., W. K. Godsoe, S. Tank, J. B. Yoder, and O. Pellmyr. 2008. Distinguishing coevolution from covicariance in an obligate pollination mutualism: Asynchronous divergence in Joshua tree and its pollinators. *Evolution* 62:2676–87.

Smith, C. I., S. Tank, W. Godsoe, J. Levenick, E. Strand, T. Esque, and O. Pellmyr. 2011. Comparative phylogeography of a coevolved community: Concerted population expansions in Joshua trees and four yucca moths. *PLoS One* 6:e25628.

Smith, F. A., S. K. Lyons, S. K. M. Ernest, et al. (9 authors). 2003. Body mass of late Quaternary mammals. *Ecological Archives* E084-094.

Smith, S. D., R. K. Monson, and J. E. Anderson. 1997. *Physiological Ecology of North American Desert Plants*. Springer-Verlag, Berlin.

Smith Woodward, A. 1900. On some remains of *Grypotherium* (*Neomylodon*) *listai* and associated mammals from a cavern near Consuelo Cove, Last Hope Inlet, Patagonia. *Proceedings of the Zoological Society of London* 1900:64–79.

Soibelzon, L. H., and B. W. Schubert. 2011. The largest known bear, *Arctotherium angustidens*, from the early Pleistocene Pampean region of Argentina: With a discussion of size and diet trends in bears. *Journal of Paleontology* 85:69–75.

Soibelzon, L. H., and V. B. Tarantini. 2009. Estimación de la masa corporal de las especies de osos fósiles y actuales (Ursidae, Tremarctinae) de América del Sur. *Revista del Museo Argentino de Ciencias Naturales* 11:243–54.

Sokolov, V. E. 1974. *Saiga tatarica*. *Mammalian Species* 38:1–4.

Sorkin, B. 2006. Ecomorphology of the giant short-faced bears *Agriotherium* and *Arctodus*. *Historical Biology* 18:1–20.

Southon, J. R., and R. E. Taylor. 2002. Brief comments on "Terrestrial evidence of a nuclear catastrophe in Paleoindian times" by Richard B. Firestone and William Topping. *Mammoth Trumpet* 17 (2): 14–17.

Spaulding, W. G. 1985. *Vegetation and Climates of the Last 45,000 Years in the Vicinity of the Nevada Test Site, South-Central Nevada*. US Geological Survey Professional Paper 1329.

——. 1994. *Paleohydrologic Investigations in the Vicinity of Yucca Mountain: Late Quaternary Paleobotanical and Palynological Records*. NWPO-TR-022-94. Dames and Moore, Las Vegas, NV.

Spaulding, W. G., and P. S. Martin. 1979. Ground sloth dung of the Guadalupe Mountains. In *Biological Investigations in the Guadalupe Mountains National Park, Texas*, edited by H. H. Genoways and R. J. Baker, 259–69. National Park Service Proceedings and Transactions Series 4.

Springer, K., J. C. Sagebiel, E. Scott, C. Manker, and C. Austin. 2005. Additions to the late Pleistocene vertebrate paleontology of the Las Vegas Formation, Clark County, Nevada. http://www.sbcounty.gov/museum/discover/divisions/geo/pdf/SVP_2005_LasVegas.pdf.

Springer, K., E. Scott, J. C. Sagebiel, and C. R. Manker. 2010. The Tule Springs Local Fauna: Late Pleistocene vertebrates from the Upper Las Vegas Wash, Clark County, Nevada. *Geological Society of America Abstracts with Programs* 42 (5): 250.

Spurr, J. E. 1903. *Descriptive Geology of Nevada South of the Fortieth Parallel and Adjacent Portions of California*. US Geological Survey Bulletin 208.

Stafford, T. W., Jr., A. J. T. Jull, K. Brendel, R. C. Duhamel, and D. Donahue. 1987. Study of bone radiocarbon dating accuracy at the University of Arizona NSF Accelerator Facility for Radiocarbon Analysis. *Radiocarbon* 29:24–44.

Stafford, T. W., Jr., H. A. Semken Jr., R. W. Graham, W. F. Klippel, A. Markova, N. G. Smirnov, and J. Southon. 1999. First accelerator mass spectrometry ^{14}C dates documenting contemporaneity of nonanalog species in late Pleistocene mammal communities. *Geology* 27:903–6.

Stanford, D. J. 1991. Clovis origins and adaptations: An introductory perspective. In *Clovis: Origins and Adaptations*, edited by R. Bonnichsen and K. L. Turnmire, 1–13. Center for the Study of the First Americans, Corvallis, OR.

———. 1999. Paleoindian archaeology and late Pleistocene environments in the plains and southwestern United States. In *Ice Age Peoples of North America: Environments, Origins, and Adaptations*, edited by R. Bonnichsen and K. L. Turnmire, 281–339. Oregon State University Press, Corvallis.

Stanford, D. J., and B. A. Bradley. 2012. *Across Atlantic Ice: The Origin of America's Clovis Culture*. University of California Press, Berkeley.

Starr, T. N., K. E. Gadek, J. B. Yoder, R. Flatz, and C. I. Smith. 2013. Asymmetric hybridization and gene flow between Joshua trees (Agavaceae: *Yucca*) reflect differences in pollinator host specificity. *Molecular Ecology* 22:427–49.

Steadman, D. W. 1989. Extinction of birds in eastern Polynesia: A review of the record, and comparisons with other Pacific Island groups. *Journal of Archaeological Science* 16:177–205.

Steadman, D. W., J. Arroyo-Cabrales, E. Johnson, and A. Fabiola Guzman. 1994. New information on the late Pleistocene birds from San Josecito Cave, Nuevo León, Mexico. *Condor* 96:577–89.

Steadman, D. W., and Ó. Carranza-Castañeda. 2006. Early Pliocene to Early Pleistocene birds from central Mexico. In *Advances in late Tertiary Vertebrate Paleontology in Mexico and the Great American Biotic Interchange*, edited by Ó. Carranza-Castañeda and E. H. Lindsay, 61–71. Universidad Nacional Autónoma de México, Instituto de Geología and Centro de Geociencias, Publicación Especial 4.

Steadman, D. W., and J. I. Mead. 2010. A late Pleistocene bird community at the northern edge of the tropics in Sonora, Mexico. *American Midland Naturalist* 163:423–41.

Steadman, D. W., and N. G. Miller. 1987. California Condor associated with spruce–jack pine woodland in the late Pleistocene of New York. *Quaternary Research* 28:415–26.

Steffensen, J. P., K. K. Andersen, M. Bigler, et al. (20 authors). 2008. High-resolution Greenland ice core data show abrupt climate change happens in few years. *Science* 321:680–84.

Stein, B. R. 1997. Annie M. Alexander: Extraordinary patron. *Journal of the History of Biology* 30:243–66.

Stephens, L. D. 1982. *Joseph LeConte: Gentle Prophet of Evolution*. Louisiana State University Press, Baton Rouge.

Sternberg, C. H. 1881. The Pliocene beds of southern Oregon. *Kansas City Review of Science and Industry* 4:600–601.

———. 1884. The fossil field of southern Oregon. *Kansas City Review of Science and Industry* 7:596–99.

———. 1898. Pliocene man. *Popular Science News* 32:82.

———. 1903. Experiences with early man in America. *Transactions of the Kansas Academy of Science* 18:89–93.

———. 1909. *The Life of a Fossil Hunter*. Henry Holt, New York.

Stock, C. 1917. Structure of the pes in *Mylodon harlani*. *University of California Publications, Bulletin of the Department of Geology* 10 (16): 267–86.

———. 1920. Origin of the supposed human footprints of Carson City, Nevada. *Science* 51:514.

———. 1925. *Cenozoic Gravigrade Edentates of Western North America*. Carnegie Institution of Washington Publication 331.

———. 1930a. Problems of antiquity presented in Gypsum Cave, Nevada. *Science* 72:405.

———. 1930b. *Quaternary Antelope Remains from a Second Cave Deposit in the Organ Mountains, New Mexico*. Los Angeles Museum Publication 2.

———. 1931. Problems of antiquity presented in Gypsum Cave, Nevada. *Scientific Monthly* 32:22–32.

———. 1932. *A Further Study of the Quaternary Antelopes of Shelter Cave, New Mexico*. Los Angeles Museum Publication 3.

———. 1933. Researches by Chester Stock. *Carnegie Institution of Washington Year Book* 32:329–30.

———. 1936a. A new mountain goat from the Quaternary of Smith Creek Cave, Nevada. *Bulletin of the Southern California Academy of Sciences* 35:149–53.

———. 1936b. Sloth tracks in the Carson Prison. *Westways* 28 (7): 26–27.

———. 1956. *Rancho La Brea: A Record of Pleistocene Life in California*. Los Angeles County Museum of Natural History Science Series 20, Paleontology 11.

Stock, C., and E. L. Furlong. 1927. Skull and skeletal remains of a ruminant of the *Preptoceras-Euceratherium* group from the McKittrick Pleistocene, California. *University of California Publications, Bulletin of the Department of Geology* 16:409–34.

Storer, R. W. 1989. The Pleistocene Western Grebe *Aechmophorus* (Aves, Podicipedidae) from Fossil Lake, Oregon: A comparison with recent material. *University of Michigan Contributions from the Museum of Paleontology* 27 (12): 321–26.

Storer, T. I. 1972. Loye Holmes Miller: Un hombre muy simpatico. *Condor* 74:231–36.

Stout, B. 1986. Discovery and C14 dating of the Black Rock Desert Mammoth. *Nevada Archaeologist* 5 (2): 21–23.

Straus, L. G., D. J. Meltzer, and T. Goebel. 2005. Ice Age Atlantis? Exploring the Solutrean-Clovis "connection." *World Archaeology* 37:507–32.

Stuart, A. J., P. A. Kosintsev, T. F. G. Higham, and A. M. Lister. 2004. Pleistocene to Holocene extinction dynamics in giant deer and woolly mammoth. *Nature* 431:684–89.

Stuart, A. J., and A. M. Lister. 2007. Patterns of Late Quaternary megafaunal extinctions in Europe and northern Asia. *Courier Forschungsinstitut Senckenberg* 259:287–97.

———. 2011. Extinction chronology of the cave lion *Panthera spelaea*. *Quaternary Science Reviews* 30:2329–40.

———. 2012. Extinction chronology of the woolly rhinoceros *Coelodonta antiquitatis* in the context of late Quaternary megafaunal extinctions in northern Eurasia. *Quaternary Science Reviews* 51:1–17.

———. 2014. New radiocarbon evidence on the extirpation of the spotted hyaena (*Crocuta crocuta* (Erxl.)) in northern Eurasia. *Quaternary Science Reviews* 96:108–16.

Taulman, J. F., and L. W. Robbins. 1996. Recent range expansion and distributional limits of the nine-banded armadillo (*Dasypus novemcinctus*) in the United States. *Journal of Biogeography* 23:635–48.

Teale, C. L., and N. G. Miller. 2012. Mastodon herbivory in mid-latitude late-Pleistocene boreal forests of eastern North America. *Quaternary Research* 78:72–81.

Tennyson, A., and P. Martinson. 2006. *Extinct Birds of New Zealand*. Te Papa Press, Wellington, New Zealand.

Thalmann, O., B. Shapiro, P. Cui, et al. (31 authors). 2013. Complete mitochondrial genomes of ancient canids suggest a European origin of domestic dogs. *Science* 342:871–74.

Thomas, D. H., ed. 1985. *The Archaeology of Hidden Cave, Nevada*. Anthropological Papers of the American Museum of Natural History 61 (1).

Thompson, R. S. 1984. Late Pleistocene and Holocene environments in the Great Basin. PhD diss., University of Arizona, Tucson.

———. 1985. The age and environment of the Mount Moriah (Lake Mohave) occupation at Smith Creek Cave, Nevada. In *Environments and Extinctions: Man in Late Glacial North America*, edited by J. I. Mead and D. J. Meltzer, 111–19. Center for the Study of Early Man, Orono, ME.

———. 1992. Late Quaternary environments in Ruby Valley, Nevada. *Quaternary Research* 37:1–15.

Thompson, R. S., and K. H. Anderson. 2000. Biomes of western North America at 18,000, 6000 and 0 ^{14}C yr BP reconstructed from pollen and packrat midden data. *Journal of Biogeography* 27:555–84.

Thompson, R. S., T. R. Van Devender, P. S. Martin, T. Foppe, and A. Long. 1980. Shasta ground sloth (*Nothrotheriops shastense*) at Shelter Cave, New Mexico: Environment, diet, and extinction. *Quaternary Research* 14:360–76.

Thomson, K. 2008. *A Passion for Nature: Thomas Jefferson and Natural History*. Thomas Jefferson Foundation, Charlottesville, NC.

———. 2011. The "great-claw" and the science of Thomas Jefferson. *Proceedings of the American Philosophical Society* 155:394–403.

Tisdale, E. W. 1961. Ecologic changes in the Palouse. *Northwest Science* 35:134–38.

Titov, V. V., and V. N. Logvynenko. 2006. Early *Paracamelus* (Mammalia, Tylopoda) in eastern Europe. *Acta Zoologica Cracoviensia* 19A:163–78.

Tollrian, R., and C. D. Harvell. 1999. The evolution of inducible defenses: Current ideas. In *The Ecology and Evolution of Inducible Defenses*, edited by R. Tollrian and C. D. Harvell, 306–21. Princeton University Press, Princeton, NJ.

Tsong, N. 2010. Mountain goat kills man in Olympic National Park. *Seattle Times*, October 17.

Tuohy, D. R. 1979. Kachina Cave. In *The Archaeology of Smith Creek Canyon, Eastern Nevada*, edited by D. R. Tuohy and D. L. Rendall, 1–88. Nevada State Museum Anthropological Papers 17.

Turner, A., and M. Antón. 1997. *The Big Cats and Their Fossil Relatives*. Columbia University Press, New York.

Turner, R. M., J. E. Bowers, and T. L. Burgess. 1995. *Sonoran Desert Plants: An Ecological Atlas*. University of Arizona Press, Tucson.

Twain, M. 1885. Carson footprints, Mark Twain clears up the mystery. *Sacramento Daily Record-Union* 53 (27), March 25.

———. (1872) 1981. *Roughing It*. Penguin Books, New York.

Tylor, E. B. 1885. American aspects of anthropology. *Popular Science Monthly* 26:152–68.

Ukraintseva, V. V. 2013. *Mammoths and the Environment*. Cambridge University Press, Cambridge.

University of Nevada, Las Vegas. 2012. Researchers find first evidence of ice age wolves in Nevada. http://phys.org/news/2012-12-evidence-ice-age-wolves-nevada.html.

USDA NRCS (Natural Resources Conservation Service). 2014. The PLANTS Database. National Plant Data Team, Greensboro, NC. http://plants.usda.gov.

US Geological Survey. 2013. Digital representations of tree species range maps from "Atlas of United States Trees" by Elbert L. Little Jr. http://gec.cr.usgs.gov/data/little/.

Utah Division of Wildlife Resources. 2013a. Licenses and permits. http://wildlife.utah.gov/license-permit.html.

———. 2013b. *Utah Mountain Goat Statewide Management Plan*. http://wildlife.utah.gov/hunting/biggame/pdf/mtn_goat_plan.pdf.

Vander Wall, S. B., T. Esque, D. Haines, M. Garnett, and B. A. Waitman. 2006. Joshua tree (*Yucca brevifolia*) seeds are dispersed by seed-caching rodents. *Ecoscience* 13:539–43.

Vander Wall, S. B., K. M. Kuhn, and M. J. Beck. 2005. Seed removal, seed predation, and secondary dispersal. *Ecology* 86:801–6.

Van Devender, T. R. 1990. Late Quaternary vegetation and climate of the Sonoran Desert, United States and Mexico. In *Packrat Middens: The Last 40,000 Years of Biotic Change*, edited by J. L. Betancourt, T. R. Van Devender, and P. S. Martin, 134–65. University of Arizona Press, Tucson.

Van Geel, B., A. Aptroot, C. Baittinger, et al. (16 authors). 2008. The ecological implications of a Yakutian mammoth's last meal. *Quaternary Research* 69:361–76.

van Hoesel, A., W. Z. Hoek, G. M. Pennock, and M. R. Drury. 2014. The Younger Dryas impact hypothesis: A critical review. *Quaternary Science Reviews* 83:95–114.

van Roosmalen, M. G. M., L. Frenz, P. van Hooft, H. H. de Iongh, and H. Leirs. 2007. A new species of living peccary (Mammalia: Tayassuidae) from the Brazilian Amazon. *Bonner Zoologische Beiträge* 55 (2): 105–12.

Van Valkenburgh, B. 1988. Incidence of tooth breakage among large, predatory mammals. *American Naturalist* 131:291–302.

———. 2009. Costs of carnivory: Tooth fracture in Pleistocene and Recent carnivores. *Biological Journal of the Linnean Society* 96:68–81.

Van Valkenburgh, B., F. Grady, and B. Kurtén. 1990. The Plio-Pleistocene cheetah-like cat *Miracinonyx inexpectatus* of North America. *Journal of Vertebrate Paleontology* 10:434–54.

Van Valkenburgh, B., and F. Hertel. 1998. The decline of North American predators during the late Pleistocene. In *Quaternary Paleozoology in the Northern Hemisphere*, edited by J. J. Saunders, B. W. Styles, and G. F. Baryshnikov, 357–74. Illinois State Museum Scientific Papers 27.

Van Valkenburgh, B., T. Maddox, P. J. Funston, M. G. L. Mills, G. F. Grether, and C. Carbone. 2009. Sociality in Rancho La Brea *Smilodon*: Arguments favour "evidence" over "coincidence." *Biology Letters* 5:563–64.

Van Valkenburgh, B., and T. Sacco. 2002. Sexual dimorphism, social behavior, and intrasexual competition in large Pleistocene carnivorans. *Journal of Vertebrate Paleontology* 22:164–69.

Van Valkenburgh, B., M. F. Teaford, and A. Walker. 1990. Molar microwear and diet in large carnivores: Inferences concerning diet in the sabretooth cat, *Smilodon fatalis*. *Journal of Zoology, London* 222:319–40.

Vartanyan, S. L., K. A. Arslanov, J. A. Karhu, G. Possnert, and L. D. Sulerzhitsky. 2008. Collection of radiocarbon dates on the mammoths (*Mammuthus primigenius*) and other genera of Wrangel Island, northeast Siberia, Russia. *Quaternary Research* 70:51–59.

Veltre, D. W., D. R. Yesner, K. J. Crossen, R. W. Graham, and J. B. Coltrain. 2008. Patterns of faunal extinction and paleoclimatic change from mid-Holocene mammoth and polar bear remains, Pribilof Islands, Alaska. *Quaternary Research* 70:40–50.

Vereshchagin, N. K., and G. F. Baryshnikov. 1982. Paleoecology of the mammoth fauna in the Eurasian Arctic. In *Paleoecology of Beringia*, edited by D. M. Hopkins, J. V. Matthews Jr., C. E. Schweger, and S. B. Young, 267–79. Academic Press, New York.

Vetter, L., M. S. Lachniet, and S. M. Rowland. 2007. Paleoecology of Pleistocene megafauna in southern Nevada: Isotopic evidence for browsing on halophytic plants. *Geological Society of America Abstracts with Programs* 39 (6): 402.

Vizcaíno, S. F., R. E. Blanco, J. B. Bender, and N. Milne. 2011. Proportions and function of the limbs of glyptodonts. *Lethaia* 44:93–101.

Vizcaíno, S. F., R. A. Fariña, M. A. Zárate, M. S. Bargo, and P. Schultz. 2004. Paleoecological implications of the mid-Pliocene faunal turnover in the Pampean region (Argentina). *Palaeogeography, Palaeoclimatology, Palaeoecology* 213:101–13.

Vizcaíno, S. F., M. Zaráte, M. S. Bargo, and A. Dondas. 2001. Pleistocene burrows in the Mar del Plata area (Argentina) and their probable builders. *Acta Palaeontologica Polonica* 46:289–301.

Wagner, W. L., D. R. Herbst, and S. H. Sohmer. 1990. *Manual of the Flowering Plants of Hawai'i*. Bishop Museum Special Publication 83.

Waitman, B. A., S. B. Vander Wall, and T. C. Esque. 2012. Seed dispersal and seed fate in Joshua tree (*Yucca brevifolia*). *Journal of Arid Environments* 81:1–8.

Wallace, A. R. 1876. *The Geographical Distribution of Animals, with a Study of the Relations of Living and Extinct Faunas as Elucidating Past Changes of the Earth's Surface*. Vol. 1. Harper Brothers, New York.

Wallen, D. R., and J. A. Ludwig. 1978. Energy dynamics of vegetative and reproductive growth in Spanish bayonet (*Yucca baccata* Torr.). *Southwestern Naturalist* 23:409–22.

Ward, D., M. Spiegel, and D. Saltz. 1997. Gazelle herbivory and interpopulation differences in calcium oxalate content of leaves of a desert lily. *Journal of Chemical Ecology* 23:333–46.

Ward, P. D. 2000. *Rivers in Time: The Search for Clues to Earth's Mass Extinctions*. Columbia University Press, NY.

Warren, C. N., and C. Phagan. 1988. Fluted points in the Mojave Desert: Their technology and cultural context. In *Early Human Occupation in Far Western North America: The Clovis-Archaic Interface*, edited by J. A. Willig, C. M. Aikens, and J. L. Fagan, 121–30. Nevada State Museum Anthropological Papers 21.

Waters, M. R., and T. W. Stafford Jr. 2007. Redefining the age of Clovis: Implications for the peopling of the Americas. *Science* 315:1122–26.

Waters, M. R., T. W. Stafford Jr., H. G. McDonald, et al. (11 authors). 2011. Pre-Clovis mastodon hunting 13,800 years ago at the Manis site, Washington. *Science* 334:351–53.

Webb, S. D. 1965. *The Osteology of* Camelops. Bulletin of the Los Angeles County Museum, Science Bulletin 1.

———. 1973. Pliocene pronghorns of Florida. *Journal of Mammalogy* 54:203–21.

———. 1974. Pleistocene llamas of Florida, with a brief review of the Lamini. In *Pleistocene Mammals of Florida*, edited by S. D. Webb, 170–213. University Press of Florida, Gainesville.

———. 1992. A cranium of *Navahoceros* and its phylogenetic place among New World Cervidae. *Annales Zoologici Fennici* 28:401–10.

———. 2000. Evolutionary history of New World Cervidae. In *Antelopes, Deer, and Relatives: Fossil Record, Behavioral Ecology, Systematics, and Classification*, edited by E. S. Vrba and G. B. Schaller, 38–64. Yale University Press, New Haven, CT.

———. 2006. The Great American Biotic Interchange: Patterns and processes. *Annals of the Missouri Botanical Garden* 93:245–57.

Webb, S. D., and F. G. Stehli. 1995. Selenodont Artiodactyla (Camelidae and Cervidae) from the Leisey Shell Pits, Hillsborough County, Florida. *Bulletin of the Florida Museum of Natural History* 37 (2): 621–43.

Weinstock, J. E., E. Willerslev, A. Sher, et al. (15 authors). 2005. Evolution, systematics, and phylogeography of Pleistocene horses in the New World: A molecular perspective. *PLoS Biology* 3 (8): 1373–79.

Weiss, J. L., and J. T. Overpeck. 2005. Is the Sonoran Desert losing its cool? *Global Change Biology* 11:2065–77.

Welker, F., M. J. Collins, J. A. Thomas, et al. (31 authors). 2015. Ancient proteins resolve the evolutionary history of Darwin's South American ungulates. *Nature* 522:81–84.

Wells, P. V. 1983. Paleobiogeography of montane islands in the Great Basin since the last glaciopluvial. *Ecological Monographs* 53:341–82.

Welsh, S. L. 2003. *Atriplex*. In *Flora of North America North of Mexico*. Vol. 4, *Magnoliophyta: Caryophyllidae, part 1*, edited by Flora of North America Editorial Committee. http://www.efloras.org/florataxon.aspx?flora_id=1&taxon_id=103110.

Welsh, S. L., N. D. Atwood, S. Goodrich, and L. C. Higgins. 2008. *A Utah Flora*. 4th ed. Monte L. Bean Life Science Museum, Brigham Young University, Provo, UT.

West, N. E., R. J. Tausch, and P. T. Tueller. 1998. *A Management-Oriented Classification of Pinyon-Juniper Woodlands of the Great Basin*. USDA Forest Service Rocky Mountain Research Station General Technical Report RMRS-GTR-12.

Wetzel, R. M., R. E. Dubos, R. L. Martin, and P. Myers. 1975. *Catagonus*, an "extinct" peccary, alive in Paraguay. *Science* 189:379–81.

Wheeler, H. T. 2011. Experimental paleontology of the scimitar-tooth and dirk-tooth killing bites. In *The Other Saber-Tooths: Scimitar-Tooth Cats of the Western Hemisphere*, edited by V. L. Naples, L. D. Martin, and J. P. Babiarz, 19–33. Johns Hopkins University Press, Baltimore.

Wheeler, H. T., and G. T. Jefferson. 2009. *Panthera atrox*: Body proportions, size, sexual dimorphism, and behavior of the cursorial lion of the North American plains. In *Papers on Geology, Vertebrate Paleontology, and Biostratigraphy in Honor of Michael O. Woodburne*, edited by L. B. Albright III, 423–44. Museum of Northern Arizona Bulletin 65.

White, R. 1986. *Dark Caves, Bright Visions: Life in Ice Age Europe*. American Museum of Natural History, New York.

White, R. S., Jr. 2008. Papago Springs Cave, pronghorns and paleontology: Red fields and the burden of proof. In *Neogene Mammals*, edited by S. G. Lucas, G. S. Morgan, J. A. Spielmann, and D. R. Prothero, 365–74. New Mexico Museum of Natural History and Science Bulletin 44.

White, R. S., Jr., and G. S. Morgan. 2011. *Capromeryx* (Artiodactyla: Antilocapridae) from the Rancholabrean Tramperos Creek Fauna, Union County, New Mexico, with a review of the occurrence and paleobiology of *Capromeryx* in the Rancholabrean of New Mexico. In *Fossil Record 3*, edited by R. M. Sullivan, S. G. Lucas, and J. A. Spielmann, 641–51. New Mexico Museum of Natural History and Science Bulletin 53.

Whittemore, A. T. 1997. *Berberis*. In *Flora of North America North of Mexico*. Vol. 3, *Magnoliophyta: Magnoliidae and Hamamelidae*, edited by Flora of North America Editorial Committee. http://www.efloras.org/florataxon.aspx?flora_id=1&taxon_id=103816.

Widga, C., T. L. Fulton, L. D. Martin, and B. Shapiro. 2012. *Homotherium serum* and *Cervalces* from the Great Lakes Region, USA: Geochronology, morphology and ancient DNA. *Boreas* 41:546–56.

Wigand, P. E., and P. J. Mehringer Jr. 1985. Pollen and seed analyses. In *The Archaeology of Hidden Cave, Nevada*, edited by D. H. Thomas, 108–24. Anthropological Papers of the American Museum of Natural History 61 (1).

Wigand, P. E., and C. L. Nowak. 1992. Dynamics of northwest Nevada plant communities during the last 30,000 years. In *The History of Water: Eastern Sierra Nevada, Owens Valley, White-Inyo Mountains*, edited by C. A. Hall Jr., V. Doyle-Jones, and B. Widawski, 40–62. White Mountain Research Station Symposium Volume 4.

Wigand, P. E., and D. Rhode. 2002. Great Basin vegetational history and aquatic systems. In *Great Basin Aquatic Systems History*, edited by R. Hershler, D. B. Madsen, and D. R. Currey, 309–67. Smithsonian Contributions to the Earth Sciences 33.

Wiggins, I. L. 1980. *Flora of Baja California*. Stanford University Press, Stanford, CA.

Wilder, B. T., R. S. Felger, and T. R. Van Devender. 2008. *Canotia holacantha* on Isla Tiburón, Gulf of California, Mexico. *Canotia* 4:1–7.

Wilke, P. J., J. J. Flenniken, and T. L. Ozbun. 1991. Clovis technology at the Anzick site, Montana. *Journal of California and Great Basin Anthropology* 13:242–72.

Willerslev, E., J. Davison, M. Moora, et al. (50 authors). 2014. Fifty thousand years of Arctic vegetation and megafaunal diet. *Nature* 506:47–51.

Williams, D. R. 2009. Small mammal faunal stasis in natural trap cave (Pleistocene-Holocene), Bighorn Mountains, Wyoming. Master's thesis, University of Kansas, Lawrence.

Willig, J. A. 1988. Paleo-Archaic adaptations and lakeside settlement patterns in the Northern Alkali Basin, Oregon. In *Early Human Occupation in Far Western North America: The Clovis-Archaic Interface*, edited by J. A. Willig, C. M. Aikens, and J. L. Fagan, 417–82. Nevada State Museum Anthropological Papers 21.

Wilmshurst, J. M., T. L. Hunt, C. P. Lipo, and A. J. Anderson. 2011. High-precision radiocarbon dating shows recent and rapid initial human colonization of East Polynesia. *Proceedings of the National Academy of Sciences* 108:1815–20.

Wilson, D. E., and D. M. Reeder. 2005. *Mammal Species of the World*. 3rd ed. Johns Hopkins University Press, Baltimore.

Wilson, M. C., and L. B. Davis. 1994. Late Quaternary stratigraphy, paleontology, and archaeology of the Sheep Rock Spring Site (24JF292), Jefferson County, Montana. *Current Research in the Pleistocene* 11:100–102.

Winans, M. C. 1989. A quantitative study of North American fossil species of the genus *Equus*. In *The Evolution of Perissodactyls*, edited by D. R. Prothero and R. M. Schoch, 262–97. Oxford University Press, New York.

Wojciechowski, M. F., and D. Isely. 2012. *P. spinosus. Jepson eFlora*. Jepson Flora Project. http://ucjeps.berkeley.edu/cgi-bin/get_IJM.pl?tid=40166.

The wonderful fossil beds of Oregon. 1877. *Western Review of Science and Industry* 1:608–10. From the *Guard* (Eugene, OR), September 29.

Wood, J. R., and J. M. Wilmshurst. 2012. Wetland soil moisture complicates the use of *Sporormiella* to trace past herbivore populations. *Journal of Quaternary Science* 27:254–59.

Wood, J. R., J. M. Wilmshurst, T. H. Worthy, and A. Cooper. 2011. *Sporormiella* as a proxy for non-mammalian herbivores in island ecosystems. *Quaternary Science Reviews* 30:915–20.

Woodburne, M. O. 2010. The Great American Biotic Interchange: Dispersals, tectonics, climate, sea level and holding pens. *Journal of Mammalian Evolution* 17:245–64.

Woodman, N., and N. B. Athfield. 2009. Post-Clovis survival of American mastodon in the southern Great Lakes region of North America. *Quaternary Research* 72: 359–63.

Woodward, A. S. 1900. On some remains of *Grypotherium* (*Neomylodori*) *listai* and associated mammals from a cavern near Consuelo Cove, Last Hope Inlet, Patagonia. *Proceedings of the Zoological Society of London* 1900:64–78.

———. 1902. Description of additional discoveries. In *Through the Heart of Patagonia*, edited by H. H. Pritchard, 315–30. D. Appleton, New York.

Woolfenden, W. B. 1996. Late Quaternary vegetation history of the southern Owens Valley region, Inyo County, California. PhD. diss., University of Arizona, Tucson.

———. 2003. A 180,000-year pollen record from Owens Lake, CA: Terrestrial vegetation change on orbital scales. *Quaternary Research* 59:430–44.

World Bird Database. http://avibase.bsc-eoc.org.

Wormington, H. M., and D. Ellis, eds. 1967. *Pleistocene Studies in Southern Nevada*. Nevada State Museum Anthropological Papers 13.

Worthy, T. H., and R. N. Holdaway. 2002. *The Lost World of the Moa: Prehistoric Life of New Zealand*. Indiana University Press, Bloomington.

Wroe, S., J. Field, M. Archer, et al. (10 authors). 2013. Climate change frames debate over the extinction of megafauna in Sahul (Pleistocene Australia–New Guinea). *Proceedings of the National Academy of Sciences* 110:8777–81.

Yerington, J. A. 1895. *Report of Nevada State Board World's Fair Commissioners*. State Printing Office, Carson City, NV.

Yoder, J. B., C. I. Smith, D. J. Rowley, R. Flatz, W. Godsoe, C. Drummond, and O. Pellmyr. 2013. Effects of gene flow on phenotype matching between two varieties of

Joshua tree (*Yucca brevifolia*; Agavaceae) and their pollinators. *Journal of Evolutionary Biology* 26:1220–33.

Young, J. K., K. M. Murray, S. Strindberg, B. Buuveibaatar, and J. Berger. 2010. Population estimates of Mongolian saiga *Saiga tatarica mongolica*: Implications for effective monitoring and population recovery. *Oryx* 44:285–92.

Young, T. P., and B. D. Okello. 1998. Relaxation of an induced defense after exclusion of herbivores: Spines on *Acacia drepanolobium*. *Oecologia* 115:508–13.

Youngman, P. M., and F. W. Schueler. 1991. *Martes nobilis* is a synonym of *Martes americana*, not an extinct Pleistocene-Holocene species. *Journal of Mammalogy* 72:567–77.

Yule, J. V., R. J. Fournier, C. X. J. Jensen, and J. Yang. 2014. A review and synthesis of late Pleistocene extinction modeling: Progress delayed by mismatches between ecological realism, interpretation, and methodological transparency. *Quarterly Review of Biology* 89:92–106.

Zacharias, E. H. 2013. *Atriplex torreyi* var. *torreyi*. *Jepson eFlora*. Jepson Flora Project. http://ucjeps.berkeley.edu/cgi-bin/get_IJM.pl?tid=71461.

Zaya, D. N., and H. F. Howe 2009. The anomalous Kentucky coffeetree: Megafaunal fruit sinking to extinction? *Oecologia* 161:221–26.

Zazula, G. D., and D. Froese. 2013. *Ice Age Old Crow*. Government of Yukon, Whitehorse.

Zazula, G. D., R. D. E. MacPhee, J. Z. Metcalfe, et al. (15 authors). 2014. American mastodon extirpation in the Arctic and Subarctic predates human colonization and terminal Pleistocene climate change. *Proceedings of the National Academy of Sciences* 111:18460–65.

Zdanowicz, C. M., G. A. Zielinski, and M. S. Germani. 1999. Mount Mazama eruption: Calendrical age verified and atmospheric impact assessed. *Geology* 27:621–24.

Zeigler, A. C. 2002. *Hawaiian Natural History, Ecology, and Evolution*. University of Hawai'i Press, Honolulu.

Zhang, Z., A. Feduccia, and H. F. James. 2012. A late Miocene accipitrid (Aves: Accipitriformes) from Nebraska and its implications for the divergence of Old World vultures. *PLoS One* 7 (11): e48842.

Zouhar, K. 2001. *Pinus monophylla*. USDA Forest Service, Rocky Mountain Research Station, Fire Sciences Laboratory. Fire Effects Information System. http://www.feis-crs.org/beta/.

Sources for the Great Basin Ice Age Mammal Distribution Maps

The distributions shown in figure 6.8 are heavily dependent on Faunmap II (http://www.ucmp.berkeley.edu/faunmap/about/data.html), updated by the additional sources listed here.

Arctodus: Schubert et al. 2010.

Aztlanolagus: Harris 2003; Harris 2014; Harris and Hearst 2012; B. Russell and Harris 1986.

Bootherium: G. Jefferson et al. 2002.

Brachyprotoma: Hockett and Dillingham 2004; Heaton 1985.

Camelops: Heaton 1985; Hockett and Dillingham 2004; Mead 2003; Mead and Phillips 1981; Pasenko and Agenbroad 2012; Springer et al. 2010.

Canis dirus: Arroyo-Cabrales and Carranza Castañeda 2009; Dundas 1999; Hodnett, Mead, and Baez 2009; G. Morgan and Lucas 2005; University of Nevada, Las Vegas 2012.

Capromeryx: Springer et al. 2010; G. Jefferson 1991a; White and Morgan 2011.

Equus: Brean 2013.

Euceratherium: G. Jefferson 1991a; Kropf, Mead, and Anderson 2007; G. Morgan and Lucas 2005.

Hemiauchenia: K. Adams and Wesnousky 1998; Harris 1993, 2003; G. Jefferson 1991; G. Jefferson, McDonald, and Livingston 2004; G. Jefferson et al. 2002; G. Morgan and Lucas 2005; Springer et al. 2010.

Homotherium: G. Jefferson and Tejada-Flores 1993; H. McDonald 1998.

Mammut: G. Jefferson 1991; G. Jefferson, McDonald, and Livingston 2004; G. Jefferson et al. 2002; Lucas and Morgan 2005; Pasenko 2011; Agenbroad et al. 2013.

Mammuthus: Agenbroad et al. 2013.

Megalonyx: Gillette, McDonald, and Hayden 1999.

Miracinonyx: Hockett and Dillingham 2004; M. Wilson and Davis 1994.

Navahoceros: Harris 1985, 1993; G. Morgan and Lucas 2006.

Nothrotheriops: H. McDonald 2003; H. McDonald and Jefferson 2008; H. McDonald and Morgan 2011; Springer et al. 2010.

Oreamnos harringtoni: Campos, Willerslev, Sher, et al. 2010; Mead and Lawler 1994; Mead, Czaplewski, and Agenbroad 2005.

Paramylodon: G. Jefferson 1991; G. Jefferson et al. 2002; H. McDonald 1998, 2003; H. McDonald and Morgan 2011; G. Morgan and Lucas 2005; Springer et al. 2010.

Platygonus: G. Jefferson 1991; H. McDonald 2002; G. Morgan and Lucas 2005; Murray et al. 2005; Springer et al. 2010.

Smilodon: G. Jefferson 1991; Livingston 1992; G. Morgan and Lucas 2005; H. McDonald 1998; San Bernardino County Museum 2012.

Sources for Plant Distribution Maps

Acacia greggii, Canotia holacantha, Carnegiea gigantea, Condalia globosa, Koeberlinia spinosa, Olneya tesota, Parkinsonia microphylla, Prosopis glandulosa, Psorothamnus spinosus, Robinia neomexicana, Shepherdia argentea, Yucca brevifolia: US Geological Survey 2013.

Atriplex torreyi: Welsh 2003; Zacharias 2013.

Fouquieria splendens: Henrickson 1972.

Lycium andersonii: Benson and Darrow 1981; Albee, Shultz, and Goodrich 1988; Kartesz 1988; University of Nevada, Reno, Herbarium, http://www.cabnr.unr.edu/reno_herbarium/; Calflora, http://www.calflora.org/entry/dgrid.html?crn=5230.

Mahonia fremontii: Whittemore 1997.

Ribes niveum: Intermountain Region Herbarium Network, http://intermountainbiota.org/portal/taxa/index.php?taxon=98387.

Rosa woodsii: Ertter 2012; *Encyclopedia of Life*, http://eol.org/pages/229741/overview.

INDEX

Page numbers in *italics* refer to figures and tables.

sloths in, 63; Joshua trees in, 249; mechanically defended plants in, 265; pronghorn in, 113; rabbits in, 87; seed dispersing in, 249

Southwest Museum, 202, 203

Spanish bayonet (*Yucca harrimaniae*), 254

species: as animal study focus, 45, 54; defined, 45; individual histories of, building, 290–93, 294

spectacled bear (*Tremarctos ornatus*), 56, 69, 70, 163–64

Spencer, Lee, 192, 193

spider monkeys (*Brachyteles*), 55

spineless koa (*Acacia koa*), 256, 258

spinescence, 256

Spizaetus pliogryps, 161

Spizaetus willetti, 161, 217

Sporormiella, 287

Spring Mountains, 65

spruce (*Picea*), 129, 136, 142, 269

Spruce Mountain, 23

Spurr, J.E., 135

squirrels, 53

squirreltail (*Elymus elymoides*), 247

Stafford, Tom, 173

stag-moose (*Cervalces*), 105–7, *106*

stag-moose (*Cervalces scotti*), 107

Starr, Tyler, 253

Star Trek (television series), 131, 276–77

Steens Mountain, 26

Stegomastodon (genus), 54, 55

Stehli, Frank, 101

stemmed points, 239–43, *240, 241*

Sternberg, Charles H.: Fossil Lake collecting done by, 187–88, 194; human-animal association theories disputed by, 189, 190, 191–92, 193, 222; persons working for, 203

stilt-legged horses (hemiones), 91, 92

Stock, Chester: bird remains made available by, 194; fossil remains dating challenges faced by, 166; ground sloth (*Mylodon harlani*) foot structure studied by, 12, *12*; Gypsum Cave findings reported by, 207; Harrington's mountain goat defined by, 154, 217,

231; Shasta ground sloth skull identified by, 203; sloth diet, interest in, 209, 213; Smith Creek Cave findings reported by, 217

Stockoceros (genus). *See* pronghorn (*Stockoceros*)

Stockoceros conklingi, 113

stone tools, 218, 237

storks, 155–56, 158

stout-legged llama (*Palaeolama mirifica*), 100, *101*

stratigraphy, 231, 232

structural defenses (plants), 256–57, 259

strychnine, 256

subalpine conifers, 23, 37, 39, 269–72

subalpine fir (*Abies lasiocarpa*), 22, 23, 269

Sulphur Spring Range, 224

sumac (*Rhus*), 122

Sunshine Locality, 183

supernova explosions, 282

Symbos (defined), 124

Tafimys, 51

taiga vole (*Microtus xanthognathus*), 289

tall plants: armed, 260, 264–65, *265*, 271; defined, 259; mechanically defended, 299–300; unarmed, 264–65, *265*

tapirs (*Tapirus*): Bering Land Bridge, migration across, 89; distribution of, *90*; extinction and survival of, 56; in Great Basin, possible, 273; Ice Age animals related to, 52–53; knowledge of, increasing, 166; overview of, 88; species of, 45; survival of, 163–64

Tapirus terrestris, *90*

tayassuids, 33, 93–99, *94*, 259, 274

Taylor, Erv, 282

Tegeticula antithetica, 252–53

Tegeticula synthetica, 252–53

Térapa, Mexico, 127

Teratornithidae (teratorns), 158–59, 161–62, 217

Teratornis incredibilis, 217

Teratornis merriami, 159, 217, 225

Tetracerus quadricornis, 117–18

Tetrameryx shuleri, 113–15, *115, 116*